The Mutual Construction of Statistics and Society

Metropolitan College of NY
Library - 7th Floor
60 West Street
New York, NY 10006

Routledge Advances in Research Methods

1. E-Research
Transformation in Scholarly Practice
Edited by Nicholas W. Jankowski

2. The Mutual Construction of Statistics and Society
Edited by Ann Rudinow Sætnan, Heidi Mork Lomell, and Svein Hammer

The Mutual Construction of Statistics and Society

Edited by Ann Rudinow Sætnan,
Heidi Mork Lomell, and Svein Hammer

Routledge
Taylor & Francis Group
New York   London

First published 2011
by Routledge
711 Third Avenue, New York, NY 10017

Simultaneously published in the UK
by Routledge
2 Park Square, Milton Park, Abingdon, Oxon OX14 4RN

Routledge is an imprint of the Taylor & Francis Group, an informa business

First issued in paperback 2012

© 2011 Taylor & Francis

The right of Ann Rudinow Sætnan, Heidi Mork Lomell and Svein Hammer to be identified as the authors of the editorial material, and of the authors for their individual chapters, has been asserted by them in accordance with sections 77 and 78 of the Copyright, Designs and Patents Act 1988.

Typeset in Sabon by IBT Global.

All rights reserved. No part of this book may be reprinted or reproduced or utilised in any form or by any electronic, mechanical, or other means, now known or hereafter invented, including photocopying and recording, or in any information storage or retrieval system, without permission in writing from the publishers.

Trademark Notice: Product or corporate names may be trademarks or registered trademarks, and are used only for identification and explanation without intent to infringe.

Library of Congress Cataloging-in-Publication Data

The mutual construction of statistics and society / edited by Ann Rudinow Sætnan, Heidi Mork Lomell, and Svein Hammer. — 1st ed.
 p. cm. — (Routledge advances in research methods ; v. 2)
 Includes bibliographical references and index.
 1. Statistics. 2. Public administration. I. Saetnan, Ann Rudinow. II. Lomell, Heidi Mork. III. Hammer, Svein.
 HA29.M88 2010
 001.4'33—dc22
 2010004307

ISBN13: 978-0-415-87370-3 (hbk)
ISBN13: 978-0-203-84661-2 (ebk)
ISBN13: 978-0-415-81105-7 (pbk)

Contents

List of Figures	ix
List of Tables	xi
Acknowledgments	xiii

Introduction: By the Very Act of Counting: The Mutual Construction of Statistics and Society 1
ANN RUDINOW SÆTNAN, HEIDI MORK LOMELL AND SVEIN HAMMER

PART I
Overarching Themes and Approaches

1 Numbers: Their Relation to Power and Organization 21
JON HOVLAND

2 Words and Numbers: For a Sociology of the Statistical Argument 41
ALAIN DESROSIÈRES

3 Sociology in the Making: Statistics as a Mediator between the Social Sciences, Practice, and the State 64
CHRISTOPHER KULLENBERG

4 Governing by Indicators and Outcomes: A Neo-Liberal Governmentality? 79
SVEIN HAMMER

PART II
Visibility, Invisibility and Transparency

5 Ethnicity: Differences and Measurements 99
ELLEN BALKA AND KJETIL RODJE

vi *Contents*

6 Seeing Like Citizens: Unofficial Understandings of Official
Racial Categories in a Brazilian University   117
LUISA FARAH SCHWARTZMAN

7 Ideas in Action: "Human Development" and "Capability" as
Intellectual Boundary Objects   136
ASUNCIÓN LERA ST. CLAIR

PART III
Accountability and Manageability

8 Labeling and Tracking the Criminal in Mid-Nineteenth Century
England and Wales: The Relationship between Governmental
Structures and Creating Official Numbers   157
CHRIS WILLIAMS

9 From Categorization to Public Policy: The Multiple Roles of
Electronic Triage   172
ELLEN BALKA

10 Making Sense of Numbers: The Presentation of Crime Statistics
in the Oslo Police Annual Reports 1950–2008   191
HEIDI MORK LOMELL

11 Statistics on a Website: Governing Schools by Numbers   207
SVEIN HAMMER AND SIGRUNN TVEDTEN

12 Locating the Worths of Performance Indicators: Performing
Transparencies and Accountabilities in Health Care   224
SONJA JERAK-ZUIDERENT AND ROLAND BAL

PART IV
Reporting and Acts of Resistance

13 Constructing Mirrors, Constructing Patients—with High Stakes
Statistics   245
GUNHILD TØNDEL

14 GIS in Practice: Domestication of Statistics in Policing   263
HELENE I. GUNDHUS

Contributors	279
Name Index	283
Subject Index	289

Figures

1.1	OLS regression from TBU (2005: 58), municipal income and production.	29
3.1	Survey forms, postcards, and telephone calls are assembled as programs of action.	74
4.1	An analytical framework for studies of "governmentalities".	83
9.1	The multiple domains of electronic triage.	180
12.1	Working document of quality management team for collecting the data on performance indicators requested by the Health Inspectorate.	232

Tables

2.1	The State, the Market and Statistics	45
2.2	Poverty in England at the End of the Ninteenth Century: Three Policies and Their (Three) Instruments	48
6.1	Classification for Quota Purposes as % of Classification in Socio–economic Questionnaire	122
8.1	Identifications from the Register, 1870–1873	166
13.1	Physicians' Strategies and Their Characteristics	256

Acknowledgments

Just as our galley proofs for this book arrived, so too did the news of Susan Leigh Star's untimely death. Leigh was a major source of inspiration for this book. The book consists almost exclusively of articles developed from papers first delivered at the workshop "Statistics as a Boundary Object between Science and the State," already the title reflecting our indebtedness to Leigh. Leigh graciously agreed to attend the workshop, where she not only made a keynote lecture reclaiming her *boundary object* concept from the interpretative drift it had been subjected to over time, but also attended paper presentation sessions, contributing encouraging and constructive comments. We had hoped she might review the book, reclaiming again her concept form any misinterpretations we might have perpetrated. We were hoping soon to see a book of her own, elaborating the *boundary object* concept. Instead, with both sorrow and gratitude, we dedicate this book to Leigh.

While some authors have further acknowledgments relevant to their respective chapters, we wish here to add a few more acknowledgments to individuals and institutions that contributed to the project and the book as a whole. First of all, thanks to the Norwegian Research Council (NFR) and to the Norwegian University of Science and Technology (NTNU). Together they funded the projects "For Whom the Bell Curves" and "Health by Numbers."

Besides providing research funding to many of the contributors to this volume (Ann Rudinow Sætnan as head of the projects; Heidi Mork Lomell and Svein Hammer as post-doctoral fellows; Jon Hovland, Luisa Farah Schwartzman, and Gunhild Tøndel as PhD fellows), the projects also funded the aforementioned workshop where most of the chapters in this volume (and many not included here, but reflected indirectly as sources of inspiration) were first presented and discussed. Thanks go also to the European Association for the Study of Science and Technology (EASST) which provided funding for graduate student travel grants to the workshop. And warm thanks to all the workshop participants for constructive contributions, both of other papers and of comments on those included here.

xiv *Acknowledgments*

Last but not least, thanks to editors, anonymous referees, and others in the long list of production functions we have encountered and learned from through the publishing process at Routledge. The book is the better for your help!

Introduction
By the Very Act of Counting—The Mutual Construction of Statistics and Society

Ann Rudinow Sætnan, Heidi Mork Lomell, and Svein Hammer

How does the act of counting affect the world? How does it change the objects counted, change the lives of those who count (double entendre intended)? Some say that only "what's counted counts" (Waring 1989, Anderson and Fienberg 1999). Others (purportedly Albert Einstein among them) say the opposite: "Not everything that can be counted counts, and not everything that counts can be counted" (The Quotations Page). Our argument, briefly stated, is that *society and the statistics that measure and describe it are mutually constructed*.[1] This argument addresses two counterarguments from seemingly opposite directions. On the one hand, we oppose the notion that statistics are simple, straightforward, objective descriptions of society, gathered from nonparticipant points of observation. We claim, and demonstrate, that in the act of counting we do not stand neutrally outside the "object" we count, but rather (to some extent) enter into it, redefine it, change the stakes that affect it, and thus even change the numbers we count to represent it. This is not only because all knowledge is produced from some position or other with a partial and fragmentary view of the world, not only because the so-called "god trick" (Haraway 1991: 191)—the trick of being all-seeing, as from a point both everywhere and nowhere—is a dangerous illusion, but also because the very act of counting is a specific form of viewing. Like all other specific forms of viewing, it is a social act. Counting acts in and upon the social world. Of course, this also means that *not* counting has an effect on the aspects of the world we (do and/or don't) count. What we choose to count, what we choose not to count, who does the counting, and the categories and values we choose to apply when counting are matters that matter (see, e.g., Anderson and Fienberg 1999, Waring 1989). Thus users (producers, readers, interpreters, deployers) of statistics are "users that matter" (Oudshoorn and Pinch 2003).

On the other hand, we also oppose the notion that statistics and/or society are mere fictions, to be invented at will. In other words, we do not argue for the abandonment of counting. Just because we cannot fulfill the god trick by counting (or, for that matter, by any other form of viewing and describing the world; see for instance Becker 2007), does not mean that counting is a meaningless exercise. Again, counting is a specific means of viewing and describing. Numeric statements about the world are tickets into specific discourse forums and forms. Once we enter these for(u)ms, we make ourselves susceptible to certain types of counterstatements, trials, and tests. And while these too cannot fulfill the god trick, they can nevertheless be a fruitful means of disciplining[2] our own and others' thinking about the world. Furthermore, counting has proven a powerful way of influencing the social world.[3] So rather than abandoning or rejecting statistics, and certainly rather than being controlled by them, perhaps we need to tame them—become more aware of their power and their limitations, decide what place we are willing to allow them in our lives, and learn to keep them in that place. If so, then much like learning to drive a car or shoot a gun, learning statistics needs to be about more than simply mastering the techniques of using the tool; it needs to be also about learning the power—sometimes even the danger—of that tool and learning to control it within social and ethical bounds, including learning when not to use it. In the case of statistics, the dangers lie in the routines through which statistics are applied; the discourses from which they emerge and into which they are deployed; the power relations created by those discourses; and the assumptions, meanings, and categories (on this, see Bowker and Star 1999) statistics carry with them in those discourses. Our book will sensitize stakeholders—those who count, those who are counted, and those who use the accounts—to these dangers.

Forms of government, concepts of citizenship, and statistical practices closely interact like chips of colored glass in a kaleidoscope, constantly nudging one another and creating new colors where they meet. In this interaction, statistics are not merely free-floating elements, but also play a role like kaleidoscope mirrors, reflecting and linking the shifting relationships of the elements. And of course, in the modern democratic State[4], the State and its citizens also use these statistical mirrors to peer into as self-reflecting actors, learning from reflected policy results and adjusting policies to perfect their mutual relationships according to current values and standards.

Statistics play this key role in the kaleidoscope because *to count something or someone is to make it/them count in the policy sphere*, and by corollary, only that/those which already count tend to get counted (Waring 1989, Anderson and Fienberg 1999). Categories of actions and actors are both counted and shown to be countable, i.e., cognitively separated out from the infinite universe, at the same moment by the act of counting. And not least, in that same moment they are also shown to be politically relevant, which in turn serves to further define the State as interested in them. This phenomenon is known as mutual (or co-) construction (as defined and exemplarized in Oudshoorn and Pinch 2003). Briefly stated, *the act of counting its citizens, territories, resources, problems, and so on, is one of the acts by which the State participates in creating both itself, its citizens, and the policies, rights, expectations, services, and so on, that bind them together.*

No wonder then that data collection and categorization are activities at times surrounded with struggles (Bowker and Star op. cit.)—struggles to be seen or not seen and struggles over how one is defined. These struggles, however, are not constant. While no settlement is necessarily permanent, categories do emerge that are thereafter taken for granted for long periods of time. But since these categories emerge from social negotiations, they may not be identical over sociogeographical space. There will always be *centripetal* forces drawing toward standardization of categories, for instance influences stemming from global scientific communities, international statistical organizations, national and international governing bodies' report requirements, and so on. But there will always at the same time be *centrifugal* forces pulling categories apart, e.g., local values and traditions, or grassroots opposition by minority groups against how they are defined by the majority.

HISTORICAL PERSPECTIVES

Our argument builds from a growing literature. Histories of statistics (e.g., Desrosières 1998; Hacking 1990; MacKenzie 1981; Porter 1986, 1995) demonstrate that statistics as a knowledge system and set of practices[5] has evolved in interaction with political contingencies. We will not repeat here all the centuries covered by their analyses, but rather, with their conclusions as a starting point, situate statistics in a current, historically contingent context.

Desrosières' history of statistics (op. cit.) analyzes the co-constructive interaction between, on the one hand, scientific processes of description, coding, categorizing, measurement, and analysis, and, on the other hand, the administrative and political world of action, decision making, intervention, and improvement. An important theme of the book is how phenomena like unemployment, inflation, poverty, and so on, are measured by statistics, and then used in descriptions, discussions, and justifications. In other words, "they are inscribed in routinized practices that, by providing a stable and widely accepted language to give voice to the debate, help to establish the reality of the picture described" (ibid.: 1). At the same time, statistics was not constructed in a vacuum. Rather, the forms of public action predominant in a given country or era made up a space where the themes of the surveys, the questions asked, and the nomenclatures used were formed (ibid.: 248). In the discourses of government, however, this co-constructive effect is not made visible. Instead, the distinction between the *scientific* and *politico-administrative* spheres is maintained: "It is because the moment of objectification can be made autonomous that the moment of action can be based on firmly established objects" (ibid.: 9).

Desrosières shows how different actors, tools, techniques, structures, events, actions, and so on, contribute to the establishment of a new "regime of truth." Recognizable from the seventeenth century, this regime reaches a position of hegemony from about 1940–1950 until the 1970s. The analysis can be read as a linear, progressive history culminating in a foreordained achievement, as if to say "At last, now statistics is what it should be and will remain." This, however, is an unfortunate cultural bias we as readers bring to historical analyses. It is explicitly not Desrosières' intent:

> Statistical information did not fall from the sky like some pure reflection of a pre-existing "reality." Quite the contrary; it can be seen as the provisional and fragile crowning of a series of conventions of equivalence between entities that a host of disorderly forces is continually trying to differentiate and disconnect (ibid.: 325).

Porter's (op. cit.) concern is to explain the political power of numbers in modern societies. Received wisdom has long been that quantitative methodologies won a place in the social sciences and in governance thanks to their demonstrated effectiveness within the natural sciences, and that their effectiveness there is due in large part to the natural ability of numbers to imitate and describe nature. Though not entirely without merit, there are a number of flaws in this argument. One is that mathematics is a language, no better or worse than any other language at describing or imitating nature. Next, even if mathematics were exceptional among languages in its ability to describe stars or molecules or chemical reactions, it does not automatically follow that it is equally exceptional in its suitability for describing human society.

Porter finds that, historically, there is at least as much merit in turning the received argument around: In a time when natural sciences were still a target of contention, numbers were already being gathered to describe human society. It may well have been the political success of those number-gathering efforts and of their applications toward description and analysis of society that formed a basis for their success in raising the social status of the natural sciences. The quality of numbers as a language that came to be a key in both of these successes was that mathematics came to be seen as more stringent and rule-based than other languages. Mathematics came to be seen as a language of distance, disinterestedness, and impartiality—a status it (ironically) first achieved as a political tool. And though it is now widely known that one can as easily lie with numbers as with other forms of speech, this is still a knowledge that we must remind ourselves of, whereas the presumed impartiality and objectivity of numbers remains our first assumption and response. Porter emphasizes that "objectivity" in this context is not a question of being true to nature, but of withholding judgment, resisting subjectivities (ibid.: ix). The drift from objectivity in this sense toward a perception of objectivity as truth is thus a social and historical achievement.

Hacking's (op. cit.) question is not the power of numbers in general, but of probability in particular. Even after numbers in general came to be perceived as in some ways superior to other forms of language, certain types of mathematics were still viewed with suspicion. Specifically, calculations of risk and probability were seen as the purview of gamblers and speculators. Science, meanwhile, was seen as the pursuit of certainty. The promise of scientific governance was to make policy outcomes certain, not merely predictable within some margin of error or chance. However, the masses of data and calculations flowing from postrevolutionary France, not least in light of the arguments of Quetelet, eventually convinced a critical mass of researchers and policymakers that probability could provide the sought-for regularities—even certainties—that determinist efforts had failed to produce.

Taken together, the three works just discussed paint a convincing picture of how statistics achieved the central role it holds to this day as a tool of knowledge-based governance. And yet, although statistics remains hegemonically powerful, the practices as of the 1970s described by Desrosières, Porter, and Hacking, are not precisely those we can observe in the twenty-first century. In (possibly premature) retrospect, the 1970s appear to have been yet another turning point for political climate in ways that are relevant to public statistics practices. From the 1980s onward, the faith in social planning that had characterized many Western governments since the interwar years (Thomassen 1997) seems to have begun fading away in favor of renewed liberalism and an individualist ideology (ibid.: 553–555).[6]

Ironically, during that same time period our technical capacity to store, retrieve, analyze, and distribute large quantities of data has increased exponentially. In a further irony, that same technological capacity has also fed into individualist values by promising us new means of personal freedom of choice, e.g., choosing where and when to work, when and with whom to communicate, and so on. And in yet another twist of the ironic screw, we have, perhaps by the same token of loss of faith in planning, seen an increasing focus on all manner of "risks" (Beck 1992, Adam et al. 2000, Skolbekken 1995)—a focus that finds one expression in the generation and consumption of large amounts of statistics, especially in the areas of health/disease and security/crime. At the same time, combined with liberalist ideology, risk focus finds another expression in a fear of high-tech, personalized surveillance via these data, particularly by the State. To guard against unwanted surveillance, we have also set up "firewalls" in the form of personal data protection laws and agencies.

STATISTICS AS TECHNOLOGIES OF KNOWLEDGE PRODUCTION

Like many of the aforementioned historical studies, we have taken inspiration from social studies of science and technology. We do so on the basis that statistics can be viewed as both a knowledge claim-producing tool (i.e., a form of science) and a tool of governance (i.e., a technology in a fairly traditional sense—an object designed for purposeful use).

Viewing statistics as a technology actualizes the whole panoply of theories about technology and society. The interdisciplinary field of Science and Technology Studies (STS) has distanced itself from two simple causal models—society creates technologies (social determinism) and technologies create society (technological determinism; Bowden 1995, Pinch and Bijker 1987). Each of these remains popular in the world at large, and each carries some grain of truth. Obviously, technologies do not invent themselves. Humans create technologies, often with the clear intent of changing society by deploying them. Equally obviously, looking around us, technologies have effects on society, not all of which were intended by their inventors. But when we examine processes of sociotechnical change in detail, we soon see that they are neither socially determined nor technologically determined nor a simple sum of the two. Rather we see that technology and society simultaneously (re)create one another through complex interactions with sometimes surprising outcomes. The term "co-construction" (Oudshoorn and Pinch op. cit.) is used to describe this mutual simultaneous causation process.

Translation Theory or Actor-Network Theory (Callon 1986, Latour 1987) highlights the role of the engineer as co-constructive entrepreneur. The engineer not only translates the characteristics of nonhuman elements, weaving them together into a new technology, but also weaves in human elements. In building this network, the engineer is not entirely bound by previously acknowledged properties and interests of the various elements. Rather, the successful engineer takes control over the new network by deconstructing older properties and alignments and ascribing new ones (Latour 1987). For instance, to succeed in inventing the compact camera, Kodak ascribed new properties to silver oxide by linking it with paper and cellophane foil rather than glass and created whole new social categories: the amateur photographer, the corner store photo lab, and so on (Latour 1987: 131).

In Translation Theory, the engineer-entrepreneur is a very powerful figure. Translation Theory also brings nonhuman actors into sociological analysis by showing how they can be understood as having "interests," "voices," and social roles. The co-construction concept builds on this, but also on other strands of STS research, in which technologies' end users and nonusers are highlighted. Already Cowan (1987) points out that consumers are not passive putty in engineers' hands, but active, selective, and creative in whether and how they take technologies into use. The metaphor "domestication" (Silverstone and Hirsch 1992, Lie and Sørensen 1996) has been proposed as one way of understanding this creative process. In domesticating a pet, both pet and owner adjust their behaviors toward one another. Similarly with a new technology, we both train it and are trained by it. For instance, when a public office implements a new statistics system, in learning to operate the system the office staff also "tames" it by "teaching" it categories and comparisons they find relevant to their work.

Akrich (1992) develops another metaphor—"script"—into a whole vocabulary for exploring the interactions of engineers, nonhuman actors, and users. Engineers and marketers seek to guide users' actions by "inscribing" user roles into technologies as material form (as when a green button tells you "push here"), user handbooks, help hotlines, and so on. But no matter how thoroughly inscribed, technologies remain interpretatively flexible (Bijker 1987). Thus, prospective users still have to "read" technologies' scripts and remain active, selective, and creative in whether and how they enact them. For instance, a test driver reports how he experienced a new car with built-in safety features (*Dagbladet* 2008). One feature is a warning signal when the car crosses a painted line without having signaled a lane change. The test driver reports being amused when he first heard the signal—the first three notes of the William Tell Overture. His passengers recognized the tune and hummed its continuation. Realizing what had triggered the tune, the driver crossed the lane divider again, this time for six notes' worth. This became a musical game, crossing and recrossing the lane divider until the car stopped playing William Tell and sent instead a more insistent "pling" while flashing a coffee cup symbol on the dashboard—time for a coffee break. The engineers' script is clear: *Drivers must follow traffic rules—signal before changing lanes. If the car crosses a lane divider without a signal, presumably the driver is not fully alert. A gentle reminder will help the driver wake up and drive safely. If the car continues to weave across the divider, presumably the driver is seriously drowsy. A louder signal will wake the driver up so s/he can pull off the road for a needed break.* But in this case, the test driver creates a very different reading of the script—a user script (Gjøen and Hård 1996) or anti-program (Latour 1992)—turning the car into a musical toy. The test driver was lucky. A myriad of human and nonhuman actors—the weight of the vehicle, the tires, the road surface, the weather, the banks of the curves, other drivers on the road, etc.—could have made their voices heard, showing how dangerous a car is when played as a musical toy. Luckily for driver and passengers, the other actors were cooperative this time.

The Concept of "Mutual Construction"

The concept of mutual (or co-) construction invites us to see the power (however great or small) of users to interpret and (re)shape the technologies designers and producers offer them, and simultaneously, the ways users are themselves (re)shaped by their use (or nonuse!) of the technologies around them (Oudshoorn and Pinch op. cit.).

Co-construction is a concept used mainly in the technology branch of STS. STS is a field that brings together several perspectives on both science (truth claims and the social institutions and practices that produce them) and technologies (a melding of materials, techniques, organizations, knowledge, and routines aimed at producing intended results). Since statistics are a tool for learning and governing via production of (claims of) knowledge, both the science and technology aspects here are relevant. We would argue that co-construction is equally relevant when viewing statistics through a science lens of STS. After all, STS' technology perspectives are explicitly inspired by perspectives on science (Bijker 1987).

Like technologies (or rather, the other way around) data are human products of social processes, but not without reference to nonhuman actors. The ways we construe the world, as evidenced in part by how we parse it up into data, have consequences for how we act within and upon the world, thus (re-)shaping it. And not least, data are also interpretatively flexible, placing the fate of a scientific text and its truth claims in the hands of its readers (Latour 1987).

Beginning with interpretative flexibility, this does not mean that we are free to establish whatever truths we wish. Latour, for one, is not a relativist. Remember that he acknowledges nonhuman actors as having a voice—not always audible or intelligible to us humans, but occasionally recognizable, especially when they "veto" our would-be claims. Foucault (1972) points out, in effect, that truth claims become nonhuman actors in their own right: Truth claims and other statements have histories. These histories are cumulative, claim layered upon claim in "sediments" that solidify over time in ways that discipline what claims we may make in the future, and what qualifications we must hold to make them, if we expect to be taken seriously. Nevertheless, some scope for interpretative flexibility—of the data (broadly conceived) on which we base claims, and of the claims as texts—will always remain.

The claims we make—the forms we give them, the ways we interpret them, and the ways we choose to act on them—all affect the world those claims reference. Using statistics as our example, it matters that we turn our observations into numbers. Numbers have specific qualities. They are, as Latour points out, highly mobile (Latour 1987), though we will dispute whether they are as "immutable" as Latour seems to claim. Nevertheless, being mobile and at least *appearing* immutable, they enable flows of information, and with it power, from what thereby becomes a "periphery" to wherever they are gathered, which thereby becomes a "center of calculation." It is the flow of information itself which (re)creates this division of the world into center and periphery and gives the center power over the periphery.

The categories enacted by the counting process—the "units" to which measurements are ascribed, the characteristics measured, the scales of the measurements—also impose shapes onto the world as we see it. And, as Bowker and Star (1999) point out, these shapes are far from insignificant. They are (often literally) bloody serious.

And then, when these statistics become a basis for action (e.g., administrative action by the State), they quite obviously become tools to (re)shape the world. Here the co-construction concept reminds us, however, that even as statistics (re)shape the sociomaterial world, they are also constantly (re)shaped by it. Statistics do not render those deploying them almighty. Even when deployed in the interests of governance, statistics remain interpretatively flexible and can encounter various forms of resistance such as the user scripts and anti-programs discussed previously.

The Concept of Boundary Object

Another fruitful concept in our endeavor is "boundary object." Star and Griesemer (1989) describe boundary objects as having a number of properties that make them useful in demarcating and bridging boundaries between different but interacting social worlds:

> Boundary objects are both plastic enough to adapt to local needs and constraints of the several parties employing them, yet robust enough to maintain a common identity across sites. They are weakly structured in common use, and become strongly structured in individual-site use. They may be abstract or concrete. They have different meanings in different social worlds but their structure is common enough to more than one world to make them recognizable means of translation. The creation and management of boundary objects is key in developing and maintaining coherence across intersecting social worlds (Star and Griesemer 1989: 393).

This concept reminds us that the sociomaterial world in which statistics are constructed and deployed is not a homogenous and harmonious one. Statistics are meant to move "across sites." In doing so, they will encounter different interpretations and interests. Statistics may serve to translate interests across boundaries, unidirectionally or multidirectionally. Statistics may also serve to mark out boundaries by highlighting differences between sites and/or to bridge boundaries by becoming a meeting point around which different groups may collaborate in spite of ascribing different meanings to the shared object. What roles statistics play at these boundaries remains to be seen, perhaps best by examining local statistics practices.

STATISTICS AS TECHNOLOGIES OF GOVERNMENT AND GOVERNANCE

More specifically than knowledge production in general, we will focus on the production of knowledge as a tool for governance.[7] The chapters in this book will contribute theoretically and empirically to understanding the dynamic interactions of statistical practices with the aforementioned historical changes in the roles and governance styles of the State. The book will focus on fields where knowledge and power, description and decision, *there is* and *we must*, meet and interact. At these meeting points, government is not just a context surrounding statistical practices. Rather, our analyses explicitly discuss how practices and techniques related to coding, registration, measurement, and use of the numbers functions as part of governance—and, vice versa: how the practices of planning, decision, assessment, and action are formed and transformed in relation to statistics. In short, we focus on the co-constructive play between, on the one hand, the production of standardized knowledge, and on the other hand, attempts to make the world accountable, manageable, and governable.

This perspective of ongoing co-productions can be linked to the analytical field of *governmentality* (Foucault 1991, Hammer this volume). When using this perspective, relations between statistics and concepts such as "accountability," "transparency," and "management" can be seen as part of governmental discourses, of rationalities and technologies that make some ways of thinking and acting more relevant than their alternatives. From this point of view, there seem to be some historical displacements in the mechanisms that hold the public sector accountable. That is, an adjustment from rules and professional codes to comparisons, competitions, and other techniques transferred from the market and the economy. These tendencies toward more active, decentralized use of statistics can be related to the introduction of new techniques of transparency and accountability—techniques that more or less transform the art of government, and through that constitute new ways of understanding and new kinds of subjectivities.

Accountability is a governance principle of increasing relevance. The concept can be related to more or less synonymous words such as answerability, responsibility, and liability—and through this it can be defined as an aspect of "good" (democratic, effective, efficient) governance, most particularly those aspects of governance involved with justification of decisions and actions of management, implying taking responsibility and explaining the results achieved. In this way the concept is related to *transparency*, that is, the possibility to see, reflect, and communicate on what is being done "of, by, and for" (Lincoln 1863) and to the governed. In short, through making government transparent, the practices and effects of governance are expected to become accountable.

Given this short sketch of a complex landscape of concepts, *where*, *when*, and *how* do indicators, measurements, and numbers play into this field—and *with what effects*? To a degree the answer to these questions seems relatively given: To achieve both efficacy and legitimacy in governance, needs for knowledge have always been articulated. Standardized, quantitative measurements represent an opportunity to observe, in a seemingly neutral and objective way, what is done, with what input factors, through which processes, and with what outcomes. Numbers represent a possibility to compare, assess, problematize, and discuss the state of State—and, in extension of this, they are inscribed into systematic, continuous processes where the operations of the public sector are planned, coordinated, directed, and controlled. In this way, measurements contribute to accountability and manageability, to well-grounded and explainable decisions and actions directed at avoiding what is deemed undesirable and achieving what is deemed desirable.

Taking the perspective of this book seriously, we try to avoid discussions on whether governance or statistics are the most fundamental. Rather we hold on to a co-constructive perspective, where the practices of statistics cannot be separated from related political, administrative, professional, and social processes. Every new statistic carries a potential to transform the field it operates as part of. In this way it is possible to see *statistics* both as a way of representing the real, and as an important mechanism in the production of this reality. And, tuning this perspective in the opposite direction, the *fields of governance* can be seen as (in part) constructed through indicators and numbers. And yet, we are not claiming that this construction is all there is; we are not taking a radical relativist position. The descriptions given by statistics are valid only to the extent that the objects they exhibit can be seen as consistent with their descriptions, that is, part of "the real." In this way the main perspective of the book can be described as a sociomaterial constructivism.

FROM MACRO- AND MESO- TO MICROPERSPECTIVES

Our book also parallels a trend that can be observed across all three areas discussed earlier—the historical literature on statistics, approaches to understanding science and technology, and changes in understandings of governance—namely a move from macro- to microperspectives.

This trend is least prominent in historical studies, where analysis of documents often requires one to look for large-scale trends and turning points. At this macro level, O'Malley (2004) has extended the aforementioned studies of Desrosières, Porter, and Hacking into the present day, while also pointing out a need for micro-level studies. O'Malley emphasizes a point the three share, namely that statistical practices vary from place to place and time to time, reflecting not least differences in governmental rationalities. O'Malley tracks variations in intentions, interpretations, and (at the macro level) practices of statistics in relation to three forms of liberalism—classical, social, and neoliberalism. These three forms value, manage, and distribute risk and uncertainty in somewhat different ways. O'Malley shows how the categories of risk and uncertainty are formative parts of governmental discourses: Where risk is seen as a calculable danger, one that can be dealt with through specific, even individualized precautionary behaviors and contractual arrangements, uncertainty can be seen as an incalculable possibility that must be dealt with through a generalized foresight (e.g., economic thrift and cautiousness of action). These are not so much real differences as differences of perception, where O'Malley focuses on perceptions at the level of rationalities and technologies of governance. In a gross oversimplification, one might see risk as associated with numerical computations while uncertainty is managed more descriptively. O'Malley warns against such oversimplification. He sees both uncertainty and risk and the technologies for managing them as complex assemblages. The differences are not so blatant as the use or nonuse of statistics, but more the manners in which statistics and other descriptive and analytical tools are used.

Also, several meso-level studies, institutional histories charting the roles and practices of national statistics bureaus and other statistics-producing agencies (e.g., Anderson and Fienberg 1999, Haggerty 2001, Lie and Roll-Hansen 2001, De Michelis and Chantraine 2003), probe developments closer to our time. For instance, Lie and Roll-Hansen's (op. cit.) history of Statistics Norway extends into the 1990s, focusing there on public and political fears regarding large data registers and on the opportunities such registers provide for large-scale prognostic analyses in state planning and evaluation contexts. Eurostat has also published its own history (De Michelis and Chantraine op. cit.), emphasizing the need for standardization of statistics, primarily economic indicators, as a basis for negotiating and monitoring equitable trade agreements and working conditions. Igo (2007) analyzes private meso-level institutions, such as the Gallup Institute, and their roles in creating the political and commercial worlds we know today. In our view, these and other meso-analyses, lend further support to the understanding of statistics as an ongoing and changing product of, and at the same time a contributor to, social contentions and negotiations closely related to our conceptions of the State.

Statistics have come to play a key role in knowledge-based governance. However, it is important to recognize that the role they play is not a natural given but a social achievement. The historical studies discussed above are all macro- and meso-level studies of the role of numbers and statistical calculations in modern governance. That is to say, they view governmental styles and their use of statistical tools from above and/or from the level of relatively large institutions—from the perspective of the State, the monarch, the administration, the parliament, the census bureau. Citizens are implied as masses for whom the States' applications of these tools have consequences, but whose adaptations to them are not subject to close analysis. Nor are (sub-)meso- and micro-level adaptations—the day-to-day practices of lower level administrators as they gather, codify, store, retrieve, analyze and apply data—the primary targets of this literature.

Recognizing that macro-level intentions are not an adequate description of meso- and micro-level practices, O'Malley reminds us that intentions underlying the construction and deployment of a technology of governance, such as statistics, are in themselves real and worthy of study:

> [T]he existence of a plan, or of the design of a governmental technology, is quite distinct from whether or not it is translated into practice. [. . .] Government is always an *attempt* [emphasis in original] to create subjects of a certain kind, or achieve results of a certain form. Failure, judged internally or by external critics, is routine. To some degree, such "failures" to translate the blueprints of rule into practice are part of the genealogies of new technologies and rationalities (O'Malley op. cit: 26).

The STS field has taken a strong turn from macro toward micro through the growing emphasis on the co-productive role of technologies' and knowledge claims' users. And the governance field has taken a double turn from macro toward micro—both theoretically in emphasizing the co-production of governance by the (self-)governed, and politically/pragmatically at the level of (ideologies of) practice where belief in planning and steering has faded in favor of belief in local adaptation and the "market" force of citizens as "consumers" of public services. Whether responding to O'Malley's (or Foucault's, Cowan's, and others') call or simply as citizens of our time, we have chosen to focus on micro-level practices, where statistics are defined, gathered, and used.

This book will address the micro level of statistical practices and their relationships to macro-level discourses. Thus we go beyond the current state of the art in the study of the governance of numbers as we add a further and until today neglected dimension to this highly relevant field of research by describing how, when, where, by which means and to what ends public, private, and political actors construct, shape, understand, and deploy statistical entities. In addition to a top-down approach, we add a focus on the bottom-up process of negotiating the adequacy and usefulness of statistics to be used for good (or not so good) governance.

STRUCTURE OF THE BOOK

In this Introduction we have presented our general argument and positioned that argument within the literature. The rest of the book is structured in four parts. In Part I, issues of overarching theoretical interest are discussed further. The chapters, while written to stand each on its own, can also be read as part of a narrative whole: Hovland discusses the rationality that is constituted through the numbers and their use; he proposes a vocabulary that parses up the processes of producing and using statistics, thus rendering more visible how those processes imbue statistics with power. Desrosières gives an analysis of numerical practices in relation to various configurations of statistics and State, reviewing and updating his already classic historicization of statistics as co-productive with modernity. Kullenberg exemplifies one of these configurations through the analysis of two centers of calculation where statistics, social science, and the Swedish welfare state meet. Providing one means of summing up Part I, Hammer presents a perspective on indicators and numbers as part of a neoliberal governmentality.

The remaining parts will together provide support for our main argument by empirically exploring three sub-themes: In Part II we look at how statistics are used to construct, reveal, and conceal group identities. The focus is on the use of statistics about, and sometimes instances by, "minority" "ethnic" groups (scare quotes used here to signify that we question both of these categories). These cases help us explore issues of the role of statistics in forming, presenting, and representing identities. Balka and Rodje discuss the rationale for recording citizens' "ethnicity" and relating this to health statuses: While health statuses can be shown to vary substantially across "ethnic" categories, and while this situation is clearly a challenge to a State that aspires to egalitarian democracy, the authors show that for most understandings of the possible causal and/or governable relationships between "ethnicity" and health, data on "ethnic" categories are simply irrelevant. In relation to an affirmative action policy in Brazil, Farah Schwartzman explores the flexibility of "racial" self-identities as they interact with body self-images, family histories, experiences of discrimination, class consciousness, political programs, and categories imposed by public agencies and/or significant others. Lera St. Clair wraps up Part II by discussing issues of categorizations and egalitarian policies at the macro level: How do concepts such as "human development," "capability," and "poverty" travel across policy contexts and development theories?

Moving from the co-construction of identities of statistics and citizen groups to the co-construction of statistics and relationships between citizens and states, Part III discusses statistics in contexts of changing paradigms of government. Williams studies the labeling and tracking of "the criminal" in the nineteenth century, in an analysis of the relationship between governmental structures and the creation of official numbers. Balka presents data from hospital emergency rooms, where she studies the electronic trace(s) of "triaging activity" from categorization to public policy: the data produced during the triaging process, the places those data go, the other data with which triage data are combined. Lomell focuses on the presentation of crime statistics in Oslo police annual reports through fifty-eight years. Looking historically at how the presentation and interpretation of the statistics have varied, we get a better understanding of how various modes of governnance constitute and affect how the police read meaning into the numbers. Hammer and Tvedten present a Norwegian website where performance indicators on school quality are published as part of a national quality assessment system in education, and analyze the interplay between understandings of quality, the statistics presented, and the guidelines for using the numbers. Lastly, Jerak-Zuiderent and Bal analyze what they conceptualize as "performing transparencies and accountabilities in health care," by following performance indicators and questioning what they "actually" do. In sum, these chapters give a multiple picture of the interplay between acts of counting on the one hand, and on the other hand governmental acts of constructing, making sense of, and using the numbers.

Part IV takes the theme of governance by statistics a step further by focusing on the "production site." What happens to the numbers produced in so-called "high stakes" statistics routines, i.e., what happens to the production of numbers when the numbers produced have predictable and serious consequences for those who produce them? Based on micro-level studies of how end users produce, report, and relate to aggregated data, the two chapters focus on how counting affects the subjects that are simultaneously counting and counted. While Gundhus demonstrates how police officers actively resist using aggregated crime data in their daily work, Tøndel shows how physicians' awareness of how their work is used as management information (re)shapes the practice of diagnosing. Together with earlier chapters (Farah Schwartzman, Jerak-Zuiderent, and Bal), Tøndel and Gundhus' chapters convincingly show that those at the bottom of the counting hierarchy, the "counters" and the "counted," are not passive objects of governance, but active and acting subjects that count.

NOTES

1. Alternatively, we could use the term "co-construction" as do Oudshoorn and Pinch (2003).
2. For discussions on the history of statistics as a discipline and how statistics have disciplined scientific and administrative thinking over the years, see for instance Desrosières 1998, Hacking 1990, Porter 1995. For a discussion of how discourses more generally discipline thinking, see Foucault 1971.
3. As witness the title "The Power of Large Numbers" (Cole 2000), that book's argument on the social functions of statistics, or the many other deployments of nearly identical phrases on the Internet.
4. Throughout the book we have attempted to be consistent in two rather pesky linguistic differentiations. We use "State" when referring to abstract or general principles of nationhood or territorial governance and "state(s)" when referring to specific instances thereof. We use "statistics" as a singular when referring to an overall body of mathematical theories and practices and as a plural when referring to specific collections and presentations of numerical accounts of the world.
5. We use the phrase "knowledge system and set of practices" to include not only mathematical formulae, but also routines for gathering data, assessments of what phenomena can and should be measured and recorded, understandings of the meanings of these data, and practices for their distribution, interpretation, and application. Each of these facets of statistics has been shown to be responsive to political contingencies.
6. The financial crisis of fall 2008 may nudge the pendulum back towards state planning and regulation again, but the chapters in this book were for the most part written before October 2008.
7. We use the terms "governance" and "government" as more or less synonymous, but where government also evokes images of institutional arrangements and governing bodies, governance emphasizes the dynamics of their practices.

REFERENCES

Adam, B., U. Beck, and J. van Loon (2000) *The Risk Society and Beyond: Critical Issues for Social Theory*. London: Sage.

Akrich, M. (1992) "The De-Scription of Technical Objects," in *Shaping Technology, Building Society: Studies in Sociotechnical Change*, ed. W. Bijker and J. Law, 205–224. Cambridge, MA: MIT Press.

Anderson, M.J., and S.E. Fienberg (1999) *Who Counts? The Politics of Census-Taking in Contemporary America*. New York: Russell Sage Foundation.

Beck, U. (1992) *Risk Society. Towards a New Modernity*. London: Sage.

Becker, H. (2007) *Telling About Society*. Chicago, IL: University of Chicago Press.

Bijker, W.E. (1987) "The Social Construction of Bakelite: Toward a Theory of Invention," in *The Social Construction of Technological Systems*, eds. W. Bijker, T. Parke, and T. Pinch, 159–187. Cambridge, MA: MIT Press.

Bowden, G. (1995) "Coming of Age in STS," in *Handbook of Science and Technology Studies*, eds. S. Jasanoff, G. Markle, J. Petersen, and T. Pinch, 64–79. Thousand Oaks, CA: Sage Publications.

Bowker, G.C., and S.L. Star (1999) *Sorting Things Out: Classification and Its Consequences*. Cambridge, MA: MIT Press.

Callon, M. (1986) "Some Elements of a Sociology of Translation: Domestication of the Scallops od Fishermen of St Brieua Bay," in *Power, Action and Belief: A New Sociology of Knowledge*, ed. J. Law, 196–223. London: Routledge & Kegan Paul.

Cole, J. (2000) *The Power of Large Numbers: Population, Politics, and Gender in Nineteenth-Century France*. Ithaca, NY: Cornell University Press.

Cowan, R. S.(1987) "The Consumption Junction: A Proposal for Research Strategies in the Sociology of Technology," in *The Social Construction of Technological Systems: New Directions in the Sociology and History of Technology*, eds. W. Bijker, T.P. Hughes, and T. Pinch, 261–280. Cambridge, MA: MIT Press.

Dagbladet (2008) "Sikrere enn framtida," November 18, 2008, available at http://www.dagbladet.no/tekstarkiv/artikkel.php?id=5001080089322&tag=/tem.words=Wilhelm%3BTell%3Bbil%3BWilhelm+Tell (accessed March 30, 2010).

De Michelis, A., and Chantraine, A. (2003) "Memoirs of Eurostat: Fifty Years Serving Europe." Luxembourg: European Communities. http://epp.eurostat.cec.eu.int/pls/portal/docs/PAGE/PGP_ADM_FILES/PGE_ADM_FILES_DS/HISTORY.PDF (accessed March 30, 2010).

Desrosières, A. (1998) *The Politics of Large Numbers: A History of Statistical Reasoning* (Translated from French). Cambridge, MA: Harvard University Press.

Foucault, M. (1971) *L'Ordre du discourse*. Paris: Gallimard.

———. (1972) *The Archaeology of Knowledge & the Discourse on Language*. Tavistock.

———. (1991) "Governmentality," in *The Foucault Effect*, eds. G. Burchell et al., 87–104. Chicago, IL: Chicago University Press.

Gjøen, H., and M. Hård (1996) "Challenging the Engineering Script: The Mutual Appropriation of the Electric Vehicle and Its Drivers," Paper presented at the joint annual meeting of 4S and EASST, Bielefeld, October 1996.

Hacking, I. (1990) *The Taming of Chance*. Cambridge, UK: Cambridge Univeristy Press.

Haggerty, K.D. (2001) *Making Crime Count*. Toronto: University of Toronto Press.

Haraway, D. (1991) *Simians, Cyborgs, and Women: The Reinvention of Nature*. London: Free Association Books.

Igo, S.E. (2007) *The Averaged American: Surveys, Citizens, and the Making of a Mass Public*, Cambridge: Harvard Univeristy Press.
Latour, B. (1987) *Science in Action: How to Follow Scientists and Engineers through Society*. Cambridge, MA: Harvard University Press.
———. (1992) "A Note on Socio-Technical Graphs," *Social Studies of Science* 22(1): 33–57.
Lie, E., and H. Roll-Hansen (2001) *Faktisk talt. Statistikkens historie i Norge*, Oslo: Universitetsforlaget.
Lie, M., and K.H. Sørensen (1996) "Making Technology Our Own? Domesticating Technology into Everyday Life," in *Making technology our own? Domesticating technology into everyday life*, eds. M. Lie and K.H. Sørensen, 1–30. Oslo: Scandinavian University Press.
Lincoln, A. (1863) "The Gettysburg address" http://showcase.netins.net/web/creative/lincoln/speeches/gettysburg.htm (accessed January 18, 2010).
MacKenzie, D.A. (1981) *Statistics in Britain 1865–2930: The Social Construction of Scientific Knowledge*. Edinburgh, UK: Edinburgh University Press.
O'Malley, P. (2004) *Risk, Uncertainty and Government*. London: GlassHouse Press.
Oudshoorn, N., and T. Pinch, eds. (2003) *How Users Matter. The Co-Construction of Users and Technology*. Cambridge, MA: MIT Press.
Pinch, T.J., and W.E. Bijker (1987) "The Social Construction of Facts and Artifacts: Or How the Sociology of Science and the Sociology of Technology Might Benefit Each Other," in *The Social Construction of Technological Systems: New Directions in the Sociology and History of Technology*, eds. W.E. Bijker, T.P. Hughes, and T.J. Pinch, 17–50. Cambridge, MA: MIT Press.
Porter, T.M. (1986) *The Rise of Statistical Thinking 1820–1900*. Princeton, NJ: Princeton University Press.
———. (1995) *Trust in Numbers. The Pursuit of Objectivity in Science and Public Life*. Princeton, NJ: Princeton University Press.
Quotations Page, The. Quote Details: Albert Einstein: Not everything that can ... http://www.quotationspage.com/quote/26950.html (Accessed August 21, 2008).
Star, S.L. and J.R. Griesemer (1989) "Institutional Ecology, 'Translations' and Boundary Objects: Amateurs and Professionals in Berkeley's Museum of Vertebrate Zoology, 1907–39," *Social Studies of Science* 19(3): 387–420.
Silverstone, R., and E. Hirsch (1992) *Consuming Technologies: Media and Information in Domestic Spaces*. London: Routledge.
Skolbekken, J.-A. (1995) "The Risk Epidemic in Medical Journals," *Social Science in Medicine* 40: 291–305.
Thomassen, Ø. (1997) *Herlege Tider. Norsk fysisk planlegging ca. 1930–1965*, Dr. art. dissertation, Trondheim: Senter for teknologi og samfunn.
Waring, M. (1990) *If Women Counted: A New Feminist Economist*. San Francisco, CA: Harper Collins Publishers.

Part I
Overarching Themes and Approaches

1 Numbers
Their Relation to Power and Organization
Jon Hovland

Numbers and measurements are central devices in organizational planning and execution. The search for symbols of precision, accuracy, and accountability often leads to numbers. The bureaucratic model, with laws and regulations to be followed in both decisions and execution, implies the use of numbers to quantify organizational behavior. In New Public Management (NPM) and, more to the point, NPM tools such as Balanced Scorecard, proponents argue that the use of numbers is not merely a symbol, but the only way to both produce a secure coherence from goals through strategies to organizational outputs, and also to control the relation between resource input and production output (Carter 1989, 1991; Corporater.com; Mayston 1985; McNamara 2008).

In studying the relationship between numbers and governance, concepts such as organization and governing tend to come to the forefront. As social scientists, it intuitively makes more sense to study the social entity. Isolating numbers as object of study is a dangerous path that may lead to unrewarded determinism, as numbers are arguably social. Yet, this is what I do in this chapter. I first asked what happens when an organization observes itself and its relation entities through numbers. I also asked what happens when they communicate through the language of numbers: Does it matter, and if so, how?

However, for or against numbers is a debate that does not work, because numbers are symbols, and their meanings only exist through the processes they are part of. So I turned to the processes, and found numbers to dissolve into four fundamentally different debates (and corresponding dilemmas). Expanding on these made it possible to discuss numbers in relation to and also as representing power. Then it is possible to move forward to see these processes again as a whole. They are not a whole per se, but "use of numbers" is seen as one uniform process; through this fallacy we can understand how the use of numbers becomes part of processes of power (and feelings of powerlessness), and partakes in production of apparently self-supporting systems of decision and control that sometimes seem to make no sense at all.

The first section of the chapter expands on the meaning of numbers and organization, and important research on this. The second expands on the four different dilemmas I found, thereby providing names and language as a new framework for understanding the use of numbers in organizations, and how the uses of numbers relate to power. The last section is a short summary and conclusion.

THE GHOST OF NUMBERS: HOW NUMBERS ARE (NOT) SEEN, AND WHY IT'S ALL SO DIFFICULT

Looking into organizational use of numbers, it is plain to see that they produce anomalies. We all know examples of supposedly rational decisions based on objective facts (that is, those available) that to many concerned seem just silly—one spectacular example being how the Financial Accounting Standards Board arrived at the conclusion that 9/11 was not an extraordinary event, according to their accounting procedures (Liesman 2001)[1]. Another example could be the blindered focus on journal publications and similar accounting audits to measure research quality in universities (Fillitz 2000).

And yet, such measures are widespread and influential. This implies that we are observing a system of knowledge processes that are complex, powerful, and against which there is little relevant language and much feeling of powerlessness. Searching for the reason behind these decisions that seem so strange, we may find practices that should be named "bad statistics," where simple lecturing would seemingly make the difference. But this is odd, because those doing the calculations are generally far more skilled than this author—and still argue for pushing the limits of the methods toward a situation where precision and accuracy are words referring only to the numbers and not what they represent. This is bad statistics, but there are two ways of saying that: First, from inside the logic of statistical methodology, one would be able to find some judgments less based on mathematical proof than others, the handicraft is more or less well accomplished. Desrosières (2001) brings this kind of argument further, stating that a "realistic" approach toward statistics may be seen as three different approaches: statistics as pure measurement (metrological), as pragmatic accounting, or as database material for argumentative purposes. Alternatively, we could choose a situational rather than a methodological contextualization, asking whether the output is reasonable or not (reasonable as opposed to rational). Desrosières' accordingly suggests a fourth approach: [statistics as] "definition and coding of the measured variables are 'constructed,' conventional, and arrived at through negotiation" (ibid.: 340). The technician would argue that this kind of evaluation is part of the methodology, from the method-logic this is one of several validations of the accuracy of the analysis. From a context-logic viewpoint, it *is* the analysis, and so we have different terms and different room for thinking anew.

This leads to a different, and maybe even stranger anomaly: the observation that those officials most closely connected to processes of calculation and analysis of numbers also seem to be, if not the most critical, at least those with the strongest arguments against their own practice. One could expect the opposite, that they were merely technocrats, and that they rarely reflected upon their work as more than calculation and measurement; but my observation is the opposite. This is parallel to Latour's description of the Janus face of science; on the one hand, there is "science-in-the-making," and on the other, there is "ready-made science" (Latour 1987). Although science is presented as ready-made, packaged knowledge, the participants creating this science will still work on this knowledge as something in-the-making, with shortcomings that seriously weaken the knowledge in question. Hence it is fully possible for those managing and operating the numbers to be those best equipped and even willing to criticize the very numbers they operate.

All these anomalies tell us something about the role of numbers in organizations. They are blurry, contested, important, invisible, taken-for-granted, created, creating, compromise agreements. In short, they are very similar to what Star and Griesemer (1989) name "boundary objects." One basic feature of boundary objects is that they are observed from different viewpoints at the same time, represented by different kinds of membership orientation. For one, this creates anomalies, because different points of observation imply the possibility, even probability, of observing different objects. There is no single understanding of how numbers work, nor uniform understanding of the relationship between the different viewpoints. Further, this means that the debate we enter into when discussing the use of numbers is not a clarified one. So we are at the risk of being sidetracked, or even fragmented, into various questions not initially posed.

Organizations are, as most of the entities of the present analysis, problematic concepts. In the words of Eco (1990: 59): " ... in the natural sciences the conjecture has to try only the law, since the Result is under the eyes of everybody, while in textual interpretation only the discovery of a 'good' Law makes the Result acceptable." Terms such as organization, technology, and governing are all results of (con)textual interpretation, and are accordingly only acceptable when observed through a "good" perspective, theory, argument, or "Law." Following this, it is difficult to discern what is *organization* from what is *governing*, or for that matter what is *number* from what is *organization*. This may hamper arguments, but has not prevented scholars from approaching these kinds of relations, and attempting to name constituents and processes at some level of abstraction.

In his study of performance measurement systems, Power (2004) names them *counting*, *control*, and *calculation*; however, he does not follow up on this distinction. Rather, he fronts two basic processes involved in these practices: first, the inherent reductionism we find in counting practices; second, the relationship observed between measurement and technologies of monitoring and control. Again, the focus is on the observation of the relationship, and not so much on how there can be such a relationship. Hood's (2007) approach is to distinguish *targets, rankings,* and *intelligence*[2] as different ways of operating performance measurement. Here the focus is entirely on the different effects of these kinds of measuring practices. Written from a management point of view, he concludes that the more standardized "target" measures are more useful for creating transparency in organizations. This is a different description from Desrosières' (2006) distinction between quantification (in general) as, on the one hand, a process of comparing the incomparable and how this phenomenon has been debated, and, on the other hand, a process of simultaneous convention and measurement and how these two processes together make quantification possible. This distinction is traced back to the debates of Quetelet and Cournot. Lampland and Star (2009) take yet another tack, crafting a new umbrella concept of formalizing practices under which we can observe standardization, quantification and formal representation as different ways of designing social practices.

These studies provide important insights into the different ways of numbers and different roles of numbers in organizations, but not the big "Why." Why are numbers so useful in organizations, and why do they relate to power? Is this possible to answer, given the different meanings of organization we encounter through these analyses? This chapter concludes that this is possible, provided that we differentiate different debates (and corresponding dilemmas) concerning number practices. Through expanding on these dilemmas, I show how numbers can be described in relation to power and as representing power.

A MODEL FOR CRITIQUE OF MODELS

To better discuss practices concerning use of numbers, I suggest the terms *standardization, quantification, modeling,* and *accountable communication,* which I will expand on here. Standardization refers to the creation of formal norms or conventions for objectifying a process or object, rendering it possible to categorize and measure. Quantification refers to turning objects, standardized or not, into numbers. Modeling is the process of mediating reality into graspable mechanisms; concerning statistics, it is the process of using numbers to infer correlations and patterns of causality. Accountable communication refers to the phenomenon of personal communicative relationships being replaced with communicative relationships based on standards and numbers (e.g., through practices such as auditing). Each part is introduced with one or two examples to clarify the meaning of the concept.

This is an inquiry into the processes where numbers are used and about what it means to use numbers. Is it relevant, the semi-articulated feeling of power through numbers? I argue it is, and that to see it, the language employed has to separate these essentially different processes and debates, as is done here, to make it manageable.[3] I do not start at the notion of governance; I depart from the idea of numbers and statistics, and argue my way toward more observable kinds of governing.

Standardization—Power to See Through Creating Equals

Mari Storstein suffers from spinal muscular atrophy. Her muscles do not develop, so she is in a wheel chair and in need of assistance. Through a fairly recent reform, all Norwegians who receive governmental assistance are to be registered in an accounting system named IPLOS, where degree of functionality over a spectrum of parameters is scored from 1 to 5. The intention of this register is, according to the Norwegian Directorate of Health (Helsedirektoratet 2009), to promote transparency and control toward service units and to provide data for research. Some municipalities also experiment with using accumulated IPLOS data to estimate nursing load, for allocation purposes. Storstein was shocked to realize that she was also registered, that somewhere there was a number for her capability of defecation and intimate hygiene. Through her televised documentary (Storstein and Storstein 2007), we are allowed to follow her struggle to find out her scores. Finally, after two years, she can lift the closed envelope and ironically state: "This is my life." Tears swell up as she finds not only her scores for defecation, but that she has a low score on "socially acceptable clothing" and a bottom score on "eating in a culturally acceptable manner." These do not represent her self-perception. So starts the struggle to be erased from the register, or at least to adjust the scores she resists.[4]

If we are not happy with regarding the outer world as an infinite mass of undifferentiable experience, we will have to find ways of dividing it up into entities and putting things back together. Instead of a world of new and unanticipated experiences, we walk through a world of categories, based on our knowledge and those virtues in an object or process we choose to emphasize. If we agree on these categories and how to identify and evaluate them, the categories have turned into standards. Standardization is when we agree on how long a meter is, how to define production at the University, or how disabled a disabled person is.

Standardization opens up the boundaries of inside information, while running the risk of silencing essential tacit knowledge or even more explicit knowledge. Porter (1995) describes how relationships to powerful outsiders and reduced public trust accelerate processes of standardization. When a representative of an organization of knowledge and information engages in information exchange with other systems, trust, power, and relatedness are instructive to the degree of standardization he needs to follow.

Bowker and Star (1999) study what happens to the objects that are standardized and categorized, which is a slightly different perspective on the same phenomenon. How can we trace the process of individual phenomena, objects, or actors turning into one class, and what are the consequences? They argue that this is where it starts making sense to use the concept of "boundary objects," because these classified pieces of information can move between subsystems of the organization with less friction (ibid.: 152, 296–98).

Three central questions occur: First, why does it seem that standards are something we can trust in situations of mistrust? Second, is the power of the outsider that pushes for these standards one that is desirable? Third, in what way is power a part of this process of standardization in itself? From how we choose to answer these questions, we may derive the pros and cons of standardization.

There are many ways of approaching the trust in numbers; one that I will come back to through a discussion of Luhmann's theory of communicative systems (1995) is that it is not really a matter of trust, but of the possibility of communication. The more direct answer would be that it is actually a matter of trust in numbers' ability to create transparency between separate systems. This is a belief we find, for example, in Hood (2007). Standards, he claims, create ways of knowing that are disconnected from knowledge systems and inside information, and are thereby neutral and predictable. The problem with this lies in the notion of transparency[5]—a taken-for-granted concept that, examined more closely, contains an antagonism: What is it to be able to see through? From a point of view regarding the outside world as an objective reality ready to be reported on, this question is not relevant. However, from a more constructivist point of view it is probably necessary to contrast transparency with visibility. For a phenomenon or object that exists within a system of knowledge to be visible, it has to be viewed with reference to the relevant system; in another system, it is a different object. The transparency we assume in standardization applies an outside reference system, and the object or phenomena observed may or may not be the same as the one observed with reference to the logic of the system itself.

From this point of view, the influence of the powerful outsider becomes political. There is a potential of replacing visibility with a representation that serves other interests. On the other hand, it provides the possibility of reducing power inside systems that develop in undesired ways. An example of the latter would be audit systems that could have exposed the Enron scam before it was too late. An example of the first would be audit systems in academia that restrict the faculty's possibilities for independent research (Strathern 2000). Taking the former view, IPLOS' standards could penetrate local knowledge frontiers, so that an accountant unfamiliar with the singular case could state that Storstein was receiving too little nursing resources according to her deficiency score. In the latter view, Storstein's score undermines seeing her as a social, productive, critical person, perhaps more well-functioning than many unregistered citizens. By now it should be clear that this debate is very much a struggle concerning power. Standards are a way of overcoming monopolies of knowledge and truth. This may create possibilities for open communication, predictability, and visibility, or may create a form of control that disrupts positive potentials found within that very monopoly.

Quantification—Possibilities of Knowledge, Power of Convention

The Three Strikes Law in California (California Penal Code § 667, subdivisions (b)–(i)) has resulted in personal tragedies of Kafka-like disproportionality. Lampland (2009) provides us with a review of The People of the State of California v. Vernell Gillard, who was injured at work in 1993. While explaining at his first medical examination that he had been in a car accident six years earlier, he did not mention this at the insurance company's medical examination. As the doctor who did the initial examination failed to send relevant information to the insurer's doctor before the examination, the insurance company was not informed of this. Mr. Gillard was convicted on three counts of insurance fraud and one count of perjury, a felony which carries penalties ranging from one year in county jail up to five years in state prison. As Mr. Gillard had two prior felony convictions, he was, according to Section 667, sentenced to 25 years to life. He was released on parole after eight and a half years, after a trial judge dismissed one of his earlier convictions (Lampland 2009).

A Norwegian municipality decided to intensify initiatives against victimization in schools. One school counselor explained in a research interview how she decided which schools were advised to change their approaches to victimization: They summed up pupils' self-reported victimization, estimated the mean, and all schools above the mean were considered to have a problem and advised to review their practices.

Numbers work in mysterious ways; standardization is one aspect of how they move and also a presumption for accountable communication. When we are doing numbertalk we do more than this. One obvious part of numbertalk is quantification—essentially summing up classified objects, such as felony convictions or victimized pupils, into numbers. While standardized information may or may not be represented by a number, it definitely turns into numbers when quantified.

Quantification seems not so complex a process as the others discussed here. There are strong reasons to quantify, as it makes it possible to sum up large amounts of knowledge and observation into small packets of information that are more easily handled, which presumably is why ancient traders invented measures and even why revenue authorities invented numbers in the first place (Flegg 1983). And yet, this seemingly simple process also seems somehow a magical transformation. Indeed, measurement has been surrounded by superstition since ancient times: "He was the author of weights and measures, an innovation that changed the world of innocent and noble simplicity, in which people had hitherto lived without such systems, into one forever filled with dishonesty" (Flavius Josephus, cited in Kula 1986). Josephus is referring to Cain, pointing out the ancient roots of numbers' slightly dodgy reputation, a reputation maintained up to our day, for instance when a merchant is suspected of fixing his weights. I'll go as far as to say that this is inherent in our collective cultural knowledge; it's a spinal reflex. Assuming this, it is not difficult to understand why criticism of measurement systems so often is brushed aside as old-fashioned, as a reaction without reflection.

As Desrosières (2006) points out, quantification is more than measurement; measurement in itself is nothing. Quantification is applied convention, and measurement based on this convention. In other words, we can quantify anything that we can create a convention for. We can quantify love if we agree on how to measure it; we can quantify production in municipalities if we know what it is for a municipality to produce.

Now we see that quantification is not so simple after all. The immediate reaction toward measurements revolves around possibilities for and experiences with cheating. If that were the entire diagnosis, a more thorough audit would be the medicine. However, most of us are not quite sure what love is, and we do not really know what it is for a municipality to produce; this is a matter of debate and construction of a convention that was not there before. As soon as this convention is applied to a form of quantification through measurements we will necessarily observe a cementation of contested norms. This means that however open and auditable a process, the construction and stabilization of the convention is a matter of power to define norms in a very explicit way.

The measurement (as opposed to the convention) is a logic rather than a praxis. Yes, we may well argue that general measurement systems, like arithmetic, are social constructs, yet then this construction is part of the convention. A systems' logic is never right or wrong, it is always one out of many communicative systems. Following Luhmann's definition of social systems (1995), systems are found through their differentiating boundaries of communication. Through communication, systems define which communications are internal and which cross their borders. Systems make decisions accordingly, and are reproduced through this self-reference. One system of logic, based on one system of communicative codes (e.g., measurement) cannot communicate with other systems without translating into different codes or basic assumptions for arguments, thereby altering the content of what is communicated. My point is therefore that apart from implying a socially constructed convention that may be contested, quantification also induces a logic of measurement. As a logic, it cannot be contested on its own terms; it can only be contested through contrasting systems of logic, systems powerful enough to contest it.

In the example of Mr. Gillard it is obviously the legal system that is in the forefront; the judge's decision is based on whether the actions are legal or not, and which sections of the law that apply. Then a measurement with the specific benchmark (three - the number of offences) is applied and interferes with the legal system. In the municipal counselor's procedure, good school/bad school becomes a benchmark on a shifting scale of averages, where above is bad and below is good. The problems are obvious: Self-reported victimization may as well be an indicator of pupils' awareness as of their victimization; a school doing a good job of sensitizing children to their rights may score poorly. Furthermore, the benchmark seen in isolation denotes some level of victimization (anything less than average) as acceptable—a position the municipal counselor explicitly opposes elsewhere.

Modeling—Mediators for Understanding Complexity

The Bell Curve (Herrnstein and Murray 1994) struck American public debate hard, and became one of the America's best-selling popular science books of the 1990s (Wikipedia.org 2009). The volume concerns questions relating to education, welfare, and population groups. The researchers found a significant coherence between IQ and social status and welfare. Further, they found minority groups to generally score lower on IQ. Based on this, they concluded that public welfare policies were missing the point as IQ was decisive, and minority groups were worse off because they had low IQ.

The Norwegian Bureau of Technical Calculation for Municipal and County Administration Economy (TBU)[6] provides annual statistical reports on the state of affairs around the country. In some of the reports we also find regression models of the current relation between income and production (Figure 1.1).

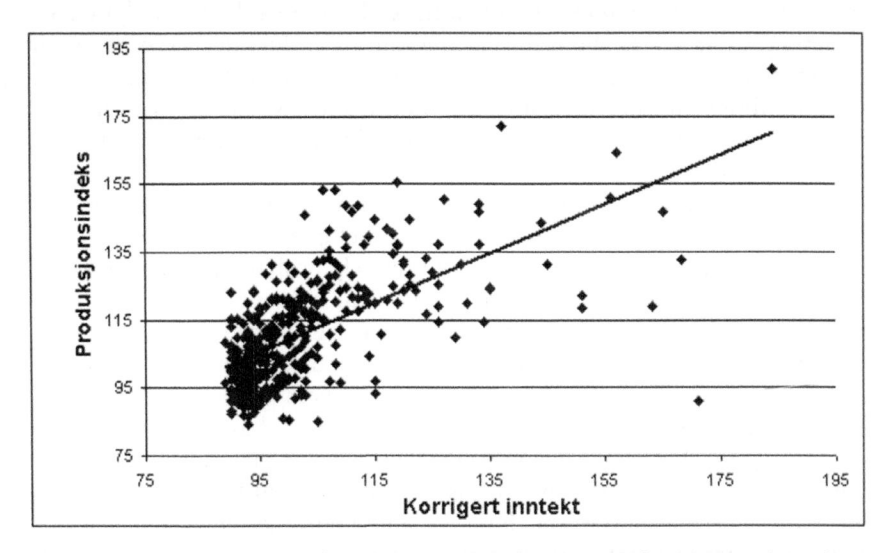

Figure 1.1 OLS regression from TBU (2005: 58), municipal income and production. 'Produksjonsindeks' means 'production index'; 'Korrigert inntekt' is 'income corrected for costs of capital and earmarked grants'.

The production index is a summary of different municipal activities, basically everything quantified. The regression model is used as an advisory scheme for municipalities to see if their production level is less than should be expected.

We always use models. Even the most surrealist artist will use models for understanding and communicating insights. The difference lies not in using models in themselves, but whether or not the models are a static or reflexive part of the process of understanding, something in which also the anomalies are of interest as showing something the model is not able to allow for. This chapter is most definitely based on a model. Models can be viewed very broadly, as autonomous mediators (Morrison and Morgan 1999) that consist of ordered objects, relations, and operations on these objects (Suppes 1961 in ibid.). It may also be worth describing models as idealized systems described as mechanical texts (Giere 1988 in ibid.). Following a phenomenological point of view, modeling is a way of organizing the world around us, a way of making sense and creating ideas of causality to be able to rationalize our actions and make decisions. Being a sociologist using models is maybe just as much human as it is scientific. As C. Wright Mills pointed out (1959), modeling through sociological terms is something everybody does to orient oneself in the everyday world. The science comes from the systematic use of sociological imagination and inquiry.

In the context of this chapter, we can discuss modeling in a broad sense or more specifically. The broad sense would be the way systems of performance measurement are based on different kinds of models. The specific point is made clearer when we look at the algorithms for inference analysis. When we are talking about use of statistics from outside the analytic construction, we talk less about the calculation itself, the brief moment in any report, analysis, or article that we may term the statistical moment. Usually we talk about and maybe criticize standardizations, the ideas of normal distribution, or how a given statistic is used, but rarely the actual inference that is carried out.

In numbertalk, modeling is about statistical inference and its logic of modeling. The magnificent advance of the development of statistical modeling is our increased ability to estimate precisely probability, causality, and explanatory power. The evolution of statistical inference is truly a revolution, and our ability to investigate (and create) relations between phenomena has reached a level one could only dream of a century ago.

Most inference models related to social numbers (be it in social science or administrative statistics) are based on some kind of analysis of variance statistics (e.g., ANOVA and other residual-based tests). There are many other branches of statistics, but this is one we meet regularly and that I will focus on. The one important thing we all (should) know about ANOVA analysis is that residual analysis is analysis of the relationship between values as the model predicts them and values as they are found in the data. In other words, it is a test of to what degree the values in the data fit into the model, as opposed to a test of the model's ability to describe the data (Hovland 2007).

This is also the main argument in Freedman's (1987) critique of path analysis, as advocated by Blalock, Duncan, and Blau (Blalock 1964; Blau and Duncan 1967; Duncan 1966). Abbott (2001) maintains that path analysis struck sociology hard in its day, and argues this was because these young mathematical sociologists claimed causality with a previously almost unheard-of self-assurance, backed with very complex methodology. As Blalock states in his preface (1964: p.vii), "Causal interpretations will undoubtedly be extremely useful if not necessary for a long time to come." But, however new and complex their methods were, the assumptions were not. Through these assumptions, Freedman (1987) deducts some sophisticated descriptions of the models of the world that are assumed if one accepts that path analysis provides pictures of causality. He concludes, to put it bluntly: It's crazy to believe path analysis displays causality.

Following this, it is not a very far stretch to state that ANOVA analysis is inherently conservative, as it only tests how closely the world as we observe it fits the ideas we already have. The principle is basically the same as when we apply theoretical models, but the mechanism becomes more evident and visible in these kinds of automated scripts.

This conservativeness may transform into a tool for governance. Evaluation of social and educational programs changed during the Bush II administration from including a wide spectrum of methodologies to a situation where, according to House, only randomized experimental designs were accepted as sources of truth (House 2006). House connects this with processes of fundamentalism, and following this, processes of control. The fundamentalism has to do with the anxiety that only one way of evaluating provides a real source of truth, and that the watershed of qualitative methods has diluted scientific authority and should be curbed. As we have noted, these kinds of experiments do not produce ideas; they test ideas. This is where we again encounter the matter of control and power, that is, control through maintaining scientific authority in the evaluation programs, but also through controlling the evaluations. As shown, models, such as randomized experimental designs, may more easily leave stakeholders outside, as the questions asked and experiments designed are (usually) molded by authority.

Returning to *The Bell Curve* (Herrnstein and Murray 1994), the main argument rests on the model of analysis. There are not likely to be problems with the data or the testing, but the model assumes one out of many possible causalities by defining IQ as an independent variable, which it arguably is not. The regression figure from the TBU report outlined earlier assumes that income is the independent variable toward production. But what is municipal production? The causality arrow points in an intuitively sensible direction, but other coherences may be just as important.

Both quantification and modeling are processes through which numbers are produced and mediated, but they differ both in the principles of the processing and the kinds of dilemmas we have to engage in and debate. Modeling is a mediator. The more powerful the mediator, the more evident its results, and the more the model (re-)produces itself. The great danger of self-reproducing results comes when the model is no longer seen as a part of the process and is taken for granted, whether this concerns a statistical or a theoretical model.

Accountable Communication—From Human (Dis)Trust to Numbers

How is the accounting made by the International Monetary Fund (IMF) made relevant in national accounting? The economy analysts at IMF have a "particular interest in the circumstances surrounding the emergence of financial imbalances [. . .], the policies to overcome such imbalances, and the corrective policy criteria for making loans. This involves going on missions to the country in question" (Harper 2000: 25). Harper provides a wonderful ethnography of how an IMF mission's work was carried through in a country nicknamed Arcadia: During the first weeks after arrival the team worked intensively at registering tables of economic activity, gathering knowledge of the economy, and also controlling them toward international standards and reviews. Central in this process was to "sort out facts from the facts," "build up a picture," and talk to local sources to work out what the figures "really meant." The turning point in the ethnography is where the IMF mission arranges an official meeting with the Central Bank Governor. Through this five hour long meeting, tables are presented, while the Central Bank officials run through and interpret the mission's tables. The conclusion of the meeting is that they all agree to these tables, with some modifications. If this had not been the case, the mission would have faced new weeks of investigations, calculations, and adjustments. As the accounting was sanctioned and verified by the Central Bank it would now be possible to carry on working with the numbers in question, to structure the kinds of suggestions IMF would forward to the country in question, and also to be prepared for what kinds of demands IMF would impose, were the country to ask for financial support. In the words of one of the Arcadians the mission had been working with: "We've been waiting for you [the mission] to come back and solve these policy dilemmas. You should come back more often" (ibid.: 50).

In one Norwegian municipality the administration recently had reorganized into a 2–3 level hierarchy, and introduced Balanced Scorecard software[7] as a governing tool. In a research interview, two chief officers explained why this was done:

> City Manager: Lines of communication are [now] short from bottom to top, [you may call it] freedom of information—part regulation, part

communication. Throughout the system towards leaders, the communication in that direction has been somehow problematic. Why do this? It concerns making it easier for the leaders to get hold of information, critical signals that are sent out by our employees.

Executive officer: With many parallel units, the span of control is much wider. You may say—it involves a whole other range of demands on leadership. Then you want this in person, each unit leader wants to speak with the chief officer, but the chief officer cannot discuss with fifty leaders, and then you need to find new ways to deal with such things. The flat organization is much harder to govern politically, you are provided with less political insight and governance.[8]

Accountable communication is comprised by processes whereby performance indicators, patient satisfaction queries, accounting practices and so forth are incorporated into administrative practices and routines so that they replace other relations as the link between (sub)systems, such as different divisions of an organization. This observation was in many ways the starting point of the study that inspired this chapter: What does it mean when municipal administrations embrace Balanced Scorecard-like tools into their organization? What kinds of processes does it derive from, and what does it result in? Is it about improved precision of efforts, and increased justice to those not otherwise heard? Or is it a matter of control? At least, if it is a matter of control, it is so deeply embedded in positions that it is problematic to speak of such at an individual level. It seems to be a matter of reducing responsibility and simultaneously staying in control.

The whole idea of these kinds of accountable communication rests upon a silenced debate and seeming consensus of standardization, with the arguments of transparency and predictability as central mantras. Also, accountable communication is necessarily tied up with processes of quantification and modeling, and thereby the debates related to these. In other words, we could have made accountable communication the object of this study, and presented it as the sum of standardization, quantification, and modeling. All these seem in themselves connected to statistics and society, and I have already shown how they are connected to power issues. However, we must be sure to distinguish the two contexts of scientific argument: in society and inside science (Desrosières 1997). In society, statistics is, in addition to the methods in use, about a mutual reliance on processes of power and governing. These are processes through which relations and structures in the organization are made formally reducible to agreed routines, tests, and actions—accountable communication. On the one hand, this is not necessarily about statistics in the sense of numbers on a page; it is a question of organizational structures that infuse those numbers with power. On the other hand, this is all about statistics, as accountable communication depends on standardization and trust in methods and vice versa: The power of the numbers is dependent on relationships taking on the form of accountable communication, and the rituals of verification (Power 1997) preceding and accompanying this form. Therefore we must observe and try to understand this as a separate phenomenon and debate, although it is one that is more difficult to capture. The IMF mission did not introduce accounting at the Arcadian Central Bank; numbers were obviously already at the core of their activity. IMF's representatives worked their way toward their own accounting of the national economy, and through an identifiable ritual of verification (which could have turned out otherwise—nonverification) this account was turned into a meeting point both parties could work on. The municipal example is not all that different. In the everyday work of a unit leader, such as a headmaster or nursing home director, numbers and accountings are clearly important. Their primary job however, is to manage a good school or nursing home, not to provide accountable figures. Through the Balanced Scorecard, the numbers become a medium of communication between unit leaders and the chief officer, an accountable communication through which the outcomes also should be expectable. If parents' satisfaction with the school drops below the mean, explanations and initiatives are expected, perhaps dependent on some ritual of verification.

I have defined this process of accountable communication as the replacement of personal relations with numbers. Of course this does not mean that human interaction is abolished, but that the measurements, models, and standards, or in Hood's (2007) terms, the targets, rankings, and intelligence, make up the axis of the interaction (e.g., between the unit leaders and the city manager or the manager's staff). We may talk about objects that operate in the interface between subsystems of the organization, and as previously argued, in this context it may make sense to talk of these as boundary objects. The question all along has been why the numbers have such a strong ability to occupy this position, where actors from different systems agree that these are neutral and possible to use (although one of the major points with boundary objects is that they are read through, observed, and created inside a subsystem). We could answer this very simply, by stating that they are connected with standards and thereby with transparency and predictability. And yet, "everybody knows" that statistics lie, that numbers deceive, and that measures only show the measurable part of the picture.

To understand this anomaly, we have to ask: What's in a system? Until now, this chapter has not explicitly discussed this. Talk of system has been implicitly linked toward formal, organizational systems and subsystems. We have also—implicitly—offered ideas inspired by Luhmann's concept of communicative codes. These are different ideas of system. Formal systems are observable activities in daily life, but the tools for observing them are not obvious. Luhmann's concept is an abstraction, a way of observing formal systems (but also of challenging their boundaries) as communicative fields, defined by the code of communication (Luhmann 1995, 2000;

Luhmann et al. 2007). These codes of communication create systems not made up by actors or members, but by communications, and by referring to its own code, they become self-referential and self-maintaining (autopoiesis). Examples could be politics (system) and power/not power (code); economy (system) and profitable/not (code); media (system) and newsworthy/not (code); law (system) and legal/illegal (code), and so forth. Within a system any statement will be considered through the code. If it is not possible to decide through the code, the statement is irrelevant for that system, but a different system might consider the same statement relevant.

These examples illustrate that this argument does not rest on the assumption that numbers are introduced for the first time, or even that we are essentially talking about numbers. But let's say we are, as this chapter is dedicated to numbers. If we accept that quantifications are made up of a convention and a measurement, and that the convention is a social construct while the measurement is a logic, then Luhmann's understanding of system becomes relevant. There are two simultaneous ways of defining this logic as a code: One way—measurable or not—concerns the relationship between the number and the phenomenon it is claimed to represent. The other—position on a numerical scale—concerns the size of that number. Both codes are attached to the numbers wherever they are observed. This is the twist: *If numbers operate as boundary objects between organizational subsystems, then by virtue of being numbers, they introduce communicative codes across these boundaries. Numbers are Trojan horses.*

THE CHICKEN AND THE EGG, AND OTHER CONCLUSIONS

What came first? The need to employ numbers or the power the numbers gained from being used? The question is as backward as the chicken/egg question. The chicken/egg question has a good Darwinist answer: The egg came first, laid as a mutation by a pre-chicken. The question is what's wrong. The right question is: What processes of variation and selection turned the pre-chicken to chicken? Similarly for numbers: The numbers gain power from their positions. The power of numbers makes them adequate for organizations in need of trust and visions of accountability. However, this power is not straightforward, it's not in the numbers but in their processes, and it is far more than one uniform process. Through this paper I have shown four fundamental practices of numbers, and how these represent dilemmas and power.

Setting the Standard

As we have seen, there is a very direct power in this, and the struggle over standards is very visible and important, both because it is important to imprint one's own outlook in what will appear in the standardized observation, but also because standards open a door into knowledge systems: The standards and the indicators are frequently a tool for those outside a system of knowledge, with an urge to control something on its inside.

Calling the Count

Quantification is representation through measurement. Measurement implies activating a convention for the measure, and utilizing a scale. Hence quantification is not a neutral practice; quantification is selecting one convention over others and also introducing the logic of numbers and scale to a phenomenon. "What gets measured gets done. What gets paid for gets done more," is one of many punch lines of "management guru" Tom Peters (1998: 284). To quantify is to activate power.

Limiting Reason

Modeling with numbers is not essentially different from other modeling: It is a mediator. Models provide ways of depicting causality and coherence, through mediating information about the world as observed, but the model is also dependent on the knowledge we already employ, this employment arguably being a form of power. Modeling through numbers is a mediator of those ideas and knowledges we have accumulated from standards and from quantifications.

Amplification of Thought

Simplifying to the absolute extreme: Systems of accountable communication, where numbers and counts replace other kinds of relations, produce organizations permeated by numbers on the boundaries of subsystems. However, if we see numbers as representatives of a communicative system (not as organizational subsystems), they are not on the boundaries, they define the system in themselves, and therefore the process of accountable communication (as defined here) is a way of colonizing subsystems with the logic of numbers and measures.

Is this really true? I think so, but there are a number of important questions to be answered: Do these structures of situational power work in these ways? Are actors and members of organizations inclined to be governed in these ways? Are these processes of organizational implementation of knowledge shaped in other ways than by the very act of counting?

Answers to these questions have tended to be slightly desperate, reflecting the feeling of debility that overwhelms participants in systems of counting, when ones only means of offloading frustration is through mocking humor. In the jokes and one-liners we see the stupid and silly manager or statistician. But they are neither silly nor stupid. What we really see is the (unrecognized?) observation of (unassailable?) power. Unreconizability and unassailability may be because there is no language through which to debate this apparatus of power—a situational power that seems to sail under one flag, but which at a closer look fragments into a circle of processes. I am certain it is possible to find better languages than the one I suggest here. What is important is to gain a language through which this object becomes visible, and that this is a language that makes the essential differences and connections visible, and also that expresses the strengths and important, constructive virtues of these processes. This way we can study numbers as bureaucratic tools, or as tools of New Public Management or liberal arts of governing, with essentially the same framework. It would not be surprising to find differences in what kinds of number/power relations are fundamental in different regimes. We would also be likely to find that the numbers are often presented as good ideas and tools, friends, yet that they sometimes enable us to build monsters of systematic stupidity and ruthlessness.

NOTES

1. This example is also referred to in Lampland and Star (2009), a volume abundant with wonderful examples of accounting and standardizing practices that, "silly" or not, do imply social consequences.
2. Targets are standards of performance or change in performance to be achieved over a period of time; rankings are data allowing comparison among a set of (constructedly) rival units; intelligence is background information (Hood 2007: 96).
3. This may also be read as a mechanism. We may argue that standardization is a presumption for accountable communication, while modeling and quantification are different processes related to processes of accountable communication. This way they are part of the same object of study, namely a general perspective on use of statistics and numerization in organization and power relations. I choose not to emphasize this for now, as it is not my intention to present a mechanism, a unitary model that would restrict the room for interpretation and negotiation, but rather to name a language through which it is possible debate these processes. Although I choose not to walk in that direction, I would like to thank Ellen Balka for pointing out to me the relationship between the concepts.
4. It is clear through this documentary that Storstein was subject to a number of injustices, morally, formally, and juridically. First of all, the general rule was that the scores were to be set in cooperation with the user where possible. In this and many similar cases, this was not done. Although the documentary did not invoke general debate, some debate has occurred and changes have followed. At the Ministry's homepages you will now find information and forms for "You who wish to see your IPLOS registration" and "You who wish to change your IPLOS registration" (Helsedirektoratet 2009b). Also, the general regulations have been revised in early 2009, including a stronger emphasis on user participation in registration (Helsedirektoratet 2009b). A whole other issue is whether this may redirect IPLOS from a standardization regime to a pastoral confession regime (Foucault et al. 2007 [1977–1978]).

5. Inspired by Jonathan Kahn's comments at the workshop "Statistics as a Boundary Object between Science and the State," Trondheim May 2007. For further discussion of the notion of transparency, see Hovland (2008).
6. TBU stands for Teknisk beregningsutvalg for kommunal og fylkeskommunal økonomi.
7. The main idea behind "Balanced Scorecards" is to provide an information grid where all activity, not only economy numbers, but also quality measurements in their different guises, is included. An example of Balanced Scorecard software can be found at www.corporater.com.
8. My translation.

REFERENCES

Abbott, A. (2001) *Time Matters: On Theory and Method.* Chicago, IL: University of Chicago Press.
Blalock, H.M. (1964) *Causal Inferences in Nonexperimental Research.* Chapel Hill, NC: University of North Carolina Press.
Blau, P.M., and O.D. Duncan (1967) *The American Occupational Structure.* New York: Wiley.
Bowker, G.C., and S.L. Star (1999) *Sorting Things Out: Classification and Its Consequences.* Cambridge, MA: MIT Press.
Carter, N. (1989) "Performance Indicators: 'Backseat Driving' or 'Hands Off' Control?" *Policy & Politics* 17(2): 131–139.
———. (1991) "Learning to Measure Performance: The Use of Indicators in Organizations," *Public Administration* 69(Spring): 85–101.
Corporater.com (undated) "Enterprise Performance Management Suite: Measuring Business Performance Just Got Easier," http://www.corporater.com/products/corporater-enterprise.html (accessed October 12, 2008).
Desrosières, A. (1997) "Du singulier au general. L'argument statistique entre la science et l'État," in *Cognition et information en societè*, ed. I. B. C. L. Thévenot, Paris: Éditions de l'École des Hautes Études en Sciences Sociales, Translated by Linda Sangolt.
———. (2001) "How Real Are Statistics? Four Possible Attitudes," *Social Research* 68(2): 339–355.
———. (2006) "From Cournot to Public Policy Evaluation: Paradoxes and Controversies involving Quantification," *Prisme* 4(7).
Duncan, O.D. (1966) "Path Analysis: Sociological Examples," *American Journal of Sociology* 72(1): 1–16.
Eco, U. (1990) *The Limits of Interpretation.* Bloomington, IN: Indiana University Press.
Fillitz, T. (2000) "Academia: Same Pressures, Same Conditions for Work?" in *Audit Cultures. Anthropological studies in accountability, ethics and the academy*, ed. M. Strathern, 236–255. London: Routledge.
Flegg, G. (1983) *Numbers: Their History and Meaning.* New York: Schocken Books.
Foucault, M., M. Senellart, F. Ewald, and A. Fontana (2007) *Security, Territory, Population: Lectures at the Collège de France, 1977–78.* Basingstoke, UK: Palgrave Macmillan.
Freedman, D.A. (1987) "As Others See Us: A Case Study in Path Analysis," *Journal of Educational Statistics* 12(2): 101–128.
Harper, R. (2000) "The Social Organization of the IMF's Mission Work," in *Audit Cultures: Anthropological Studies in Accountability, Ethics and the Academy*, ed. M. Strathern, 21–54. London: Routledge.

Helsedirektoratet (2009a) "Høringsuttalelser til revidert IPLOS-veileder (høring under behandling)," http://www.helsedirektoratet.no/publikasjoner/horinger/h_ringsuttalelser_til_revidert_iplos_veileder__h_ring_under_behandling__314394 (accessed March 11, 2009).
———. (2009b) "Til deg som ønsker å vite hva som står om deg i IPLOS-registeret," http://www.helsedirektoratet.no/publikasjoner/skjema/til_deg_som__nsker___vite_hva_som_st_r_om_deg_i_iplos_registeret__bokm_l_og_nynorsk__108504 (accessed March 11, 2009).
Herrnstein, R.J., and C.A. Murray (1994) *The Bell Curve: Intelligence and Class Structure in American Life*. New York: Free Press.
Hood, C. (2007) "Public Service Management by Numbers: Why Does it Vary? Where Has it Come From? What Are the Gaps and Puzzles?" *Public Money and Management* 27(2): 95–102.
House, E. (2006) "Qualitative Evaluation and Changing Social Policy," in *Qualitative Inquiry and the Conservative Challenge: Confronting Methodological Fundamentalism*, ed. N.K. Denzin and M.D. Giardina, 93–108. Walnut Creek, CA: Left Coast Press.
Hovland, J. (2007) "Et farlig redskap?" in *Kvantitativ analyse med SPSS*, eds. T.A. Eikemo and T.H. Clausen, 312–324. Trondheim: tapir Akademisk Forlag.
———. (2008) "Fashion Report from 4S Annual Meeting in Montreal: The Transgressing Use of Transparency," *EASST Review* 27(1): 5–7.
Kula, W. (1986) *Measures and Men*. Princeton, NJ: Princeton University Press.
Lampland, M. (2009) "The People of the State of California v. Vernell Gillard," in *Standards and Their Stories. How Quantifying, Classifying, and Formalizing Practices Shape Everyday Life*, eds. M. Lampland and S.L. Star, 115–118. London: Cornell University Press.
Lampland, M., and S.L. Star, eds. (2009) *Standards and Their Stories: How Quantifying, Classifying, and Formalizing Practices Shape Everyday Life*. London: Cornell University Press.
Latour, B. (1987) *Science in Action: How to Follow Scientists and Engineers through Society*. Milton Keynes, UK: Open University Press.
Liesman, S. (2001) "Accountants Say Attack Costs Aren't 'Extrordinary' Items," *The Wall Street Journal*, Oct 1, 2001 New York.
Luhmann, N. (1995) *Social Systems*. Stanford, CA: Stanford University Press.
———. (2000) *Sociale systemer: grundrids til en almen teori*. København: Hans Reitzel.
Luhmann, N., C. Madsen, D. Baecker, and B. Christensen (2007) *Indføring i systemteorien*. København: Unge Pædagoger.
Mayston, D.J. (1985) "Non-Profit Performance Indicators in the Public Sector," *Financial Accountability & Management* 1(1): 51–74.
McNamara, C. (2008) "Performance Management—Basic Concepts," http://www.managementhelp.org/perf_mng/perf_mng.htm (accessed October 12, 2008).
Mills, C.W. (1959) *The Sociological Imagination*. New York: Grove Press.
Morrison, M., and M.S. Morgan (1999) *Models as Mediators: Perspectives on Natural and Social Science*. New York: Cambridge University Press.
Peters, T.J. (1998) *Circle of Innovation: You Can't Shrink Your Way to Greatness*. New York: Knopf.
Porter, T.M. (1995) *Trust in Numbers: The Pursuit of Objectivity in Science and Public Life*. Princeton, NJ: Princeton University Press.
Power, M. (1997) *The Audit Society: Rituals of Verification*. Oxford, UK: Oxford University Press.
———. (2004) *The Risk Management of Everything: Rethinking the Politics of Uncertainty*. London: Demos.

Star, S.L., and J.R. Griesemer (1989) "Institutional Ecology, 'Translations' and Boundary Objects: Amateurs and Professionals in Berkeley's Museum of Vertebrate Zoology, 1907–39," *Social Studies of Science* 19(3): 387–420.

Storstein, A., and M. Storstein (2007) "Jakten på Sylvia B," Kristiansand: Kristiansand interkommunale AVsentral.

Strathern, M. (2000) *Audit Cultures: Anthropological Studies in Accountability, Ethics and the Academy.* London: Routledge.

TBU (2005) "Rapport fra Det tekniske beregningsutvalg for kommunal og fylkeskommunal økonomi, desember 2005," in K.-o. regionaldepartementet. (ed), Vol. Publikasjonskode H-2184.

Wikipedia.org "The Bell Curve," http://en.wikipedia.org/wiki/The_Bell_Curve (accessed May 12, 2009).

2 Words and Numbers
For a Sociology of the Statistical Argument
Alain Desrosières

"For Whom the Bell Curves"[1]: This witty title evokes a fetish object of statistics. The bell curve (also known as the "normal law") describes the way in which a number of mutually unrelated events, which nonetheless result from a *constant common cause,* can cumulate under the effect of the law of large numbers and can be represented by a frequency curve shaped like a bell. This curve was successively used in different sociotechnical configurations: the measurement errors of the astronomers Gauss and Laplace, the macrosocial regularities of Quetelet, Galton's aptitude scales. Formulated in 1738 by Abraham de Moivre as the limit of a binomial distribution of drawn lots (heads and tails game), it was first used in astronomy to describe the distribution of observation errors (thus its other name of Laplace-Gauss Law). Then, in the nineteenth century, Adolphe Quetelet realized that the bell curve fit the distribution of body heights measured in a population of conscripts and, more generally the diverse behaviors of human beings, leading to the projection of an *average individual* with stable and enduring attributes. Later, this property was again utilized by the English eugenicists, Francis Galton and Karl Pearson, in analyzing the hierarchy of human beings' abilities. It was they who, at the end of the nineteenth century, gave this curve its current name of "normal law."

Yet, playing on the image of the village bell with its regular pealing which regulates and coordinates the life of the community, this title with its double entendre reminds us that statistics is not only, as a branch of mathematics, a tool of proof, but that it is also a tool of governance, ordering and coordinating many social activities and serving as a guide for public action. As a general rule, the two aspects are handled by people of different specializations, whose backgrounds and interests are far apart. Thus, mathematicians develop the formalisms based on the probability theory and on inferential statistics, while the political scientist and sociologist are interested in the applications of statistics for public action, and there are even some who speak of "governing by indicators" (Lascoumes and Le Galès 2004). The two areas of interest are rarely dealt with jointly. The phrase "For Whom the Bell Curves" invites us to historicize and to sociologize the study of statistical tools, their technical aspects and their utilization as arguments. Often perceived by the social scientists as unwieldy (albeit indispensable), these tools are generally subcontracted to well-respected methodologists who are requested to deliver incontrovertible, reliable, ready-made algorithms.

The history of these methods and of these formalisms is intricately linked with the history of issues placed on the agenda for official decisions which themselves subsume: (1) ways of conceptualizing society and the economy, (2) modes of public action, and (3) different forms of statistics and of their treatment. Based on different examples, we shall address the question of how quantification and statistical algorithms contribute to shaping the social world into a variety of configurations and arrangements whose constituent parts are mutually complementary.

We shall draw on earlier works to present some of these configurations. First of all, at a macrohistoric level, we will sketch in the relation between five different conceptions of the role of the State and the statistics corresponding to each. Then we will "zoom in" on three ways of describing and dealing with the issue of poverty in England at the end of the nineteenth century. Finally, some connections between statistical tools, the type of argumentation and the nature of the problem concerned will be presented to illustrate the sampling survey, the exploratory statistics used by sociologists, and performance indicators, borrowed from business management and used in public policies. It is not the purpose of this chapter to present a complete model of the relation between statistical methods and algorithms on the one hand, and the social issues supported by these instruments on the other, but only to indicate with a few examples a possible way to bring these tools into public view, to open up their "black boxes" by historicizing them—but without relativizing, which sociology has been accused by some of trying to do. The examples will not be developed in detail; they are all taken from various works on the history and sociology of quantification.

FIVE FORMS OF INTERACTION BETWEEN STATE, MARKET, AND STATISTICS

The rationalization of public activities, which, beginning with Max Weber, has been presented as a function of modern states, is often characterized summarily by a few, supposedly univocal traits: the anonymization and standardization of management of the social sector, the development of bureaucracies, the growing importance of the role of technicians and engineers. Initially, rationalization was seen as coming from the outside, a contribution from a distinct "somewhere else," from science, from technology, then from the social sciences or from economics, as a phenomenon of progress. Its presumed history is thus, at best, linear, internalistic, cumulative, nonstratified. To historicize this development requires that we "re-endogenize" the discourse of rationalization along the lines opened up by the modern-day sociology of science. This implies not only reconstituting this history, taken in the traditional sense mentioned earlier, but also bringing to light its diversity, its contradictions, controversies, and points of fracture: for the history of the tools of rationalization, despite what the rationalizers sometimes claim, is as turbulent and nonlinear as the history of the ways of viewing society and that of the policies adopted to act upon it. These three themes can be perceived as being co-constructed within coherent and interrelated configurations. We propose this hypothesis in order to describe the relation between the history of statistics, that of economic theories and that of economic policies.

The history of economics has been punctuated since the eighteenth century by controversies about the relation between State and market; doctrines and policies, closely or less closely interrelated, succeeded each other. Their interactions are analyzed from the vantage point of the institutional ideas and practices associated with several stylized historic configurations: mercantilism, the planned economy, liberalism, the welfare state, Keynesianism, neoliberalism. Whatever the dominant configuration, statistical systems of observation have usually been presented as inexorable and practically univocal progress, having little connection with the evolution of the yet so many diverse theories and practices having to do with the role of the State in running or guiding the economy. The historiography of economic thought, specifically the works which deal with the reciprocal interactions between the State and the science of economics, pays little notice to the differences between modes of statistical description specific to the various forms that the relations between the State and the market have taken throughout history. In short, the history of economic policies and that of statistics are rarely presented, let alone analyzed, *jointly*.

The reason for this deficiency in economic historiography is simple. Historically, statistics was perceived to be an instrument, a subordinate methodology, a technical tool offering empirical validation for economic research and its political applications. According to this linear view of scientific progress and of its applications, statistics (understood both as a producer of information and as a mathematical method of analysis) can develop only if it keeps its autonomy among economic doctrines and practices. It is for this reason that history works dealing with economic thought and practice paid little attention to statistics, never seeing this oversight as being in any way a problem or a possible contradiction; nor did they consider it a subject worthy of specific historic treatment. By "statistics" we mean here the body of knowledge consisting of the registration and presentation of quantitative data and the algorithms for its analysis such as series, indicators, econometric models, and many other methods available nowadays in databanks and data-processing packages.

A lead for an analysis of the relation between statistics as a tool and its social and cognitive context is furnished by the history of the various conceptions of the State's role in economic affairs. Set out next are five typical configurations in simplified and stylized form (Table 2.1). They do not follow a historically chronological succession and they are not mutually exclusive; in fact, they intertwine in many concrete cases. Thus, they have been stylized only for the sake of proposing a grid for a differentiated interpretation of the history of the statistical tools used in each case. While the inception of each of these configurations will be dated, in the case of the last three it is obvious that their inception corresponds to a period of profound questioning about the role of the State subsequent to three dramatic world economic crises: respectively, the crisis at the end of the nineteenth century which gave rise to the welfare state, the crisis of the 1930s producing the Keynesian State, and the crisis at the end of the twentieth century, which saw the emergence of the neoliberal State.

1. *Direct intervention* occurs in a broad variety of settings ranging from mercantilism and Colbertism (seventeenth century) to the Socialist planned economy. The French-style *engineer State* is another of its modalities, whose statistics are comparable to those of a large enterprise organizing its departments, or those of an army unit managing logistics. Its key components are the demographic census and the flow of goods in physical quantities.
2. In contradistinction, the *classical liberal State* (end of the eighteenth century) keeps such intervention at a minimum and advocates the free operation of market forces. Its statistics, where they exist, aim at bringing the real markets into consonance with theory: complete and identical information for all stakeholders, especially in the matter of prices. The agricultural surveys which have been carried out in the United States for over a century now, represent one of its models. The dream of a Stateless "liberal and libertarian" society, founded solely on market mechanisms where prices embody all the information that is needed, is the symmetric counterpart of the dream of a perfect "engineer" State.
3. The *welfare state* (end of the nineteenth century) seeks to safeguard wage-earning workers against the consequences of an extension of the market philosophy to employment itself by establishing protection schemes for the family against unemployment, work accidents, and illness. It relies in particular on sampling surveys of the workers' employment, of their needs, incomes, and household budgets, and of consumer price indexes related to the workers' consumer spending. Those are the items on which the official statistics of this period concentrate, as shown by Lucien March in France, Ernst Engel in Germany, Caroll Wright in the United States, or Anders Kiær in Norway (Lie 2002).
4. *Keynesianism* assigns a responsibility to the State for the macroeconomic piloting of society whose mercantile nature it does not, however, challenge (1930s). Its main tool is national accounting (Vanoli

Table 2.1 The State, the Market, and Statistics

	Conceptualization of Society and of the Economy	Mode of Action	Forms of Statistics
Engineer State Production and people (since the seventeenth century)	Hierarchically structured institution, rationally organized France, from Colbert to De Gaulle; USSR.	Optimization under constraint. Reduction of costs. Planning. Technocracy. Major work projects. Long term vision.	Demography. Production in physical quantity. Input-output table. Material balance.
Liberal State Trade and prices (since the eighteenth century)	Physiocracy. An extensive market. Free competition	Fight against corporatism. Free-trade philosophy. Anti-trust laws protecting competition.	Statistics promoting market transparency: (e.g., American agriculture). Measurement of possible dominant position. Market shares.
Welfare State Waged work and its protection (since the end of the nineteenth century)	The labor market is not a market like any other, it has to be protected.	Laws on working hours, accidents, unemployment, and retirement benefits. Compulsory insurance systems ensuring social rights.	Labor statistics. Wage, employment, unemployment. Sampling surveys of workers' household budgets. Consumer price indexes.
Keynesian State Global demand and its components (since the 1940's)	The market cannot function on its own without generating crises. It must be regulated at a global level.	Supervising and managing the occasional gap between global supply and demand through monetary and budgetary policies.	National accounting. Analysis of the economic situation. Economic budgets.
Neoliberal State Polycentrism, incentives; Benchmarking (since the 1990's)	An extensive market. Free and undistorted competition Financialization Distributing the decision-making centers into a network.	Moving from rights to incentives: e.g., bonus-malus, polluting-rights market. Turning administrations into agencies. Contractualization Coordination by emulation: e.g., the European OMC	Objectification of new areas of equivalence Objectification of statistics. Construction and use of indicators to evaluate and classify performance. Benchmarking supplements, or replaces, directives and regulations.

2005) and public statistical systems are reorganized to that end. Here, the figures on consumption and the price index to quantify inflation apply to *the entire population* and no longer only to blue-collar workers. Macroeconomic models, like those of Ragnar Frisch, Jan Tinbergen, or Lawrence Klein, guide policies dealing with aggregates by comparing global supply and demand.

5. Finally, the *neoliberal State* relies on microeconomic dynamics, leading them if need be in a certain direction with the help of incentive systems and by accepting the main hypotheses of the rational expectations theory (1980s). Benchmarking, that is the evaluation, the classification, and the list of best performance, is one of its key instruments. The microeconometric models of logistic regression make it possible to separate and to isolate the "specific effects" of variables or tools used in public activities affecting their performance, and thus to improve the "target variables" of policies which are conceived in terms of incentives (especially fiscal) and of the behavior of individuals. The emulation concomitant to the use of different methods makes it possible to identify "best practices."

The evaluation of incentive-based procedures is obtained from individual data studies or from quasi experiments (microsimulations) intended to modelize the behavior of stakeholders, *including that of public authorities*. This represents an important difference between the neoliberal State and all the others mentioned previously; it follows from the rational expectations theory, which posits that the failure of public policies can be attributed to the fact that people integrate into the information which guides their behavior the effects they expect these policies to have. Seen in this way, none of the players, least of all the State, remains outside the game. The State is in fact subdivided into a number of more or less autonomous decision-making centers, or "agencies" which are run almost like a business enterprise; these are players among others, and lend themselves to the same form of modelization as any other microeconomic players.

The idea of endogenizing the construction of the statistical tool according to the historical description of different forms of State is consistent with what precedes, even if this reflexive historicization of statistical practice is not part of the toolbox of theoreticians of neoclassical economics. Actually, in a *realistic* conception, statistics are merely a measuring tool, external to a preexisting "reality," just as the State, criticized by the partisans of the rational expectations theory, is in their view external to society. However, since the production of statistical information is an essential factor in running the economy, it is not surprising that the multiplication of "governing centers" and the endogenization of their activities is accompanied by a comparable proliferation and endogenization of computational centers which produce statistical data. This data *is not "a given"*; it is the outcome of a costly social process of proliferation, whose cognitive and economic components are part and parcel of the complex global society that they are supposed to describe.

A CASE STUDY: POVERTY IN ENGLAND AT THE END OF THE NINETEENTH CENTURY

However, the foregoing macrohistoric review covering the relationship between State, market, and statistics over more than two centuries is misleading in that it suggests a too perfect coherence. On a more subtle historical scale, innovations and transformations can be seen as resulting from particular contingencies and events specific to a given country or even to small groups of individuals who often embarked on projects of which the ultimate result turned out to be completely different from what they had expected. A case in point is the example of the English eugenicists, Francis Galton and Karl Pearson. The initial context is the great economic and social crisis of the 1880s. The utter poverty which was then the lot of the working class provoked urban riots, causing great concern among the English bourgeoisie. Various reform movements proposed their interpretation of the situation and tried out different ways to react (Table 2.2).

The first took its inspiration from the Darwinian theories of the biological transmission of "abilities" and from the conviction that it was necessary to "ameliorate the overall ability of the nation" by the use of eugenic methods for the selection of the fittest, from which Galton's and Pearson's statistical innovations emerged. The second reformist movement focused on the observation and description of the living conditions of the poor by means of social surveys (Charles Booth, Seebohm Rowntree, Arthur Bowley), forerunners of the modern sampling survey. It was based on a finely differentiated categorization of groups of the population, linked to different forms of intervention. Finally, the third theory was an outgrowth in the wake of the Poor Law of 1835, with workhouses and a local assistance system by "Poor Law Unions" (Yule 1895). The points at issue then centered on how these assistance unions were to function and on the relative share that was to be given to home assistance (*outdoor relief*) as opposed to workhouses (*indoor relief*).

A comparison of these three policy approaches allows one to see beyond the misleading apparent structural consistency of "epistemes" described in broad strokes, without the contingent details of their workings. For in fact, the grave crisis at the time produced two major competing ways of viewing society, of describing and modelizing it, and of acting upon it: one is biological, the other socio–economic.

The philosophical, political, and technical controversies of this period produced some of the most valuable tools of modern statistics. For instance, the "bell curve" of Gauss, Laplace and Quetelet was completely

Table 2.2 Poverty in England at the End of the Nineteenth Century: Three Policies and Their (Three) Instruments

Types of Policy	Representative Authors	Nature and Source of Information	Technical Tools	Social Philosophy
Eugenicist Hereditarianist	Francis Galton Karl Pearson, 1880 to 1900	Measurement of biological traits (height), then of abilities	*Bell curve* (normal law) Correlation Regression towards the mean	Darwinism Eugenicism Selection of the fittest
Policies targeted according to subtle taxonomy	Charles Booth, Seebohm Rowntree, Arthur Bowley, 1885 to 1900	Social surveys classifying the poor in eight economic / moral categories	Differentiated explanation and treatment of poverty according to an *ad hoc* taxonomy	Distinction between the "deserving" poor, who can be helped, and the "unworthy," beyond salvation
Workhouse and assistance (based on Poor Law of 1835)	Udny Yule 1895–1899	Management statistics of the 580 local unions of assistance to the poor, decomposed into *indoor relief* and *outdoor relief*	Contingency table Regression Adjustment by the least squares method.	Assistance maintains and increases poverty (Polanyi 1944)

reinterpreted by Galton. Similarly, the use of administrative sources (the assistance unions), the techniques of social surveys and of sampling surveys, and a great variety of statistical taxonomy were all debated and applied in those decades in England.

Two political and cognitive constructs competed for the conceptualization of the social crisis of that time. One was the response of Galton and Pearson, drawing on biology, heredity, and eugenics. Though it enjoyed great popularity into the 1940s, this conceptualization became discredited. Postulating a "normal" distribution of supposedly hereditary abilities, Galton argued for the naturalization of the structure of social classes, perceived as a one-dimensional scale reflecting innate abilities. He interpreted the bell curve in terms of a hierarchical distribution of biological traits and hereditary abilities, and no longer merely as stable averages, as Quetelet had done. He invented the notions of median and fractiles (deciles, percentiles, etc.). Noting that the "upper classes" were less fertile than the lower classes, the eugenicists thought there was a risk that the "global ability" of the nation would diminish. A "eugenic" policy would therefore mean limiting the fertility of the poor. In order to defend these ideas, Pearson formulated the concepts of correlation, of regression and the chi-square test. Thus, though discredited, this configuration nevertheless laid the foundations for mathematical statistics subsequently taken up into inferential statistics by Ronald Fisher, Jerzy Neyman, and Egon Pearson (the son of Karl), and into econometrics by Ragnar Frisch and Trygve Haavelmo (Morgan 1990).

The second conceptualization of the crisis was social and economic. It is less homogeneous than the former. For one thing, some social reformers attempted to formulate a detailed typology of the various strata of the poor classes, including in each case a description, an explanation, and a proposal for action (Hennock 1976, 1987). In-depth social surveys were carried out by Booth and Rowntree, which were to be the mainspring for the initial sampling surveys (Arthur Bowley in England, Anders Kiær in Norway, Alexandre Kovalevski in USSR). The method of random sampling rests on a conventional equivalence between the balls picked out from Bernouilli's urn; it was consistent with democratic egalitarianism among the citizens of the nation, and with the idea that social protection in the future would cover not only the working class, but all classes of society (Kiær 1895).

Furthermore, the old system of assistance based on the Poor Law of 1835 was still in effect and was still a subject of debate. Managed by the local unions established in 580 counties, it separated home assistance to women, children, and elders, or *outdoor relief*, from assistance to able-bodied men, obliged to work in workhouses, or *indoor relief*. The relative weight of these two forms of assistance was a controversial issue in the 1890s. Udny Yule, who was trained by Pearson but felt allergic to his biological vision of the social world, intervened in this debate by analyzing the management statistics of the 580 unions of assistance. To this end, he imported the regression method from Pearson's biometrics laboratory, and "explained" the variations in pauperism by variations in the way management was carried out. In 1897, in order to estimate a line of regression, he was the first to reemploy the method of *least squares adjustment*, formulated around 1800 by the astronomers Legendre, Gauss, and Laplace. With this transfer from biometrics, Yule anticipated what in the 1930s would come to be known as econometrics. Basically, Yule contended with questionable statistical arguments that outdoor relief contributed to maintain poverty, an idea current throughout English social history since the Speenhamland laws (1795) and until today, as witness the current disputes about activating social benefits and about *workfare*.

The three configurations summed up in Table 2.2 do not constitute a coherent whole; they are deeply marked by their own irreducible contingencies and histories and are linked to very different political philosophies, which cannot be summed up by referring simply to advancing rationalization or social engineering. The domain of techniques is not univocal nor is it unaffected by the passions of the social world. Still, certain styles of reasoning go hand in hand with certain formalisms. Linear regression, for one, with its "explanatory variables" and its "explained variables," introduced by Yule in connection with social policy in 1895, was to play an essential role in guiding and evaluating the *modus operandi* of public policies by the use econometric models in the twentieth century.

FROM MONOGRAPHS TO SAMPLING SURVEYS: TWO SOCIOTECHNICAL CONFIGURATIONS

Another example of the co-construction of statistical tools and social policy is provided by the history of socio-economic surveys of family budgets. These surveys, in use since the beginning of the nineteenth century, underwent radical changes around the middle of the twentieth century. Before, they concerned only the poorer sections of the population, mainly the working class. The purpose was to analyze the conditions for the renewal of the labor force; the idea of representativity was not addressed: the families under inquiry were judged to be "typical." Le Play and his disciples provided a model of "monographic" surveys which was widely used until the 1900s. Engel's well-known "laws" on the connection between food consumption ratios and income levels were based on the same premise. They were used until the 1930s to weight the price index of worker families' consumption in order to avoid their standard of living being negatively affected by inflation. However, it never occurred to a researcher at that time to investigate the budgets of well-to-do middle-class families.

The nature and the purpose of these surveys changed completely as of the 1940s, under the joint effect of the birth of the welfare state and, following that, of the Keynesian macroeconomic policies. From then on, surveys addressed the totality of the population in order to describe inequalities between classes, on the one hand, and, on the other, to quantify global consumption. It was in this context that the method of representative sampling appeared and progressively replaced the monographic method. This turn of events is well illustrated by the ground-breaking project of the Norwegian Anders Kiær who, in 1895, proposed to carry out a "representative enumeration" based on a "purposive selection" (but not yet "random") covering "all classes of society." He presented his project, closely associating technical, social, and political arguments, before the ISI, the International Statistical Institute, in French. The justifications he gave from the outset for his survey are indicative of the transition from a period in which relations between classes were still seen in terms of rank and relative place, and were thus incommensurable, to another period in which individuals from different classes could be compared by a common yardstick, a period in which the topic of inequality—inconceivable under the older system—became fundamental, and in which problems of poverty were no longer thought of in terms of charity and good-neighborliness but in terms of social laws enacted by parliaments. Kiær pointed out that the previous type of surveys dealt only with workers (or the poor) since it was not yet conceivable to equate the different classes within a bigger whole. He was thus one of the first to raise the problem of "social inequalities," using these very terms. It is worthy of note that this is referred to at the beginning of an account, the first by a government statistician to deal with representativity, in 1895:

> [D]etailed surveys on income, dwelling-places, and other economic or social conditions conducted in regard to the working classes have not been extended in an analogous manner to all classes of society. ... Even if we restrict consideration to the so-called working-class question alone, we should compare the workers' economic, social, and moral situation with that of the middle- and affluent-classes. In a country where the upper classes are very wealthy and the middle classes quite well-off, the demands of the working classes relative to their salaries and housing should be measured by a different gauge than in a country (or region) where a majority of those who belong to the upper classes are not rich and where the middle classes are in financial difficulties. It follows from this proposition that, in order to properly assess the circumstances of the working class, one must in addition be informed on the same aspects for the other classes. But we must go a step further and state that, since society does not consist only of the working class, social surveys must not leave out any class of society (Kiær 1895: 177).

Immediately following, Kiær explains that the survey would be useful in creating a fund for retirement and social security, promoting social equity and a statistical treatment of various risks:

> A representative census has been underway in our country, the goal of which is to elucidate various issues related to the planned creation of a general fund providing for retirement and insurance for disability and old age. This census is being conducted under the patronage of a parliamentary committee whose task it is to examine these questions, and of which I am a member (ibid.: 177).

In 1897 the debate hinged on what his "representative method" offered as compared with the "typological method" then recommended at the ISI by LePlaysian statisticians. Kiær emphasized the coverage aspect, referring to a picture of the total territory in miniature which would show not only types but also "a variety of real-life cases." He did not yet mention random selection but insisted on the verification of results by means of general statistics:

> I find that the ISI terminology, "procedures in typological studies," is not consistent with my ideas. I shall have occasion to demonstrate

the difference that exists between investigations by types and representative investigations. By representative investigations I mean a partial exploration in which observations are made over a large number of scattered localities, distributed throughout the land in such a way that the ensemble of the localities must not be chosen arbitrarily, but according to a rational grouping based on the general results of statistics; the individual bulletins we use must be arranged in such a way that the results can be checked in several respects with the help of general statistics (Kiær 1897: 180).

By confronting his method which allowed a "variety of cases" to be described with the method presenting only "typical cases," Kiær operates a shift parallel to the one that Galton and Pearson had wrought in relation to Quetelet's old statistics of averages. When they started focusing attention on the variability of individual cases by using the concepts of variance, correlation, and regression, the English eugenicists shifted statistics from a study of wholes, expressed by averages (holism), to an analysis of the distribution of individuals to be compared:

The ISI has recommended investigation by means of selected types. Without contesting the usefulness of this form of partial investigation, I think it presents certain disadvantages compared to representative investigations. Even if one knows the proportions of the different types that are part of the total, one is far from being able to reach a plausible overall result since the total includes not only the types—that is, the average ratios—but the full range of cases that occur in real life. Therefore, in order for a partial investigation to be a true miniature representation of the whole, it is necessary that we observe not only types, but all and every kind of occurrence. And this is what can be done, albeit not completely, with the help of a good, representative method that neglects neither the types nor the variations (ibid.: 181).

LOGISTIC REGRESSION OR CORRESPONDENCE ANALYSIS: TWO STATISTICAL APPROACHES[2]

The interaction between formalisms and algorithms, on the one hand, and their social utilization, on the other, is illustrated by the controversies among quantitative social science experts in the years between the 1970s and 1990s. Some, mostly economists, used the method of logistic regression which stems from econometrics. Others, mostly sociologists, used correspondence analysis (Greenacre and Blasius 1994).

Logistic regression is an extension of the old idea of "eliminating structural effects." Stated facetiously, "a variable can hide another." The problem had been addressed by Yule, through multiple regression and calculations of partial correlation, at the beginning of the century. A difficulty arises when the variables in question are *discrete*—that is, made up of categories and not of continuous quantification. The models of logistic regression (of the *logit* type) allow the use of formulas devised by econometricians. But in so doing, the reasoning is that of the natural scientist, as exemplified by Ronald Fisher in his agronomical experiments, who identifies the pure "effects" of variables which act *in a homogeneous way in the entire area under consideration*. The idea underlying this way of dealing with social variables, i.e., that laws and their effects are transposable and reproducible so long as the condition of *ceteris paribus* is respected, comes from the natural sciences.

It is not our intention here to criticize this, as has often been done by those who attempt to historicize the tools of social science. "To historicize" means to examine in a given historic context the consistency, formal and social, and the effectiveness of an assembly of definitions, tables, graphs, and calculations. Such assemblies can be understood only if we insert them into a broader panorama of reasoning and action, rather than considering them only as purveyors of additional information—one more brick in the edifice of science. The elimination of structural effects was scoffed at by Simiand who quipped: "This method leads to studying and comparing the behavior of a reindeer in the Sahara with the behavior of a camel at the North Pole." This wisecrack was taken up by those who criticized the transposition of natural science models into the social sciences. Yet, the elimination of structural effects has been considerably refined since 1980 by the use of logistic regression models (*logit*) which make it possible to separate and to subtly quantify the "pure effects" of various "explanatory" variables. How does logistic regression fit into a longer chain of argumentation in which presumably reasoning, action, and decision (and not only description) occupy a central position? The logistic regression model relies on "discrete" variables which break up the world completely into distinct, separate classes. The variables hold the main role in this play: they are the ones that act, the ones that produce effects, be they pure or blurred by the effects of competing variables. In descriptive accounts they are *the subjects of the verbs*, and are thus related to the language of the natural sciences. The classes themselves are not expressive; that is left to the variables—gender, age, education, income, class of society, region, or size of the municipality. Those who are the most attracted by the models of the natural sciences, like Karl Pearson with his biometrics, are bothered by these discontinuous variables. Age and income could, at a pinch, be brought back into the camp of "true" (i.e., continuous) variables, but the others remain under suspicion of being arbitrary, "a matter of convention," and "if the nomenclature is changed, then what?"

But the core question concerning these methods remains that of the effects that some variables have on others. This interrogation is meaningful solely from the viewpoint of *action* to change the world: What needs to be acted upon to achieve a given goal? The variable then represents a goal (a social indicator) or a means to take action *general in scope*. Variables *are made to be part of the instrument panel of men of action*. Social science is an applied experimental science, but it has to accommodate the categories that states of law have produced throughout history (administrative, wage, school, familial, fiscal categories), which differ from one country to another, thus complicating the construction of unified European statistics. For this reason, the criticism leveled at these methods falls short of its target and has had no effect; it attacked only their cognitive aspects instead of dealing with the way they are *utilized* and their *social effects*.

Correspondence analysis grew out of the factoral analysis of the psychometricians, who, at the beginning of the twentieth century, followed a procedure typical of the symptomatic metrology of the natural sciences. Charles Spearman's general intelligence (or factor "g") was a latent variable, "a mean" of n school tests taken by p pupils. It was determined as the main axis of inertia in a cloud of p points representing the pupils' performance within the space with n dimensions of the test. The unidimensionality of this cloud was subsequently criticized by Léon Thurstone who tried to explore orthogonal axes in order to more faithfully describe the complexity of the topic of abilities. Without computers, the psychometricians acquired great dexterity in performing "axis rotations" in multi-dimensional spaces. Benzecri's correspondence analysis, for its part, uses contingency tables, totalizing the rows and columns. The aim is to find the most explanatory axes in such a table, and to project row and column variables on planes, thus presenting an optimum cartography of the contents of the table.

Although both are descriptive, the French method of *correspondence analysis* and the Anglo-Saxon method—John Tukey's *data analysis*—differ. The latter presents a separate *exploratory* analysis, by means of which the examination and viewing of a file makes it possible to formulate hypotheses or to outline probabilistic models to be subsequently tested by a *corroborating analysis* based on the traditional technique of inferential mathematical statistics. Correspondence analysis, on the contrary, rejects probabilistic models; it is first and foremost a descriptive technique, and is not designed to confirm or reject a previously formulated theory.

Dealing as it does with contingency tables in which individuals are distributed according to conventional classifications, correspondence analysis is well suited to a social science concept which evolved from law and political science. The classes are distributed according to proximity systems having similar property configurations. Here, the actors in the scenario are *groups* (or *individuals*) and no longer *variables*. The groups, defined by gender, age, or social class, are the subjects of the statements in the descriptive text. They have an autonomous existence in the face of an exhaustive nomenclature (in contradistinction to logistic regression). These methods can be used for classification *a posteriori* by grouping individuals (ascending direction) or by cutting up the original entity (descending), having first defined a distance, minimized within classes and maximized between them. Statistical analysis thus creates new *conventional equivalence* formulas which can be used for action.

Logistic regression and correspondence analysis are two of the most used statistical methods today; however, they are not interchangeable, as their languages and utilization are so different. They are resorted to in distinct institutional contexts which makes a sociological comparison of their application difficult. The products of logistic regression are presented as *results of causes linked to effects* involving decontextualized variables, supposedly general in scope, similar to the way experimental sciences unfold the steps of their investigations.

Contrariwise, correspondence analysis is rarely presented as a preliminary to a "corroborating analysis" meant to verify theoretical hypotheses of which it might be one of the sources. It is rather one element among others in a multitude of historical accounts of the complexity and dimensions of the social world. The "variables" do not appear *as such*, but do so through the classes that they distinguish. It is the specific configurations of these classes and of their characteristics that become the subject of the sociologist's discourse.

Should a generalization be arrived at, it would proceed from a rhetoric different from that of the natural sciences. The two-dimensional structure of the domain of French social categories, according to Pierre Bourdieu's analysis (1987), has been confirmed by various correspondence analyses of the differences between these categories: consumption inventories, cultural activities, spatial distribution in urban neighborhoods, intermarriage, electoral behavior. These configurations are historic in that they depend on more or less well-established also historic taxonomies, and on practices whose significance is evolving.

These differences in utilization reflect the relative fragmentation of the social sciences which derive their legitimacy from a patchwork of science models. The social sciences would do well to make this mix and its sociological implications *explicit* by merging the variety of discourse with the various social practices, instead of trying to make one or another of these models prevail. History shows that such purportedly epistemological disputes usually remain inconclusive, for each of the models has its own specific social utility.

IS BENCHMARKING "SOLUBLE IN ECONOMETRICS"?

In the last thirty years, the uses made of statistical products have evolved in two different directions, amenable to a sociological description. On the one hand, business companies first, and later administrations, evolved new management methods based on *performance indicators and benchmarking* which are intended to coordinate and guide the behavior of stakeholders by offering them quantifiable criteria. On the other hand, the combined evolution of microeconomic theory, of econometric tools (in particular based on individual data), and of information technology which had become readily available to all gave an impetus to research which made it possible to test and evaluate, *ex ante* and *ex post*, sector-oriented policies. Though these two types of application may occur in association, yet, whether one deplores it or not, they are most often employed by different stakeholders, using a different rhetoric which it is important to analyze.

When it comes to determining the adequacy of public policies or their effectiveness to satisfy the needs of the users, such accounting criteria as "market share" or profitability do not apply as they do for commercial activities. The traditional concept of public service presupposes a strong commitment on the part of the members of the service who are linked by the rules of hierarchical subordination. France and Germany have long been examples of this. But since the 1980s, this civic sense of public servants has come to be seen as no longer sufficient to ensure a democratic and efficient control of activities financed by public funds. Quantified indicators were introduced, their role comparable to analytical accountancy in a business enterprise with its profit-and-loss accounts and balance sheets. The indicators are not solely monetary, since public activities (schools, public health, security) cannot be expressed in the familiar equivalence of currency. The European states and the European Union are expending considerable effort to negotiate and construct *new areas of equivalence* by agreeing upon procedures to quantify their objectives and the means available for action. To cite two examples: the French *organic law relating to financial laws* (LOLF) and the European *open method of coordination* (OMC). While the historical and political context of these two instruments for directing public policies are different, they have in common the fact that they assign a key role to *statistical indicators*. They are rarely referred to in public debates even though they provide the setting and the language which frames and structures such discussions.

The LOLF provides a new way of conceiving the state budget: it is no longer based only on allocations but also takes into account the desired objectives. This requires that the objectives be made explicit and be *quantified* so as to enable Parliament to verify whether they have been fulfilled and to evaluate the efficiency of the official departments. The need to quantify the objectives and the means to achieve them seems self-evident if Parliament is to play its constitutional role, that of ratifying the budget and monitoring its execution. However, this implies a considerable amount of work, to objectify and to create an equivalence between disparate activities. These activities must be described, identified, named, discussed, qualified, compared, classified, evaluated. It is never easy to choose the most appropriate indicator. The existing social and institutional order must suddenly be described and made explicit. This can be done, but the people concerned must really feel truly motivated. Often, the very concept of quantitative indicators arouses reticence. The procedures often lead to "comparing apples and pears." Sometimes they seem absurd, the more so as those doing the work become more deeply involved in the task. The mere fact of categorizing, which in theory is meant to simplify the world and to make it easier to understand, modifies and turns the world into something different. Having adopted a different reference system, the players are no longer the same as before since their behavior is now guided by indicators and classifications which become criteria for activities and for their evaluation.

In theory, the LOLF is intended to enable Parliament to become informed on, and to evaluate, the activities of public services, so as to establish a better balance between the legislative and the executive branches. The fact that this implied conceiving and implementing large numbers of indicators, however, did not alert commentators; they took it to be a technical matter, to be solved by technicians. And yet, increasingly pointed discussions show that this stage in the quantification process (*the actual act of quantifying*) is decisive for what is to follow and yet the issues mentioned earlier were never looked into from a general perspective. Problems and perverse effects appeared here and there, and became the subject of objections or mockery. For instance, the police and the forces responsible for road traffic safety had both chosen the positive blood alcohol test as a performance indicator. But while the former wished to measure the effectiveness of their activities by an increase of the ratio ("the higher the number of people we catch, the better the deterrent"), the other services aimed at reducing the ratio ("the lower the number of people we have caught, the better the deterrent has functioned"). There's logic on both sides! Similar effects were observed in other, though different, contexts. The centralized planning of the ex-Socialist countries failed because it was impossible to set reliable indicators for the Plan's success—this due to the perverse effects that, by retroaction, these indicators induced on the behavior of stakeholders. Indicators and classifications are simultaneously both constraints and resources which, by their very existence, change the world. Furthermore, these management tools are ambiguous and polysemic; they turn up in various areas under a partially different presentation and to fulfill partly different functions.

The open method of coordination (OMC) is used by the European Union to harmonize the social policies (employment, education, social assistance) which are outside the economic and monetary domains. The first example of this was the *European Employment Strategy* (Amsterdam

Summit, 1997). The principle of the OMC (Lisbon Summit, 2000) is that the states define for themselves at an intergovernmental level common goals expressed in terms of quantified indicators, indicators which they will then use, like a prize list, for purposes of classifying and evaluating their performance. While the benchmarking results are, in theory, only indicative, the mere fact that they are published constitutes a strong incentive to orient national policies in the direction outlined during the summits. For example, an employment rate of 70 percent was fixed as a goal in Lisbon in 2000. Thus, the LOLF and the OMC confer a key role to statistical indicators: the former for the presentation and the monitoring of the State budget, the latter for the piloting of European policies.

Although it is essential, the procedure whereby the European Union member states reach agreement on the quantification process is not well known. The work is in two stages; the political authorities select the indicators and describe them in a succinct narrative. Then, they send a quantification order to statisticians at Eurostat (the European Union's statistical office) and in the national statistical institutes, leaving it to them to take care of "details" such as, e.g., the precise definition of the concept of *employment rates*, or *income available to households*. In view of the institutional differences between countries, the statisticians cannot avoid a certain degree of vagueness in some of the important specifications of the survey procedures, of coding and of quantification (Nivière 2005) and are not able to harmonize them fully. This method is said to be "open" because it is not mandatory as it allows states to adapt it to their institutional particularities, for instance by choosing to use either direct surveys or administrative registers as the source of information.

The indicators so produced are loosely focused, i.e., not exhaustively defined. Thus, they can be utilized in different areas which previously were not in contact but which can now be compared. Ordinary language is similar in this respect: speakers do not spend their time defining their words and making their scope explicit and that is what makes communication possible. The products of public statistics, e.g., rate of unemployment, price index, GDP, are in the same case. A complete and explicit account of the way they are constructed and of their content would weaken their effectiveness as arguments not only by revealing inherent conventions or approximations of which users are not aware, but also by jeopardizing the succinctness of the discussion and of the reasoning in which these statistical arguments are put forward.

In general, all of this remains implicit, except when a controversy arises. However, this lack of precision rightfully shocks professionals who consider it important to define and standardize their products rigorously. They are caught between two contradictory requirements. On the one hand, like good engineers, they want to specify their procedures. But on the other, the negotiation procedure leads them to tolerate certain compromises without which it would be impossible to provide the indicators required for benchmarking. The balance that they consequently attempt to maintain between these two requirements has hardly ever been formalized.

The instruments of New Public Management (NPM), whose distinctive features have been highlighted in the case of LOLF and OMC, were imported into the public sector from the entrepreneurial world. The production and use of these quantitative indicators are partly beyond the expertise of both statisticians and economists. Upstream, the sources can be varied or downright disparate: administrative registers, opinion polls. Overall consistency is not ensured. For instance, in the case of percentages, the numerator and the denominator may come from different sources. It is true that, fifty years earlier, similar criticism could have been leveled at the tables in national accounts which, too, combined a variety of sources. Nevertheless, the need for consistency at the time was fulfilled by the accounting balance between resources and their utilization per economic agent, per operation. Crosschecking and verification were possible in what was in fact an interdependent system. The inventors of macroeconometric modelization, Ragnar Frisch and Jan Tinbergen, based themselves on this consistency to construct a set of equations which associated the *a priori* constraints of accounting with observed statistical regularities.

It is more difficult to detect consistency factors in the case of the NPM indicators (Armatte 2005). Upstream, the technicalities of sampling methods or of *data editing*, involved in the re-treatment of administrative files, are often bypassed by resorting to the rapid, continuous-flow production of large numbers of indicators. Furthermore, the supposed incentive effect of performance indicators automatically causes their quantification to depart from the metrological objectivity which is in principle the statistician's ideal. The fact that indicators are directly implicated in the evaluation of people's performance leads to a perennial interest in an "indicator policy," with the consequent priority concern to constantly refine and improve it. Effects such as this were among the main reasons for the failure of the Socialist countries' planning, which was based on quantified objectives.

Another problem raised by the indicator policy is the long list of indicators that are proposed. This poses problems and is constantly questioned. To begin with, only a few, "very synthetic" indicators (e.g., concerning employment) are proposed. Then, critics object: "This is far too simplistic; bearing in mind the diversity of sources (e.g., registers and surveys) what we need is a whole battery of indicators." To which counter-critics retort: "We can't do anything with such a multitude of numbers. Give us an indicator which sums up the problem!" This recurring dialogue shows how important it would be to pay attention to *the economy of statistical argumentation*, to diversifying the types of interaction and the registers of the rhetoric used.

The development of this "indicator culture" at times clashes with the culture of econometricians, who are used to treating data files for other purposes, in particular to try to modelize the relation between them, while indicators on the other hand are set out in lists, longer or shorter and more coherent or less, depending on the stage reached in the typical dialogue mentioned earlier. For one thing, econometricians feel uncomfortable faced with what appears to them to be an odd assortment. But in addition, the producers of indicators may be tempted to fill the gaps in their lists by means of an estimate of the missing data, using *proxies* created by econometric models. An interesting example of this is provided by the estimates of variables covering *small land property units*. Based on sampling surveys initially carried out on samples too small to be useful, they were supplemented by models using national regularities, typically French "Jacobinism" (high-handedness) applied to statistics. Such a procedure makes it difficult to obtain a direct evaluation of a given local policy, and, more generally, to "benchmark" the innovative efforts of the authorities of local communities. This example is an interesting case of the interaction between statistical methodology and sociopolitical culture: Under what conditions is benchmarking "soluble in econometrics"?

FOR A SOCIOLOGY OF PUBLIC STATISTICS: TWO POSSIBLE WAYS TO GO, AMONG OTHERS

Statistics can be historicized from a longer or a shorter perspective. We have proceeded first, from a macrohistoric perspective, by presenting in stylized form five types of statehood, and then, by showing various examples of more singular or local issues. In conclusion, we would like to suggest two possible lines of research to pursue these two directions. One deals with the contradictory effect of the growing reliance on quantitative indicators in the implementation of public policies. The other examines the recurrent tension in the field of public statistics (in France at any rate) between statisticians, national accountants, and econometricians.

The massive use of indicators by New Public Management has had paradoxical effects. By distributing these tools widely and by presenting them as practically self-evident and not to be questioned, these have been taken outside the exclusive purview of professionals—statisticians and economists—who had previously been solely in charge of their production and interpretation. Upstream, the construction of indicators is distributed among individuals who often have a vested interest in their definition and evaluation. Information technology makes it possible to set up gigantic *data warehouses* which can be explored by the technique of *data mining*. Downstream, their implementation is not always integrated in models which would make it possible to verify their coherence and reliability. In parallel, some professionals continue to apply sophisticated quantification techniques (complex sampling, data editing) and analysis (econometrics, data analysis). The fact that indicators occupy a growing place in public management begins to be well observed and commented on by political scientists. Nevertheless, statisticians and economists are often uncomfortable with this abundance which is partly beyond their control.

Furthermore, other points of friction arise among these professionals, caused by the confrontation of three partly distinct traditions, respectively: the statistical tradition *(stricto sensu)*, the accounting tradition, and the econometric tradition. We can compare the type of realism that characterizes each of them.

The first, that of the *true statistician* steeped in the probabilistic tradition, originates from the error theory of eighteenth century astronomy. A gap in reliability is presented in probabilistic terms. This metrology was transferred to the social sciences, in particular through the sampling method. The statistical units are homogeneous; the distributions of the variables under consideration are not too far removed from the normal, and the law of large numbers can be applied. The central idea in this transfer is that, as in the case of the distribution of astronomic observations, the calculated items (averages, variances, correlations) have a substance which *reflects an underlying macrosocial reality*, which these calculations reveal. This is the kernel of metrological realism.

Accounting realism is completely different. Accounting is by its very nature, in monetary terms, an aggregate of heterogeneous elements, some of which can be measured with certainty (e.g., the liquidities, at least where the currency is stable and reliable), while others are tinged with a degree of subjective probability. The choice of these values is guided by (possibly contradictory) considerations of caution and of communication clarity. The realism of the whole, supported by the provisional definition of the variables and by their inscription in well-ordered tables, is pragmatic rather than obeying rules of metrology. In point of fact, these two orders of realism are combined, and this combination is at the heart of a procedure for the construction and utilization of numerical data which is different from that of statisticians. National accountants have to some extent inherited this double-entry bookkeeping tradition with its constraints of balance, per agent and per operation.

Finally the econometrician's realism is yet of another type. In this case, the technical and social divisions of labor between the production of statistics and their utilization have had social and technical consequences. The databank is a black box for which the upstream and the downstream elements can be clearly distinguished, provided the upstream fulfils quality standards which are nowadays ever more explicit and trustworthy. The trust placed by the user in the data production phase is a precondition for the social effectiveness of the statistical argument. To the extent that this is the case, the proof of reality is shown, downstream, by the coherence of the results, of the constructs, and of the models whose source is the databank.

NOTES

1. "For Whom the Bell Curves" was the title of a Norwegian Research Council-funded project which hosted the workshop "Statistics as a Boundary Object between Science and the State" at which this chapter was first presented in English as a keynote address. Many thanks to Dina Levias for retranslating from a French version.
2. This paragraph is an adaptation of a part of an article published in French: Desrosières 2001.

REFERENCES

Armatte, M. (2005) Eléments pour une histoire sociale des indicateurs statistiques. Communication aux Journées "Estadistica y Sociedad", Madrid, UNED-INE-EHESS.
Bourdieu, P. (1987) *Distinction: A Social Critique of the Judgement of Taste*. Cambridge, MA: Harvard University Press.
Desrosières, A. (2001) "Entre réalisme métrologique et conventions d'équivalence: les ambiguïtés de la sociologie quantitative », *Genèses*, 43: 112–127.
Galton, F. (1909) *Essays in Eugenics*. London: Eugenics Education Society.
Greenacre, M., and Blasius J., eds. (1994) *Correspondence Analysis in the Social Sciences*. London: Academic Press.
Hennock, E.P. (1976) "Poverty and Social Theory in England: The Experience of the 1880s," *Social History*, I: 67–91.
———. (1987) "The Measurement of Poverty: From the Metropolis to the Nation, 1880–1920," *Economic History Review*, XL(2): 208–227.
Kiær, A. (1895) "Observations et expériences concernant les dénombrements representatifs," *Bulletin de l'IIS*, Vol. 9: 176–178.
———. (1897) "Sur les méthodes représentatives ou typologiques appliquées à la statistique," *Bulletin de l'IIS*, Vol. 11: 180–185.
Lascoumes, P., and Le Galés P., eds. (2004) *Gouverner par les instruments*. Paris: Presses de Sciences-Po.
Lie, E. (2002) "The Rise and Fall of Sampling Methods in Norway, 1875–1906," *Science in Context*, N°3: 385–409.
Morgan, M. (1990) *The History of Econometric Ideas*. Cambridge, UK: Cambridge University Press.
Nivière, D. (2005) "Négocier une statistique européenne: le cas de la pauvreté," *Genèses*, 58: 28–47.
Polanyi, K. (1944) *The Great Transformation: The Political and Economic Origins of Our Time*. Boston, MA: Beacon Press.
Spearman, C. (1904) "General Intelligence Objectively Determined and Measured," *American Journal of Psychology*, 15: 201–293.
Vanoli, A. (2005) *A History of National Accounting*. Amsterdam: IOS Press.

Yule, U. (1895) "On the Correlation of Total Pauperism with Proportion of Out-Relief, I: All Ages," *Economic Journal*, 5: 603–611.

3 Sociology in the Making
Statistics as a Mediator between the Social Sciences, Practice, and the State

Christopher Kullenberg

SOCIAL SCIENCES, STATISTICS, AND THE STATE

What is society composed of? This has been a question for sociology at least since the eighteenth century. This chapter aims at tracking down a few specific inquiries to what *modern* societies are composed of. These forms of knowledge are situated in the intersection of what Foucault calls the epistemological and the philosophical fields (Foucault 1970), a tension that may have been overlooked as science studies traditionally has focused on the hard sciences. While the social sciences have debated their philosophical foundations in bursts, during certain historical episodes—highlighting topics such as the nature of social life, the relationship between individual, group, society, etc.—they have also necessarily been forced to render empirical facts and measurements knowledgeable in an epistemological field. This field may have certain boundaries, limitations, and possible ways of describing society, but it must hold together in certain shapes and configurations, otherwise the social sciences would appear as nothing but mere speculation.

In this chapter I will draw attention to a certain quantitative sociology that prevailed in postwar Sweden, which relies on statistics as its main research instrument simultaneously as it constructs theories of man and society. The tension between epistemology and philosophy in the modern episteme, I will argue, is mediated by statistics through a process of quantification where "man," "society," "values," and "behavior" are turned into numbers. Following this process from categories and theories to large numbers and facts is in itself an interesting topic, but statistics and the process of quantification also convey a parallel story where the State and the social sciences are bound together. It could be argued that this relationship is self-evident, as witness the etymology of *statistics*, which from the Latin words *statisticum collegium* came to designate the State sciences in the late eighteenth century, or the English *political arithmetic*. However, etymological statements do not describe the processes through which the social sciences and the State are coproduced. For this we must do a sociology of quantification (Desrosières 1998)—in this instance a sociology of the relationship between quantitative sociology and the State in Sweden from 1946, when sociology became an independent discipline, to the 1990s. In order to show two qualitatively different modes of coproduction I will follow quantitative sociology through two examples: first, the structural functionalism of the 1940s and 1950s, where *social engineering* was the figurative shape of sociology, and second, a contemporary research institute called the SOM Institute, which was founded in 1986 and, because of different historical circumstances, operates through a mode of *epistemic authority*. As these parallel stories are opened up, connections between the genealogies of sociological traditions, the Swedish welfare state, and statistics start to emerge, and will hopefully contribute to an understanding of both scientific inquiry and the constitution of modern society through the use of statistics.

THE COPRODUCTION OF STATISTICS AND THE STATE

There is a large amount of research in the field of statistics in history, where the link to the State has been analyzed. Alain Desrosières puts statistics in a central position in *unifying* and *administrating*, especially in producing norms and standards for practical management of the functions of the State (Desrosiéres 1998: 147ff). This type of constitutive coproduction has a down-to-earth practical dimension, since it stresses statistics as a tool for knowledge about society. But I will argue that statistics has even deeper consequences. The social sciences also provide a unified cartography of the State in submitting the "ultimate matter constituted by the human population" to the "spatiotemporal framework of the State" (Deleuze and Guattari 2004: 429), thus impacting social order. This way we have to look at a broader picture than thinking of the statistical sciences merely as utilities for the rise of the Swedish welfare state. The combination of the census and statistics provide concrete strategies for solving social problems, but even more important, it promotes the construction of *social bodies* (Horn 2004), which constitute the contents of the geography[1] of the social. Patriarca (1996) argues that the rise of statistical thinking in nineteenth-century Italy was a parallel process with the rise of the modern Italian nation. She shows that through a centralized survey the numbers produced by statisticians in the newly founded national surveys produced a *power of association*. By measuring the whole Italian peninsula, the barriers between the northern and southern provinces could be torn down, thus creating associative links through numbers. As holistic investigations replaced the old regional ones, a unified map of the new nation emerged, thus securing territory through statistical spatialities. Horn (2004) also takes a turn on Italy, arguing that the foundation of Italy's Central Institute of Statistics in 1926 was intimately intertwined with the practice of government in the Mussolini regime. Using the concept *social bodies* Horn conceptualizes how the "social" was created, not as an ontological demarcation from the natural sciences, but rather as a landscape where certain problematics would emerge, such as consumption, infertile marriage, or crime. Henceforth we are thus treating statistical knowledge as constitutive both in the sense of it being a practical knowledge for State intervention, as well as providing and producing a spatiotemporal framework for the experience of populations, nations, classes, and social problems. Statistical knowledge on society makes it *whole*, or rather, as Desrosières (1991) puts it, statistics do the work of *holding together* knowledge, practice, and the State.

THE SOCIOLOGY OF STATISTICS AND THE SOCIAL SCIENCES

In the field of Science and Technology Studies (STS), the natural sciences have been seen as coproduced with society at large, but what about the social sciences? Are they not essentially "soft" sciences, saturated with politics from the very beginning? And what about the use of statistics? In one respect it would be unproblematic to think of it as an ordering device for the State, if we stick to calling it political arithmetic. However, the uses of statistics have changed dramatically in another respect: Statistics has become, at least in some forms of practice, the epistemic flagship of the modern sciences—be it in biology, physics, informatics, or sociology. Even though there is a commonsense version of criticism in the saying "anything can be proved with statistics," it would probably be an understatement to say that statistics in the modern social sciences conveys the connotations "scientific" and "objective" (Porter 1995). In an ideal form, which never exists purely in practice, although it has bearing on legwork research, statistics qualifies to be a *Royal science*, which is metric, ideal, and submits the world to constant laws (Deleuze and Guattari 2004). However, the constructivist approach requires another argument founded more securely in practice, and this is where I turn to the notion of constitutive coproduction (Jasanoff 2004), which puts scientific knowledge at the very center of social order. Statistics represents science not only by symbolizing aspects such as traditions, methods, and results, but also by standing as the material representative of institutions as "centres of calculation" (Latour 1987). Furthermore, statistics can be seen as influential in political decision making (Fridjónsdóttir 1991) as well as in inscribing perceptions of social order affecting our everyday conception of what "society" is really assembled from. And statistics is constitutive in the sense that statistical social science really requires to be made, as "legwork" research.

However, there is more to the social sciences than numbers and charts. They also contain "qualitative" categories describing the nature of man, society, language, emotions, and behavior. These are never given by statistics alone, even though they may turn out differently when quantified. In order to achieve a constructivist approach on the foundations of such categories, Foucault's notion of episteme needs to be accompanied with a few concepts from science studies. Thus, we turn to Michel Callon's notion of an actor network.

Callon (1987) argues, rather provocatively, that engineers developing an electrical car in France in the 1970s were actually better sociologists than sociologists themselves, since they were doing social analysis as a matter of "life and death" as part of their engineering project, rather than having to defend their ideas in an academic arena. In order for the engineer-sociologists to succeed, their sociological assumptions on consumer markets, social stratification, and lifestyles needed to be absolutely correct. Otherwise their technological project would fail and all investments would be in vain. In short, engineers are in Callon's view *reconstructing* society rather than merely describing and defending an interpretation of it. But even more important is Callon's notion of the actor network. Actors in this network are not limited to humans as stable entities, but instead are derived from a free association of any performative elements adding up to social reality—a method elaborated elsewhere as *generalized symmetry* (Callon 1986). This argument, transferred to quantitative sociology and statistics, enables us to look more closely into the social shaping of the social sciences. What we are looking at then, is not so much the sociologist's description of social reality, but rather how sociologists *reconstruct* social reality in order to get work done, by shaping their research instruments. By studying the uses of certain methods and techniques, we are able to give a description of the relations between the social sciences as academic disciplines, and the modes of knowledge which render possible certain social and historical contexts.

QUANTIFYING THE SWEDISH WELFARE STATE IN SOCIOLOGICAL KNOWLEDGE

Swedish sociology dates back to 1903, when the first chair in economics and sociology was established in Gothenburg. However, it took until 1947 for the subject to become an independent academic discipline, even though sociological questions had earlier been discussed in the subject of moral philosophy (Fridjónsdóttir 1987). It was heavily influenced by the American flavor of social theory, and had a strong empirical focus (Fridjónsdóttir 1991), steering the subject toward quantitative methods and a logical positivist philosophy of science. During the 1960s (Eyerman and Jamison 1992), the central components of contemporary quantitative sociology were imported: the survey method, scales and tests for measuring attitudes, and the overall theory of structural functionalism.

In the early 1940s there were no systematic, detailed statistics of Swedish society, apart from census data and sporadic studies.[2] As the government authorities ordered an expansion of the social sciences, the social scientists responded with social theory influenced by the American social survey, which was described as being able to map the structure and dynamics of social phenomena through the use of quantitative methods. In a government white paper[3] from 1946 the following quote clearly displays the relations envisioned between the Swedish State and the social sciences:

> It should consequently be stressed, that there is a significant correlation between a democratic form of government and the social sciences. It is not a coincidence that this science has flourished in two of the oldest democratic and constitutional nations, England and the United States. [. . .] [T]he social sciences are an important integrating factor in the democratic social order (SOU 1946: 74, p. 18, my translation).[4]

The social sciences thus were launched with a blessing from government agencies, the perception being that they were able to describe and solve social problems. In the 1946 white paper, statistics is also described as an important social science in its own right, one in need of development, one expected to play a crucial role in such diverse areas as economy, political science, sociology, and social policy research. Moreover, this new science promised to promote the shaping of the democratic social order. The specific importance of statistics is highlighted even further in a white paper written two years earlier:

> The methodology of statistical science is fundamental to all social scientific work, and naturally indispensable to the question of our population patterns. [. . .] The development of these methods [of statistics] is of direct value to the activity of the state (SOU 1944: 19, p. 5, my translation).

When investigating the possible uses of the social sciences, statistics works by holding them together under a unified methodology. Moreover, it is perceived to be indispensable when empirical questions, such as population patterns, are to be made knowable, and finally solved. By refining these methods (and what is at stake here is of course the future funding of statistical disciplines at the Swedish universities), a direct value for the state is anticipated in return, and this direct value is to be centralized because of its importance:

> Finally, the distribution of funds, which are to be made available for social scientific research by the state, will be centralized because of the significance of this type of research (SOU 1944: 19 p. 6, my translation).

Katrin Fridjónsdóttir argues in a historical overview of Swedish sociology that between 1947 and the early 1970s there was a strong tendency toward *social engineering* even in the self-image of sociologists, where the results of the surveys were more or less directly addressed to the authorities

(Fridjónsdóttir 1987). American sociology seemed to fit adequately into what the government agencies desired: Since it was statistical, quantitative, empirical, it was *scientific*. Becoming a social scientist in the 1940s necessarily meant becoming a scientist proper, thus "speculative" and "proto-scientific" French and German social theory was excluded.

Taking a closer look at Swedish sociology ten years later will reveal how this ideal knowledge was realized in practice. The first professor of sociology, Torgny T. Segerstedt, was a key figure in shaping the Uppsala School of Sociology. In a programmatic text which appeared in the first volume of *Acta Sociologica* (1956), Segerstedt describes the sociological program to be founded in a descriptive and relativist theory of values, combined with a deductive positivist philosophy inspired by C. G. Hempel, and a theory of society stressing the importance of a *norm system*, which produces homogeneity as well as deviant or "defective" behavior causing "social problems" in a society (Segerstedt 1956). The foundation of this knowledge was based on the relationships between *norm—norm speaker—sanction—habit*. This terminology puts society on the level of a norm system and subjectivity as one or several people, or even institutions, speaking these norms. Moreover, sanctions function to induce rewards or punishments when a norm is observed or violated. Finally, "habits" describes three forms of repeated behaviors: action, emotions, and verbal elements. However, it is not a grand theory that follows. The Uppsala School only elaborated models on the level of the *group*: "By group we mean two or more people in interaction observing social norms which can be traced to one and the same norm speaker" (ibid.: 89). Empirically, groups might be study circles, regiments, corporations, institutions. They are further divided into strata, where the status of a person varies with distance to the norm speaker. A perceived social problem in the early days of Swedish sociology was *deviance*. In the days of urbanization and heavy industrial expansion it became important to find the causes of deviant behavior in order to integrate groups that did not conform to the social project. For example, the Romani minority was perceived as being a deviant group, and the sociologists of the Uppsala School tried to find the causes for their behavior, whether it was a "defect in the norm system" or not. During the 1950s, the statistical approach to "engineering society" reached its peak, and in a government white paper from 1959 it is argued that more funds and resources should be allocated in the field of statistics:

> In a modern Western society [. . .] rational interventions must be based on knowledge of human behavior, individual as well as in groups. The trend in modern society, both in the commercial and the public sector, draws toward planning, which must be adapted to the behavior of the individual. The modern forms of statistics are more and more becoming the means to solve this double task (SOU 1959: 33 p. 13, my translation).

The development in the early 1940s is clearly coproduced with both the ideas and the practices of the Swedish welfare state. However, the *content* of sociological theory was not given, only the framework of a rational pursuit of planning modern society by means of scientific methods. The choice of structural functionalism, in local shapes such as the Uppsala School, met with the desired scientific characteristics, being both statistical and empiricist. This choice turned Sweden into a landscape of numbers, providing a map for planning. Also, American social theory seemed to fit well with the rational moral philosophers who were to become sociologists. Influenced by Swedish philosopher Axel Hägerström, the Uppsala School was founded upon the "true nature of the state and society" with a descriptive, and strongly anti-normative, theory of values. Statistics still being a young science in Sweden (in the sense that there were very few professors teaching and researching it) seemed to be the linking device between social theory, philosophy of science, and the possibility of providing the State apparatus with hard and reliable data. Nevertheless, the choice of a specific sociological tradition could have been made otherwise.

In the aftermath of 1968, Marxism impacted the homogenous era of structural functionalism, and the role of sociology as social engineering was, at least intellectually, challenged. In the late 1980s, Jörgen Westerståhl, who is described as a founding father of surveys in Gothenburg, presented the primary task of the social sciences as "presenting facts and contexts rather than providing concrete solutions" (Westerståhl 1990: 68, my translation). However, we may not ascribe "fact-based sciences" any special logic with such a statement. Rather, as we are about to see, it is precisely the *struggle for autonomization* that makes the story of the SOM Institute, which is the next case in this chapter, interesting. After the turmoil in academia in the early 1970s, the social sciences in Sweden became more pluralistic. Marxism, in its many variations, was introduced as well as a critique of positivist philosophy.

QUANTIFYING EPISTEMIC AUTHORITY—ON A LOCAL RESEARCH INSTITUTE

As we move along in the history of the Swedish social sciences, many social institutions are reconfigured. In academe, new research traditions have entered the scene, and for the statistical sciences the introduction of the desktop computer allows statistical calculations to be made with ease at lower cost. In order to see how this affects the relationships between the social sciences, the State, and statistics, we must restart our investigation with another local example.

However, the scene which we are about to enter is not a clean slate. The Uppsala School of Sociology laid down a few fundamental elements which bear strongly upon the SOM Institute. Sociology was turned into a *collective* enterprise that required epistemological cooperation between researchers. It also involved rigorous methodological training and development of the survey technique of measuring social phenomena. A key element in the production of sociological facts involved certain standardized practices of constructing surveys, which enable quantification and calculation. The focus on surveys as the empirical base of sociology was a stabilizing factor mediating between concepts and theories, scales of measurement and statistical diagrams. The Uppsala School successfully developed relations with the state apparatus and a political thrust toward a planned society. The postwar sociologists managed to translate the general interests in a rationalization of social institutions and governance, thus receiving both funds and legitimacy. This displaced sociological articulations from a U.S. context and realized them in a local practice. Theories, methods, concepts, and attitudes to scientific research were imported from any one of several traditions of American sociology. As they materialized in a Swedish setting, they changed and their outcomes became contingent but integrated parts of Swedish society. Traces of these processes are to be found some in the mid-1980s as the SOM Institute is established.

The SOM Institute

The SOM Institute is one of the largest social scientific research institutes in Sweden. They conduct surveys on national and regional levels, and each year they produce a book series presenting the results of their research. Their surveys have large statistical samples and certain questions in their questionnaires have remained intact with previous research dating back to the 1950s, hence their time series are in certain cases unique in international comparison. Before 1986 the researchers at the faculty for social sciences in Gothenburg only had access to various statistical surveys performed every third or fourth year. It is stated in books and reports from the early 1980s that the social sciences needed annual surveys, with proper epistemology, in order to achieve a longitudinal and reliable repository of statistical data. However, thinking of the establishment of the institute simply in innovational terms of performative actions will lead us away from the more important aspects of how statistics, sociology, and society are coproduced over the following twenty years. Instead we must pay careful attention to what lies before the SOM researchers in contemporary social science. Three main accounts will be elaborated here: (1) statistics emerging from a dialogue with global sociology in the shapes of scales and measurement techniques, (2) statistics functioning as an epistemic authority as the SOM researchers struggle for autonomy in the academic pursuit of knowledge, and (3) statistics used for the practice of constructing surveys and describing the social world.

Reproducing Global Sociology

Performing the institute's first survey entailed a lot of work. As statistical sociology is filled with epistemological standards such as measurement methods, scales, techniques, and response rate criteria, the researchers were faced with some hard work of packing social reality and epistemology together in order to represent the domain of the social in facts and figures. The first task was in the epistemic arrangement, which imported already available scales of measurement. When constructing the first national survey the Rokeach Value Survey and the works of Ronald Inglehart were inserted to the SOM survey. This of course enabled cross-national comparisons, which function both to strengthen the knowledge being produced and to extend the statistical map and display social patterns on a European or global level.

There are more features than just scales imported into the SOM Institute. As statistics in the Pearson-Quetelet-Spearman tradition, with independent and dependent variables, is quite plastic in the sense that it is universally applicable throughout the social sciences (and other sciences as well), they are naturally an important part in the foundation of the institute. Now, as with the scales mentioned, there is more to statistics than meets the eye. Scales and measurements necessarily limit the modes of representation while being productive in shaping a positive ground of knowledge. For example, the Rokeach Value Survey reduces the complexity of the notion of human values because it contains a predefined set of eighteen value statements. Combined with, for example, exploratory factor analysis, a very productive assemblage is formed, which allows such entities as *morality*, *sociality*, *integrity*, and *aspiration* to emerge as constitutive traits of the nature of man in relation to society. As with the Uppsala School, the SOM researchers also turned to the American tradition of social survey. However, now there was no need to do much theoretical work because the scales and measurement techniques were already preconfigured, as was the newly introduced machinery programmed for computing the social in numbers.

In the early 1980s, the SOM Institute was progressively shaped into an all-encompassing research center. It covered central social scientific topics, such as culture, values, and social trust—traits that were measured by translating and importing international models. The notion of human values and culture were, in one respect, *reproductions* of the works of, for example, Milton Rokeach and Ronald Inglehart. But reproduction is not at all a bad thing. It allowed the SOM researchers to test their findings and compare them with other surveys, and they conveyed theories and concepts such as *terminal human values* and *postmaterialism*. As they were imported into the SOM survey, they were conserved, since longitudinal studies must ask the same questions every year to be able to account for changes over time. Thus, the SOM Institute served as a local *embodiment* for the social theory of structural functionalism.

Autonomization through Practice

In order to get the yearly surveys up and running in the first place, the researchers pursued a problematization of becoming indispensable. Competing actors in the early 1980s were other institutes such as *Gallup, SIFO Research*, and *Statistics Sweden*. Indispensability was achieved by claiming the epistemological precedence of university-based social science, combined with the neutral and nonpolitical character associated with academic pursuit of knowledge (Holmberg and Petersson 1980). The arguments resemble Westerståhl's rhetoric, as mentioned earlier, where the social sciences are claimed to present "merely facts" rather than doing social engineering. As the SOM researchers spoke of epistemic authority and objective knowledge, they were speaking in the name of epistemology. By "trespassing" into the territory of epistemology through the uses of statistics, they were able to circumscribe the problematics of how to deal with the multifaceted expressions of man, behavior, and society and speak in the name of science. This is what allows them to speak of social reality in their everyday work as social scientists—whose effects are *real in practice*, even if only for a short time historically. The SOM researchers argue that other research institutes, such as Gallup and SIFO, are influential on the democratic process in a negative way, especially when publishing opinion polls close to the election dates. Moreover, their methodology is not scientific enough, since their telephone-based surveys produce statistics with "questionable" standard margins of error (ibid.).

As this problematization was mainly performed on epistemological grounds, the technical vocabulary of statistics and methodology shifted from tools to scientific rhetoric. In other words, this is where statistics is used in a dual sense, allowing the foundations of knowledge simultaneously as it is used as rhetoric in strengthening an actor network. It shifts from only "mobilizing the world" in the sense that it allows scientific facts to emerge, to playing a central role in the autonomization process, strengthening epistemic authority.

Statistics as a Tool for Response Rate Accumulation

An aspect not yet elaborated is the reconstitution of science and society through encounters with social reality. In the 1980s, the SOM Institute was very successful in constructing national surveys, approaching 70 percent response rates. But in order to achieve such numbers, which is crucial for every statistical measure, the researchers at the SOM Institute had to apply their sociological skills in another type of practice. In the early days of the institute, books were written on the nature of the survey in relation to behavioral patterns of the respondents (Ohlsson 1986). To solve the puzzle of why some people never replied to the survey, the SOM researchers themselves applied a sociology of rational choice based upon Talcott Parson's distinction between instrumental and expressive action. Combined with a model of mass

74 *Christopher Kullenberg*

communication, where the actual survey was described as a message and the respondents as the receivers, the SOM researchers were put to test. What we have then is similar to Callon's engineers: sociological skills applied to concrete innovations. Structural functionalism and theories of mass communication were used in order to "get to the world" rather than describing it.

The methodological text written by Ohlsson divides the respondents in segments such as age, profession, political interest, and so on. Then it goes on to talk about the costs and benefits of responding to the survey. It continues to describe further methodological obstacles that must be accounted for once the respondents have replied, in other words, a debate on validity. Young adults have unstable identities since they are seeking a position in society; middle-aged persons have stable identities but are busy with work and reproduction; retired people are losing their stable identity because they are feeling empty as they leave their professional lives behind (ibid.: 46ff). Thus, knowledge and practice are tightly interwoven, and when applied, there is a knowledge *of* practice. However, the trial of strength will not reside in method talk alone, but the encounter between the sociology of the SOM researchers and respondents is manifested in action. This trial may be represented in Figure 3.1.

The Sociology of Response Rates

1986: Survey + Postcard + Letter and survey + Letter and survey + Telephone call = 68%

1999: Postcard + Survey + Postcard + Letter and survey + Postcard + Telephone call + Letter and survey + Telephone call + Letter and survey + Telephone call + Letter and survey + Survey = 67%

Postcard   Telephone call

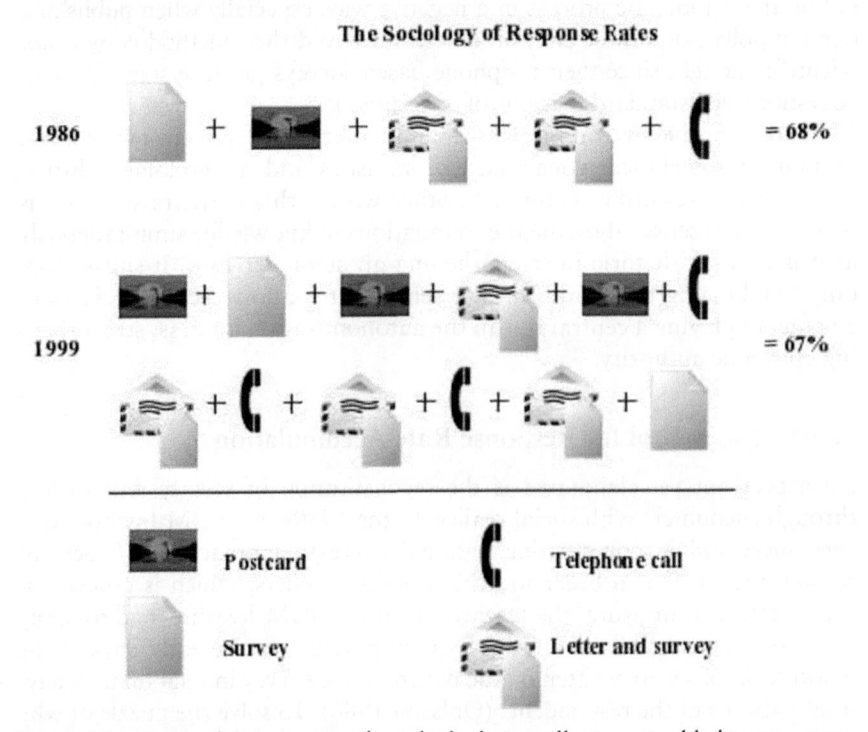

Survey   Letter and survey

Figure 3.1 Survey forms, postcards and telephone calls are assembled as programs of action. By adding elements the research instrument is strengthened, and the researchers can achieve a high response rate. However, it takes sociological skills to make assumptions about, and measurements against, what will lie in between, causing anti-programs.

By observing the composition and progression of the SOM researcher's instrument (Figure 3.1), we see the engineering skills of sociology manifested in response rates. Now we are at the level of "legwork" sociology, where there is a "direct" interaction with the study object. The survey, sent via mail, must be followed with progressive programs of action (Latour 1992), which are loaded syntagmatically by adding reminder postcards, more surveys, and finally telephone calls prompting the respondents to fill out the survey and send it back. The SOM researchers' model of mass communication is being manifested in letters and postcards, then sent to 2,500 homes, and the rational choice model is thus expressed in the flow of returned survey forms. All together, this concrete assemblage, which is neither pure theory nor merely letters and postcards, extracts a sufficient response rate for statistics to produce reliable data. The 68 percent response rate adds up to the criteria set by the tradition of quantitative social science, which had been active since 1947 in Sweden.

Over time, this assemblage is radically expanded as, in order to achieve a response rate higher than 60 percent, the SOM Institute had to load more and more actants (Latour 1992: 108) into it. Even though the 1986 survey was described as a success, the SOM researchers cannot take society for granted, and throughout the annual fieldwork the method is constantly evaluated. Two years later another book on surveys is published drawing from the experiences of the SOM fieldwork. Urban dwellers are responding less than rural people, thus telephonic reminders are concentrated to the larger Swedish cities, as well as in certain segments of age. Also, an *antiprogram* is identified among some respondents. As the SOM researchers are phoning, they discover that phone calls themselves provide a motive against replying: suspicions regarding the anonymity promised to participants (Holmberg and Weibull 1988). However, as I have stressed previously, the interpretation of this anti-program follows the epistemic arrangement, and is explained in the sociology of surveys elaborated by Ohlsson. As we see in Figure 3.1, the programs of action are progressively added up until 1999, through the dual logic of *knowing about* society and *acting in* society.

While the engineer-sociologists in Callon's story failed in linking together consumers with fuel cells and electric vehicles, the SOM researchers are quite successful according to the statistical standards of survey response rates. The abstract Parsonian sociology thus "simply works," and its *practice* does so in relation to theoretical reasoning, if we think Callon's dictum to its full extent. It would be an exposé in thinking sociology as agency, however when we start to draw things together, a more complex picture emerges, which makes a purely innovational story impossible.

CONCLUSION—STATISTICS AND SOCIAL SCIENCES IN TRANSITION

The two cases, the Uppsala School of Sociology and the SOM Institute, share similarities in respect to the theoretical foundations in structural functionalism. However, the story is not simply one of influences, traditions, or cumulative data collection. This first version is common among social scientists themselves to write and reflect upon. A second approach to describing the social sciences would be to speak about paradigms in the Kuhnian sense (see Friedrichs 1970, Burrell and Morgan 1979, Brante 1985). However, such approaches have focused more on the content of sociological theory than the social and historical framework that has made possible the emergence of such a knowledge of society in the first place.

In the case of the Uppsala School, we have seen the parallel developments of a progressive welfare state and emergent sociologists seeking knowledge of society. Statistics is the mediator making both pursuits possible: political arithmetic and sociology pass through the same *boundary object* (Star and Griesemer 1989). Through this, the social bodies, the processes of urbanization, divorces, and deviance are subjected to the language of numbers. The foundations of sociology as a practice and academic subject in Sweden I compare to the figurative of *social engineering*, not a metaphor with high ideals, but operative for the pursuit of rational and secure knowledge.

History moved on, but the Uppsala School had paved the way for certain elements, which were adopted by the SOM Institute. Social science as *collective* procedure was necessary in order to collect and organize the data that came out of the surveys, and this mode of organization still exists as central to statistical social science. Also, the methodological approach of random samples, statistics, and surveys were already there, both in terms of training, and in terms of being a privileged epistemic position. Surveys and statistics were already a *gold standard*, readily adopted.

As we then turn to the SOM Institute, the figurative of social engineering steps aside in favor of *epistemic authority*. Statistics now functions as a mediator between the SOM researchers and other research institutes in terms of autonomization, where it is used as a tool for *boundary work* (Gieryn 1999). But it also mediates between the theoretical assumptions in structural functionalist theory and respondents to the surveys. In this dual sense it quantifies both the boundary between the SOM Institute and other surveys in terms of errors of margin, criteria, and sample and response rates. But the boundary is also evidenced by negotiations through programs of action with the respondents.

Statistics may, as these two cases show, have several functions for holding social scientific research together. It also relates to the State apparatus, in that "it can function both as a basis for knowledge and as an incitement to action" (Hammer, this volume).

Even though the social sciences of today include qualitative as well as quantitative traditions, statistics will most likely appear in various forms, not as given constants, but rather as contingent historical expressions.

NOTES

1. The geographical terminology is not metaphorical. The social sciences provide in the historical examples I draw from a very concrete mapping of the "social," allowing the localization of social problems as well as social associations constituting entities such as the nation, the region, the poor, the middle class, and so on, emerging from these landscapes.
2. However, during this time Gunnar Myrdal performed what probably would count as the most famous Swedish work in social science—a study called *An American Dilemma: The Negro Problem and Modern Democracy* (1944). But it would not influence the rise of the social sciences in Sweden to any significant extent
3. In Sweden white papers are called "Statliga offentliga utredningar" (SOU), which translated into English means "Government public investigation."
4. SOU 1946: 74 is entitled "Concerning the State of the Social Sciences at Universities and Colleges." It was delegated to several professors in the social sciences, among them Torgny T. Segerstedt, who became the first professor of sociology, and Jörgen Westerståhl, a central player in the Gothenburg tradition of social surveys.

REFERENCES

Brante, T. (1985) "Paradigmteori och samhällsvetenskaperna," Häften för kritiska studier, 3(18): 6–28.

Burrell, G., and G. Morgan (1979) *Sociological Paradigms and Organizational Analysis: Elements of the Sociology of Corporate Life*. London: Heinemann.

Callon, M. (1986) "Some Elements of a Sociology of Translation: Domestication of the Scallops and the Fishermen of St Brieuc Bay," in *Power, Action and Belief: A New Sociology of Knowledge*, ed. J. Law, 196–233. London: Routledge & Kegan Paul.

———. (1987) "Society in the Making: The Study of Technology as a Tool for Sociological Analysis," in *The Social Construction of Technical Systems: New Directions in the Sociology and History of Technology*, ed. W.E. Bijker, 83–106. London: MIT Press.

Deleuze, G., and F. Guattari (2004) *A Thousand Plateaus—Capitalism and Schizophrenia*. London: Continuum.

Desrosières, A. (1991) "Social Science, Statistics and the State," in *Discourses on Society—The Shaping of the Social Science Disciplines*, ed. P. Wagner, 194–217. Dordrecht: Kluwer Academic Publishers.

———. 1998) *The Politics of Large Numbers: a History of Statistical Reasoning*. Cambridge, MA: Harvard University Press.

Eyerman, R., and Jamison, A. (1992) "Varför är svensk sociologi så amerikansk?" *Tvärsnitt*, 4: 92.

Foucault, M. (1970/1994) *The Order of Things—An Archeology of the Human Sciences*, New York: Vintage Books.

Fridjónsdóttir, K., ed. (1987) *Om Svensk sociologi*. Stockholm: Carlssons.

———. (1991) "Social Science and the 'Swedish Model': Sociology at the Service of the Welfare State," in *Discourses on Society—The Shaping of the Social Science Disciplines*, ed. P. Wagner, 247–268. Dordrecht: Kluwer Academic Publishers.

Friedrichs, R.W. (1970) *A Sociology of Sociology*. Toronto: The Free Press.

Gieryn, T. (1999) *Cultural Boundaries of Science—Credibility on the Line*. Chicago, IL: University of Chicago Press.

Holmberg, S., and O. Petersson (1980) *Inom felmarginalen: En bok om politiska opinionsundersökningar.* Stockholm: Liber Förlag.
Holmberg, S., and L. Weibull (1988) *SOM-undersökningen 1987.* Göteborg: Statsvetenskapliga institutionen.
Horn, D.G. (1994) *Social Bodies: Science, Reproduction, and Italian Modernity.* Princeton, NJ: Princeton University Press.
Jasanoff, S. (2004) "Ordering Knowledge, Ordering Society," in *States of Knowledge: The Co-Production of Science and Social Order,* ed. S. Jasanoff, 13–45. London: Routledge.
Latour, B. (1987) *Science in Action—How to Follow Scientists and Engineers through Society.* Cambridge, MA: Harvard University Press.
———. (1992) "Technology is Society Made Durable," in *A Sociology of Monsters: Essays on Power, Technology and Domination,* ed. J. Law, 225–258. London: Routledge.
Ohlsson, A. (1986) *Att svara eller inte svara—det är frågan. Empiriska studier av bortfalls- och svarskvalitetsproblem vid postenkätundersökningar.* Göteborg: Statsvetenskapliga institutionen.
Patriarca, S. (1996) *Numbers and Nationhood: Writing Statistics in Nineteenth-Century Italy.* Cambridge, UK: Cambridge University Press.
Porter, T.M. (1995) *Trust in Numbers—The Pursuit of Objectivity in Science and Public Life.* Princeton, NJ: Princeton University Press.
Segerstedt, T. (1956) "The Uppsala School of Sociology," *Acta Sociologica,* 1(1).
SOU 1944: 19 (1944) *Om inrättande av ett samhällsvetenskapligt forskningsråd: betänkande / avgivet av tillkallade sakkunniga.* Stockholm: Ivar Haeggströms Boktryckeri.
SOU 1946: 74 (1946) *Betänkande angående socialvetenskapernas ställning vid universitet och högskolor,* Stockholm: Ivar Haeggströms Boktryckeri.
SOU 1959: 33 (1959) *Organisatoriska riktlinjer för svensk statistik : utredningar och förslag / av 1956 års statistikkommitté,* Stockholm: Esselte.
Star, S.L., and J. R. Griesemer (1989) "Institutional Ecology, Translations, and Boundary Objects: Amateurs and Professionals in Berkeley's Museum of Vertebrate Zoology, 1907–1939," *Social Studies of Science,* 19: 387–420.
Westerståhl, J. (1990) "Om forskning och forskningsresultat," in *Svenska samhällsvetenskaper,* ed. K. Fridjónsdóttir, 64–70. Stockholm: Carlssons.

4 Governing by Indicators and Outcomes
A Neoliberal Governmentality?

Svein Hammer

Through five seasons of the television series *The Wire*, we follow an ongoing game between police and criminals, at the complex borderline between morality and immorality, success and failure, individual choices and collective mechanisms. One of the threads that runs through this sociological fiction of urban life is the strategies and problematizations that can be related to governing by numbers. The examples here stretch from attempts to *not* take responsibility for fourteen dead prostitutes (the risk of bad statistics) to a tough campaign to improve confidence in policing by reducing the number of crimes. In the latter case, weekly statistical reports have a double effect in the organization. On the one hand, they stimulate attempts at new ways of policing. On the other hand, and this is the more common reaction, they stimulate a practice of decentralized, creative use of statistical categories in the police reports. In this way, in a co-constructive process between police work and monitoring through numbers, new tactics and techniques are invented, implemented, and used to improve the outcomes of policing (see also Williams, Lomell, and Gundhus in this volume).

Of course, a story like this easily leads toward discussions on good or bad statistics, smart or stupid government, and so on. This chapter, however, will not problematize the quality of the numbers produced; rather, it is an analytical reflection on the co-productive processes between the indicators, measurements, and tables of statistics on the one hand, and the reflections, practices, and tactics of government on the other. The chapter will partly be theoretical and historical, partly empirical and contemporary. At one level it is a general discussion anchored in works of Michel Foucault, Alain Desrosières, and a few others, which gives insight into various rationalities and technologies of government; at a more concrete level, a discussion reflecting three empirical sources on municipal measurement and government. These sources are: participant observation in one municipality, newspaper statements on public statistics, and two governmental documents describing examples of implementation and use of the system Balanced Scorecard. The objective will not be to give any empirical conclusions, but rather to use the material to reflect on how the Foucauldian concept of "governmentality" can contribute to an analysis of statistics and its co-productive relations to various forms of government.

STATISTICS AND THE GOVERNMENT OF MAN

Statistics represents a possibility to generalize, to take us away from the individual and concrete, to give an overview of a larger landscape. At the same time, and this is perhaps not so obvious, the numbers and their distribution work as an individualizing technique; through categorizing they tell us where we belong and what factors our lives are connected to. In the discourses of the welfare state this duality was important, because it made it possible to combine holistic, long-term planning with direct intervention in face of deviations from statistical normality. What happens, then, when the dream of social engineering loses its power? Can standardized indicators and numbers have any force in a time of liberal dynamics and individual solutions? It would not be of any surprise if their position were challenged and maybe weakened. However, the duality described here seems to persist—and even increase its power—inside a partly changed, but perhaps most of all extended governmental field.

When investigating this theme I will not treat statistics as an independent unit, and I will not ask whether existing statistical knowledge is correct or wrong, good or bad. Rather my perspective is how the category of "statistics" and the variety of concepts, practices, and techniques it is related to, is part of discursive processes where the real is formed and transformed; or, in this instance, where public sector statistics are produced and used in co-productive relations to practices directed toward governing social life. Tendencies toward increasing use of quantitative techniques and tools will here be met with curiosity; the main question will be what kind of rationality this revitalizing indicates, what kind of techniques it is connected to, and whether we from this can describe the present governmental landscape as "neoliberal." The intellectual baseline for the reflections on these questions will be the writings of Michel Foucault through the 1970s.

Knowledge, Power, Government

Talking about Foucault, our thoughts easily go to concepts like "power" "discipline," and "surveillance." And, of course, his analysis of *discipline* (Foucault 1975) contributes to our understanding of the modern subject: Through practices where we learn to observe, assess, compare, and adapt, normalizing processes emerge that not only enforce homogeneity, but also individualize, making it possible to discern (and thereby create) a plurality of differences and make use of them for the good of society. Discipline, however, is not to be understood as the general form of power. In his first book on the history of sexuality, Foucault (1976) writes about *regulation of the population*; that is, techniques of power that move away from the individual bodies, that operate at a different level and use other strategies and tools. Through watching, controlling, stimulating, strengthening, increasing, and organizing the life of the population, through relating knowing and action in new ways, this power in more indirect ways contributes to improving the society, making it healthier and more productive.

Given this perspective, these two forms of power—discipline and regulation—can be said to reinforce one another, as an interplay between body and population. In both cases "the norm" is at work; the norm conditions a continuous work to define levels of quality, to measure, assess, and order every side of the life of the population. Statistics enters into this process as a central mechanism, attached especially to the establishment of connections between each man and the entirety he can be categorized and ascribed characteristics as part of. Through statistical mappings and measurements, possibilities emerge for systematic, ever increasing overview over life and its different practices and effects. In this way, the categories of discipline are transformed into something more general, something further-reaching that opens up for more subtle ways of stimulating the forces of society to strengthen themselves.

From these reflections on regulation of life and bio-power, Foucault moves into another, but still connected theme: governmentality, as discussed in *Security, Territory, Population* (1977–1978) and *The Birth of Biopolitics* (1978–1979). The discourses he then intervenes into are described as five hundred years of circulation around our questions, discussions, investigations, and solutions of the problem of "how to govern." There are many nuances in Foucault's reflections on the governmental discourse. The main point, however, seems to be that it is a constitutive interplay between the growing art of government and more or less new categories such as "population," "society," and "the economy." With the introduction of the problems and possibilities related to these concepts, the focus was turned away from discussions on the power of the governor and the governing of families toward another level of the real—which via statistics exposed new phenomena, movements, and regularities. The numbers indicated aggregations that could not be traced back to singular factors; they revealed patterns and processes that could not be understood simply as effects of the decisions of the sovereign, the regulations of law, or the life of the family. In short, there was a new acknowledgment that in society, in the life of the population, there are resources that can be mobilized and function as part of the capacities of the State. In an early phase, this opened for use of governmental practices and strategies with direct, totalizing, and paternalistic character. Over time, however, a more liberal variant crystallized: an increasing reflection on the limits of government, on the possibilities to develop indirect techniques of government, directed toward stimulating society's self-development.

The consequence was that *the population* from now on became a basic factor of government—at one and the same time an object in the hands of the governor, and a subject with needs, ambitions, and possibilities. Foucault introduces the theme of the shepherd to illustrate the implications. This analogy, with roots in the Jewish tradition, functioned as part of Christian reflections on pastoral power, and then moved into a broader governmental discourse on the relations between the one who leads and the ones who are led. In short, the shepherd's task is to collect, take care of, and lead both the crowd and its singular individuals, as purposively and systematically as possible. In this way the power of the shepherd lies neither in a stabile law nor in detailed regulations, but in the capacity to meet and take care of the changing "unity" of the crowd. In practice this implies a displacement in the art of government, from obedience in face of the law to strategic and tactical management to strengthen the population in itself.

Of course the juridical laws, the administrative rules, and the disciplinary practices do not disappear. However, they can be said to be transformed into a new form of practice, among other things characterized by a development of new techniques for production and use of knowledge on the strengths and capacities of the society. Statistics can in this way be seen as part of processes that open up society for government. It functions not only as part of the play of truth, as the tool that makes it possible to operate with distinct categories, connect things, compare, indicate probability, and so on. It also relates to processes where society is formed and transformed. Through giving changing practices and their plural effects a standardized status, measuring them, and taking the numbers into the public discourse, the complexity of life is transformed into something that can be worked with inside the logic of the numbers (see Hovland, this volume).

An Analytical Model

In an earlier article (Hammer 2008), I used this Foucauldian perspective to develop an analytical model for studies of the connections, movements, and upheavals in and between discourses of government, or "governmentalities." The idea then was that these discourses can be analyzed as a play between four components: On the one hand, an ongoing, co-productive interaction between *techniques and tactics of government* (the concrete initiatives, technologies, programs, etc., that can be connected to governing practices) and *governmental rationalities* (the more abstract processes where "the how of government" is questioned, discussed, conceptualized and written into reasonable strategies), and on the other hand, a dimension that stretches between vertical *practices of domination* ("the governing of others" through formal decisions, directives, and procedures) and more horizontal *practices of subjectivation* ("the governing of the self" through identification, interaction, assessment, and development). Figure 4.1 illustrates these two dimensions and the connections between them.

If we are to use this model in a Foucauldian spirit, it is important to not give any of the four components fundamental status; rather they should be seen as mutually constitutive inside an open, moveable tableau that is formed and transformed through time.[1]

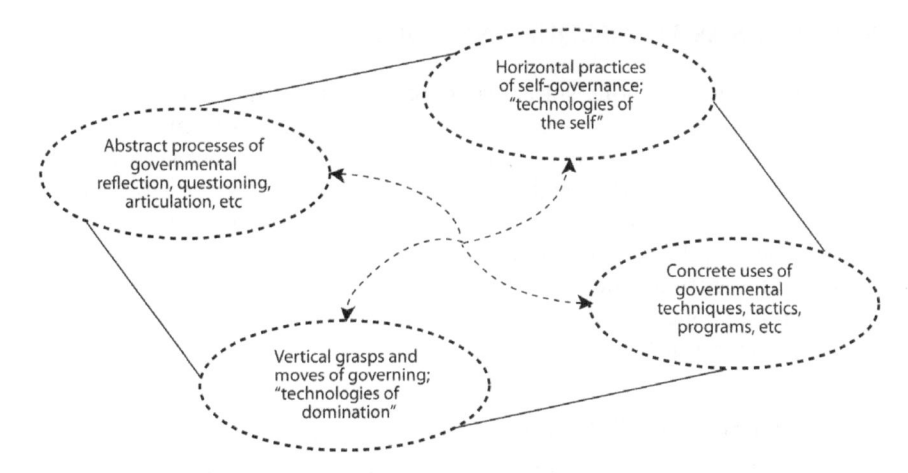

Figure 4.1 An analytical framework for studies of "governmentalities".

The first dimension directs our focus toward the ongoing interplay between (abstract) statements of knowledge and (concrete) techniques and tactics of power, which indicates that it will be wrong to reduce the concept of governmentality to a tool for idealistic analyses of ideas and rationalities; the focus is on practices and concrete techniques as much as on words and formulations (see Gordon 1991: 3). To this, the other dimension adds a similar view of the connections between (vertical) technologies of domination and (horizontal) techniques of self-governance. In a lecture Foucault says that if we want to analyze the genealogy of the subject in Western civilization, we have to take into account not only techniques of domination but also techniques of the self—or more precisely, the interaction between them. We have to take into account the points where technologies of domination of individuals over another have recourse to processes by which the individual acts upon himself. And conversely, we have to take into account the points where the techniques of the self are integrated into structures of coercion or domination: "The contact point, where the way individuals are driven by others is tied to the way they conduct themselves is what we can call, I think, government" (Foucault 1993).

With this two-dimensional model as "looking glass," we can set out to study government in the modern, liberal world—that is, we can move to the inside of empirical settings, and there trace the relations, connections, and breaks that can be seen through the optic of the model. This will be done in three analytical steps: from statistics as *concrete tool*, via statistics as part of *abstract reflections*, to statistics in the governmental play *between the vertical and the horizontal*. The intention then is to illustrate the analytical perspective of Figure 4.1 through empirical reflection.

STATISTICS AS TECHNIQUE AND TOOL

Let's start talking about statistical practices as we meet them, as concrete techniques used to collect, analyze, and mediate information. This, however, does not mean that we are going to talk just about statistics. As Desrosières (1998) so thoroughly has documented, there have always been interactions and co-productive effects between the developments and uses of statistics, and administrative and political practices directed toward action, intervention, and improvement. Our discussion will therefore be anchored in a concrete field of practice; the production and use of statistics in the governing of Norwegian municipalities.[2]

Municipal Statistics Practices

There are three governmental levels in Norway: the national state, nineteen counties, and more than four hundred municipalities. At the municipal level there have been a multitude of transformations since the 1990s (see Bukve and Offerdal 2001, Opedal et al. 2002, Spurkeland 2005, Johnsen 2007), of which we here will take a closer look at the interplay between three tendencies: first, *the nonhierarchical organization*, which in few years became "the paradigm model" of Norwegian municipalities. In most cases this implicates a structure with just two levels of authority—designed along a distinction between a strategic level that designates the long term development of the community, and an operative level with responsibility for day-to-day services (kindergartens, schools, health services, etc.). Second, a discourse on *empowering the employees*, materialized as statements on the active use of individual competences for the good of each and all—and, in extension of this, work on establishing systems characterized by ongoing, active, and flexible interactions between different employees, competences, functions, and levels.

There seem to be co-productive connections between these two tendencies and a third—the adjustment of and expansion in the municipal *production and use of numbers*. An effect of the first two transformations is a municipal reality where the distance between top and bottom is reduced. Furthermore, and only in part due to that reduction, both the sovereign authority and the structures, rules, and procedures of bureaucracy are less determinant of everyday practices than before. In the more open field of practice thus constituted, the relations between system and actors typically has been secured through contracts ("leader agreements") between the chief officer and the producers of services, specifying important outcomes and how they will be measured. Undergirding and enacting this we can identify ongoing practices where municipal indicators are defined, measurements prepared and executed, numbers collected, analyzed, and reported.

Of course, it is hardly new that the politico-administrative system produces numerical descriptions on various aspects of the municipal world.

However, there is something new going on when municipalities invest in technology and competence to use SPSS and other statistical systems. This indicates a tendency to incorporate scientific techniques into the numerical field, and through that transform and formalize municipal-statistical practices. Partly these developments can be seen as elements of departmental strategies: Since 1997, municipal activities and outputs have been registered through KOSTRA, a standardized system for statistical reporting between municipalities and the state. The numbers were among other things used as a basis for so called *efficiency networks*, in which representatives of similar municipalities were invited both to compare their results and to relate them to examples of good practice.

The aforementioned processes of encoding, measurement, and use of numbers are often understood just as technical and practical problems, steps to be solved through following recipes and working systematically. Such solutions, however, can also be described as surface effects of complex processes of negotiation, through which decisions on what to measure, how to do the measurements, and not least how to use the numbers are always about something more than statistical questions. For example, when categories such as *quality* and *efficiency* are operationalized into indicators, measured, and then used to describe and develop the municipal field, they become "inscribed in routinized practices that, by providing a stable and widely accepted language to give voice to the debate, help to establish the reality of the picture described" (Desrosières 1998: 1).

This, of course, has something to do with the magical transmutation of statistical work; a transfer from one level of reality to another, and, at the same time, from one language to another (from "this solution works well" to "the quality-index is improving"). An effect of this is that increasing use of statistics implies that more of the real is transferred into the language of numbers, into the objectifying logic of exact knowledge. Or, in other words, the reality of the statistical world extends, and thus intervenes into zones previously described, discussed, and worked on through other forms of knowing. In this way, statistical practices not only induce effects, but will be part of the construction of the real. However, in order for reliable data to be produced, the world described must already be "equipped with codes for recording and circulating elementary, standardised facts" (Desrosières 1998: 44)—and these codes are best understood through relating them to a more abstract level of governmental thinking.

STATISTICS AS PART OF GOVERNMENTAL REFLECTIONS

In the municipal field, statistics is partly used "as before," as a basis for general planning and strategic decisions. This kind of direct governing, however, has been supplemented by a more decentralized use of numbers, where especially indicators on outcomes and quality represent something new. It is possible to see this as an expansion, where the numbers not only function as an "objective" baseline for planning, but also as indirect incitements in iterative, multi-level processes of assessment, interaction, and improvement (Hammer 2004, 2008).

To understand this change we move into a much broader governmental landscape: the Foucauldian analyses of the connections between concepts such as population, society and the social, and the constitution of *regulation of life* as a form of power. From the end of the nineteenth century this opened for the establishment of a discourse on the welfare state, articulated around an idea of collective needs that could be met through a combination of administrative regulation, apparatuses of social security, and direct interventions into people's lives.

As part of this, statistics came to play a more decisive role, giving consistency to things affected by collective action (as shown in Kullenberg, this volume). Through new practices of numerical addition, description, and diagnostics, connected to standardized procedures and assessments, statistics came to occupy a privileged position as common point of reference—"a cognitive space of equivalence constructed for practical purposes" (Desrosières 1998: 17). This opened new possibilities to monitor the landscape, to categorize and explain phenomena in a stabile and objective light, to plan and develop society writ large, and to intervene in human lives when necessary—in short, "to know, programme and transform the social field" (Sending 2003: 215).

In Norway, connected to the dominance of the Social Democrats, in an epoch which stretched from 1935 to the 1970s, this discourse seemed to have established itself as the ultimate solution to the challenges of government. And yet, while political disagreements within the discourse on the welfare state revolved around questions on which techniques of government that could make society orderly and safe, this discourse constantly ran up against discourses focused on the (especially practical) bounds of government. This indicates how liberal discourses were active—in the nineteenth century as defining forces in the work on clarifying the relations between a rational subject and a democratic society, later as an opposition against social democratic practices of government, as a continuous questioning of the belief in complete overview, detailed calculations, and deductive operations.

When I here introduce categories such as "liberal" and "liberalism," it is not to signify closed ideologies, but rather open, changeable discourses. However, one family resemblance of most liberal discourses is the treating of society as independent of government, as a set of relations and movements that cannot be governed in its totality. An early formulation on this is the concept of "the invisible hand," which connotes that every attempt to govern faces a complex and self-regulating mechanism (especially the market), which can only be observed through its effects. The only legitimate way of governing, then, is to act such that "reality" develops according to the principles and mechanisms of society itself. In other words, the liberal art of government is exercised in a field of constant questioning, searching, clarifications, and adjustments, focused on how a domain beyond direct government can be governed indirectly, without bringing it completely into the field of politics.

In his lectures of 1978–1979, Foucault discusses how liberalism was transformed into neoliberalism. At least three nuances are important here. First there is a shift in the view of the relation between state and market. Earlier statements on governing as little and carefully as possible are displaced by statements on the need to govern the market to make it function well. Intuitively this seems to indicate that the mechanisms of the market are reduced in power. However, it is possible to argue the opposite, and this is the second point: The market is no longer seen as a limited sphere, but is generalized throughout social space—also to parts of society previously not governed through the logic of economic calculation.

Connected to this we can see a third effect of neoliberal rationality, namely the idea of "one's life as the enterprise of oneself" (Gordon 1991: 44). Again we can talk about a displacement, this time in the understanding of the working man: In classical liberalism "homo economicus" is the fundamental figure, but now there are new connotations connected to this figure—a movement from the capitalist who seeks to maximize his profit, to the working "entrepreneur" of everyday life who through continuous calculations and choices seeks to maximize the effects of his actions.

The effect of these displacements is that while a liberal position used to imply a claim for less government, the claim rather has become to govern differently. In other words, while liberal discourses earlier contributed to delimit public power, it now seems that neoliberal discourses function as the most important form of rationality in the public sector. In the end this constitutes a reality where government and subjective practices have melted into one common segment. As in traditional liberalism we can still identify an idea of the market as a regime of truth, which can be used to assess whether the government functions well or not. When the market is "spread out," this then involves new claims to submit the public sector to continuous evaluation, a form of ongoing, evidence-based critique, where measurements and assessments of results function as a governmental baseline.

Reflections on Municipal Government

Can these thoughts on general transformations help us to understand current statistical practices of Norwegian municipalities? To answer this, we now move away from the observations of concrete practices toward a more reflexive level of government, as can be studied in the weekly Norwegian newspaper *Kommunal Rapport* (hereafter KR, all quotes from the material are my translations from Norwegian), which reports from and discusses problems and tendencies related to the municipal level. A search in the web archive[3] indicates increasing talk (from around the year 2000) related to new ways of using statistics. There are a variety of statements on this, but a main theme seems to be how to develop and improve the outcomes of government—this in a discussion related both to national governance of the municipalities, municipalities' governance of their areas of responsibility, and the operational units' governance of their services.

The discourse here seems to circulate around some basic governmental questions: Is it best to govern a municipality through detailed rules and centralized authority, or more indirectly through delegation, measurements, and numerical reporting? Should the main focus of the organization be on outcomes, or is the production of public services too complex for such reductions? Is measurement of satisfaction a fundamental indicator of quality, or does this ignore that municipal government also is about developing local society and upholding professional and legal standards? Should the numbers be used in processes of internal assessment and development, or is it necessary to make them public and comparable?

Related to these questions, it is possible to identify a discourse focusing on the complex reality beyond the numbers, problematizing the effects of capturing this reality through standardized measurements. In the words of one mayor: "We must not use the KOSTRA numbers to make all municipalities equal. The average must never become an answer" (KR November 9, 1999). The alternative here could be to argue for contextual insight and reflexive discussions as a better governmental strategy.

Another discourse, however, does not ask whether reality is measurable, but how we can improve the reliability and use of the numbers. Here we meet statements on the importance of measuring outcomes, both to assess the quality of services and to use the numbers as tools for improvement. At the national level a project leader for example argues for the value of establishing "a standardized system, where every municipality is integrated" (KR August 13, 2003). And a school leader relates good results in his municipality to systematic work on improvement, where "the numbers from tests and investigations are used" and where "the units which improve their results, are valued" (KR January 22, 2003).

An aspect of this discourse is a critique toward lack of numbers. In the words of a journalist: "There are frighteningly few key numbers that tell anything at all about the quality of the municipal services. Until we get better numbers, we have to live in the unknown" (KR January 12, 2006). Many of the statements in the material are related to this theme of knowing. This can be talk on the necessity for more systematic and comparable knowledge, the need to know the performances and effects of services, the importance of replacing "beliefs" with "facts," the possibility to see successes and failures through systematic use of indicators. To compare the numbers (either toward a standard, or the outcomes of others, or the numbers from earlier years), represents here a systematic technique that makes differences and changes visible, and through that possible to manage. An important aspect of this is the idea that competition sharpens quality, and that benchmarking constitutes a possibility for such competition: "The results are posted on the wall [. . .] then everyone can see where on the list they stand" (KR March 9, 2005).

This discourse can also be studied in documents supporting municipal implementation of Balanced Scorecard. This result-based system was originally constructed for American business, then transferred to the public sector where it diffused throughout the governmental landscape and seemed in just a few years to become a point of reference in the municipal governing-by-numbers discourse. An interesting empirical source here is two documents published by the Ministry of Local Government and Regional Development (2002, 2004), presenting and discussing examples on how various Norwegian municipalities have implemented and used Balanced Scorecard. In the words of the ministry, these texts present "a method to gain better governing," rooted in "long term visions, plain objectives, and systematic measures of results."

The strength of the methodology is said to be making visible and explicit the areas where it is most important to achieve success (2004: 5). To achieve this, the municipality has to inscribe existing and new indicators of outcomes into a system which is about "moving the focus toward the essence; the hunt for the most important, measurable fields of performance" (2002: 20). At the same time, however, the few indicators selected are intended to give an account of *the whole*: the results of yesterday, today, and tomorrow, external and internal circumstances, hard and soft results and/or contributing factors. If a municipality manages to achieve this, the organization has got a system of government that "in a genuine way gives a basis for assessments of whether the municipality is moving in the right direction" (2002: 20). In this way, one thread of the newspaper discussions interacts with ministerial statements; together they legitimate and give direction to the ongoing work on developing, implementing, and using statistical techniques as part of municipal government.

STATISTICS BETWEEN THE VERTICAL AND THE HORIZONTAL

The discussion so far has given a picture of the first dimension of Figure 4.1; the co-productive processes between *concrete* practices of governance and measurement, and the more *abstract* reflections on "how to govern" and "how to use statistics." Given the model, what remains is to go into the other dimension; the play between *vertical* grasps and moves of government, and *horizontal* practices of self-governance. The meeting point between these two is what Foucault defines government to be about. Of course, this theme has been an implicit part of the preceding discussions, especially in the thesis on adjustments from ambitious attempts of holistic, direct governing, toward a liberal reflection on the possibilities to achieve more while (or even by) governing less.

This double act of contributing to improvements while at the same time retreating creates challenges: The coworkers who operate in the fields of governmental practices—who have to administer without administration, govern without governing—are moved into processes where new technologies of government have to be designed and implemented. The main question, then, will be how to develop the constitutive elements of individual practices in a way that also strengthens the force of government. Or, in more technical words, how to constitute processes where humans as creating entrepreneurs, calculating economists, and choosing customers are cultivated, and through that transmit situational assessments and choices into instrumental practices of optimalization. In this story, hierarchical structures move into the shadows, while continuous reflections on how to govern ourselves and others, which techniques to use, and how to use them, come into the governmental light. Or, as Nikolas Rose (1999: 69) has formulated it, "the achievement of the liberal arts of government was to begin to govern through making people free."

Exact Numbers, Situational Choices

To illustrate the vertical-horizontal dimension we can move into one of the municipalities of our study, which has been organized in two formal levels since 2001: on the one hand, a strategic level that ideally takes care of the municipality as a whole; on the other, an operative level of different service units (schools, kindergartens, culture, health, etc.). In practice, however, it is not that easy: At the strategic level we find the chief officer and his six directors, supported by a staff of about fifty coworkers. At the operative level there are about 200 units each, with one leader and a flat structure of employees. Between the directors and the staff, between the strategic and operative levels, between the leaders and their employees—in all these meeting points we can identify a structure where the vertical and the horizontal is in flux. At the strategic level, ideally everyone represents the chief officer, but this is more easily said than done when the coworkers are more or less specialists with specific knowledge and experiences, and therefore never totally identical assessments.[4] This aspect is further complicated when we include the operative level, with its multitude of functions and competences.

All in all, this constitutes a complex field of practice inside a "low and wide" structure, where the flow of facts, assessments, and choices has a dynamic and open character. The effect of this is increased because the reduction of structural levels goes hand in hand with processes toward more empowerment, that is, a system where each coworker is increasingly expected to assess the situation and make choices based on situational possibilities. How can we understand the function of statistics in this dynamic organization? Pat O'Malley (2004) here gives a distinction that can be useable: On the one hand, the technologies of *risk*, which can be understood as different techniques for aggregating systematic knowledge about life, and the use of this knowledge to make life more secure. On the other hand, the practices of *uncertainty*, which point to situations where the real cannot be aggregated or calculated, and where our choices therefore have to be based on experience and "know how."

In one way the municipal adjustments described earlier could be seen as a radical shift where *risk* has been more or less replaced by *uncertainty* as the main governmental logic. This thesis, however, becomes less likely as soon as we talk not only of open structures and empowered coworkers, but also include the expanding use of result-based systems such as Balanced Scorecard, leader agreements, and ongoing assessments of outcomes. Techniques like these seem to be rooted in *risk technology*. The stress on both defining what is most important and seeing how these fundamental factors function as part of a whole is all about trying to avoid surprises, gaining control, inscribing every movement into a systematic journey. A technical condition for this is to work out explicit criteria of success through unambiguous and measurable indicators. More important, however, is that the coworkers actually act in accordance with the indicators—that is, actively choose solutions to produce positive effects in the statistical systems.

In other words, the storyline here can be said to be about an open system where humans have to interpret, assess, and act in a steady flow of new situations—and precisely therefore need something to relate their actions toward, something that both motivates and helps direct their conduct. The paradoxical question then is whether we are observing a tendency to give humans more possibilities to act through uncertainty, but where the numbers still work as a kind of risk technology. In other words, can the adjusted use of indicators and measurements be seen as something that *partly* simplifies the complex, but at the same time *partly* closes the open—and through that works as a sophisticated, dynamic, and indirect way to regulate the frames and direction of the practices of municipal coworkers?

In the two aforementioned documents from the Ministry of Local Government and Regional Development (2002, 2004), this question can be related to statements that stress *freedom of action, firmness on objectives*—in other words, that the focus should be "less on what the employees do and more on the effects that are achieved" (2004: 77). The introduction of management by outcomes, we are told, "is primarily motivated in the wish to give increased power of development to the municipal organisation" (2004: 38), and the objective is "a system of government where attention, conduct and control are moved from administrative to service-contribution processes" (2002: 55).

Other statements deepen this perspective: to strengthen the wish for development, "the need for standardizing has to be subordinate to the necessity for more adaptation to every unit" (2004: 59). Of course every result-based strategy has to be rooted in quantitative measurements of standardized indicators, but still it is stated that one should "attach more weight on ownership than exactness, because ownership towards the interpretations also means understanding for the initiatives" (2004: 39). Not strange then that involvement is stressed:

> [I]nvolvement in the work with the scorecard and the work with outcomes gives ownership and responsibility towards initiatives that have to be implemented. It is the employees who, in the final analysis, via changes in attitudes and ways of working, in fact work out the improvements (2004: 85).

From this perspective it becomes a point in itself that every employee should be able to go into the Balanced Scorecard, and see "which critical success factors and indicators the municipality are governed after, and which fields of focus they stand in relation to" (2002: 39). It is especially important to involve "the employees that have most direct effect on the outcome," because this gives both better analysis and increased "understanding and motivation to work out initiatives, and with that improve the possibility for the initiative to be carried out with the desired result" (2002: 50).

If these statements indicate a somewhat "soft" system, other statements mediate a tougher story, one where the measurements are fed into plan and budget processes and used to encourage improvement of outcomes, and where it is important "that initiatives are attached to the results of measures, and as much as possible come after those, when one has knowledge about the situation" (2002: 31). In one municipality these strong demands for action and consequences are summed up like this:

> When the results are available they must be analysed . . . A plan for improvement or maintenance of results is made within 3–4 weeks, with clearly defined and preferential initiatives. The plan is checked out. A new, revised result-agreement is entered into (2004: 75).

The perspective seems to be that all this work with defining strategic success factors, finding measurable indicators, setting objectives on what measures that are needed, reflecting all these in agreements for leaders and developmental talks, and so on, contributes to realism, strengthens the focus on quality, and urges everyone to do a better job: "Explicit results develop more explicit and result-oriented leaders" (2002: 10). If our discussion had been inside the field of theory of knowledge, much could have been said about this inconsequent movement between, on the one hand, the argument that exactness has to be given lower priority than adaptation, involvement, and interaction—and, at the same time, the demand for using the (nonexact) numbers as the (exact) basis for initiatives and plans.

Governing by Indicators and Outcomes 93

Analytically, however, what is important is not whether the practices of municipalities live up to criteria for serious knowledge. Rather, the inconsequence in itself can be interpreted as functional—because, as indicated previously, it probably tells how the system of result measurements is practiced through intertwining processes between the technologies of *risk* and *uncertainty*: From one point of view the nonhierarchical organization and empowered employees seem to be part of a more open and dynamic reality, where the interplay between employees, competences, roles, functions, levels, and so on, is situated in time and space, beyond direct government. At the same time, from another point of view, we see something quite different: The indicators and their measurements sharpen the focus of the participants, tell how things are connected, open up for comparisons and competition, increase the possibility for prediction, indicate what should be maintained or increased—in short, they function as a sophisticated, subtle reregulation of the municipal vertical-horizontal order.

CONCLUSION

This chapter started with a short presentation of *The Wire*, intended to function as a mental picture through the following discussions. Or, from the opposite perspective, the theoretical reflections and empirical examples have outlined a context which gives analytical meaning to the series. The Wire, of course, is fiction—but this does not mean that the picture it gives is wrong or too simplified. Up to a point, the statistical practices portrayed parallel empirical realities, even in a "social democratic" country such as Norway.

The baseline of the municipal analysis was an observation of a "levelling out" of vertical structures, and increasing focus on the horizontal practices of self-governance—tendencies which were related to expanded and more flexible uses of statistical tools at all levels of government, and explained in the co-constructive interplay between the concrete techniques and tactics of governance and more abstract governmental processes of discussion, reflection, conceptualizing, and problematizing. Of course there are many nuances in this field, not least connected to the fact that the influence of the sketched neoliberal discourses is here observed in a social democratic landscape. Probably this means that there are elements that in a given situation can be articulated into alternative discourses, for example through a revitalizing of the vertical, bureaucratic components of government. This, however, can still be analyzed through Figure 4.1—at least as long as we can observe reflections and practices connected to the inclusion of more horizontal processes in the governmental game.

As typical for the Foucauldian project, the analysis of governmentality does not picture a power that represses, which says no; rather, it indicates a technology of government which transforms "freedom itself and 'the soul of the citizen', the life and life-conduct of the ethically free subject" to an object for its own capacity (Gordon 1991: 5). Through the empirical material I have indicated how this can be connected to pluralizing words such as "deregulation," "competition," and "freedom of choice." At the same time, however, my argument is that this increased freedom and competition is co-productive with a space for ongoing, technical-instrumental processes where targets are fixed, contracts are signed, indicators are defined, procedures for measurement and reporting are developed, strategic moves of improvement effected, and so on—in short, a space of intensified control through self-governing.

Inside this neoliberal governmentality the acting subject is no longer identical with the actors of the social democratic formation. Rather we meet a rational exerciser of strategic choices, an active coplayer who is increasingly integrated into the government of himself and others and who through calculations, in interplay with other calculating actors, is expected to maximize benefit. As art of government this liberation can be understood as a move in the subtle practices of governing, where aggregating the acts of individuals goes hand in hand with techniques for the framing and forming of these acts. In this play statistics has the advantage that it can function both as a basis of knowledge and an incitement to action. Related to practices where everyone, continuously and constantly, is expected to make strategic moves in constructed fields of assessment and choice, statistics contributes to a process of subtle reregulation, where the indicators, measurements, and numbers purportedly work to optimize what we do and what we achieve.

Through this chapter we have just touched on what happens in a concrete field of practice when statistical techniques are inscribed into open processes of self-governing. The intention has not been to give any empirical conclusions, but rather to contribute at a meta-level to an analysis that communicates with the other chapters of this book.

NOTES

1. The model also relates to Kullenberg's (this volume) analysis, where every performative element is understood as part of complex processes of construction, and where the productive force of the singular element is a contextual, empirical question rather than a theoretical one.
2. In the years 2001–2005, I was working in a Norwegian municipality, at an internal service with responsibility for assessment and development. This gave an opportunity to observe the strategic and operational implementation of new techniques of measurement and government, an "analytical view from the inside" which constitutes an empirical basis for my interpretation of municipal practices (see Hammer 2004).
3. The research was done through logging into the website of the paper, and then using the search terms "measure," "quality," "governing," "assessment," and "benchmarking" for the years 1999–2006.
4. At the operative level this sometimes leads to the characteristic of the staff as "a many-headed troll."

REFERENCES

Bukve, O., and A. Offerdal (2001) *Den nye kommunen: kommunal organisering i endring*. Oslo: Samlaget.
Desrosières, A. (1998) *The Politics of Large Numbers*. Cambridge, MA: Harvard University Press.
Foucault, M. (1975) *Surveiller et punir, naissance de la prison*. Paris: Gallimard.
———. (1976) *Histoire de la sexualité 1. De volonté de savoir*. Paris: Gallimard.
———. (1993) "About the Beginning of the Hermeneutics of the Self," *Political Theory*, 21(2): 198–227.
———. (2007) *Security, Territory, Population: Lectures at College de France 1977–78*. New York: Palgrave MacMillan.
———. (2008) *The Birth of Biopolitics: Lectures at College de France 1978–79*. New York: Palgrave MacMillan.
Gordon, C. (1991) "Governmental Rationality: An Introduction," in *The Foucault Effect: Studies in Governmentality*, eds. G. Burchell, C. Gordon, and P. Miller, 1–51. Chicago, IL: University of Chicago Press.
Hammer, S. (2004) "Hvordan kan vi måle dette?" ["How can we measure this?"], *Sosiologi i dag* 4/2004.
———. (2008) "Styring, statistikk, subjektivitet" ["Government, statistics, subjectivity"], *Tidsskrift for Samfunnsforskning*, 49(1): 73–106.
Johnsen, Å. (2007) *Resultatstyring i offentlig sektor*. Bergen: Fagbokforlaget.
Kommunal Rapport, http://www.kommunal-rapport.no/ (accessed February 21, 2006).
Ministry of Local Government and Regional Development (2002) "Resultat og dialog. Balansert målstyring i kommunal sektor," http://www.regjeringen.no/upload/kilde/krd/rus/2002/0005/ddd/pdfv/148013-veileder_bms.pdf (accessed April 7, 2010).
———. (2004) "Resultatledelse. Bruk av balansert målstyring og andre former for systematiske resultatmålinger i kommunal sektor," http://www.regjeringen.no/upload/kilde/krd/bro/2004/0004/ddd/pdfv/221475-veileder_resultatledelse.pdf (accessed April 7, 2010).
O'Malley, P. (2004) *Risk, Uncertainty and Government*. London: Glasshouse Press.
Opedal, S.H., I.M. Stigen, og T. Laudal (2002) *Flat struktur og resultatenheter: utfordringer og strategier for kommunal ledelse*, NIBR-rapport, Oslo: Norsk institutt for by- og regionsforskning.
Rose, N. (1999) *Powers of Freedom*. Cambridge: Cambridge University Press.
Sending, O.J. (2003) "Fattigdom og politisk rasjonalitet," in Neumann & Sending (eds.): *Regjering i Norge*, Oslo: Pax forlag.
Spurkeland, J. (2005) *Relasjonskompetanse: resultater gjennom samhandling*. Oslo: Universitetsforlaget

Part II
Visibility, Invisibility and Transparency

Metropolitan College of NY
Library - 7th Floor
60 West Street
New York, NY 10006

Part II

Visibility, Invisibility and Transparency

5 Ethnicity
Differences and Measurements

Ellen Balka and Kjetil Rodje

INTRODUCTION

The Canadian Health Act suggests that health services should be equitably and universally accessible to all Canadians.[1] In an era that emphasizes measurement and accountability (Timmermans and Berg 2003, Wiener 2000), we ought to be able to determine whether or not services are equitably distributed and universally available, and, if they are not, Canadians should be able to use the Canada Health Act to argue for more equitable service delivery.

Although interpretation of "universal and equitable delivery of services" within the Canada Health Act can be debated, the diversity of the Canadian population[2] (Statistics Canada 2009) and Canada's focus on multiculturalism suggest that equitable can be interpreted to mean access to health services across diverse ethnic groups. Although in principle this seems simple, in practice, operationalizing this idea is quite complex and raises numerous ethical issues and debates which are the subject of this chapter.

UNPACKING ETHNICITY

Ethnicity is a frequently used and often ill-defined concept. It can reflect many different factors, such as place of birth, nationality, physical features, language, ancestral origin, name, religion, identity, sense of belonging, culture, tradition, lifestyle, and so on. Multiple aspects of ethnicity are often grouped together, and interpretation of the meaning of ethnicity is often left to survey respondents and research users.

Ethnicity is hard to classify and operationalize, and the very act of using ethnicity as a metric raises complex questions about how and by whom ethnic labeling and categorization might—or should—be applied. We can assign ourselves ethnicity through self-identification, or ethnicity can be a label or category applied to individuals or groups by others (e.g., relatives, communities, governments, media, researchers). Ethnicity is often confused with race, which may be used as a proxy for characteristics often viewed as synonymous with skin color.

Ethnicity is a complex and heterogeneous constellation of many factors, and agreement is lacking about whether it should be measured, and, if it is measured, how this should be done. Holding the Canadian government accountable for universal and equitable provision of health services across ethnic groups creates a need for data about ethnicity and health which can be used to determine how accessible health services are. Without such data it is impossible to determine whether or not gaps exist in service delivery, or health status.

In this chapter, we consider issues which arise in relation to determining whether or not ethnic groups in Canada have equal access to health services. We begin with an overview of health service use and health status among ethnic groups, which sets the stage for examination of the issues which arise when conducting research about ethnicity and health in Canada. We outline some of the challenges associated with the measurement of ethnicity, and continue with a discussion of ethnicity and race. As Sætnan, Lomell, and Hammer (this volume) suggest, truth claims, including those based on statistics, have histories which are cumulative, and statistics are not simple, straightforward objective descriptions of society. This point is borne out in our review of how ethnicity has been measured in Canada, which nicely demonstrates how our understanding of the ethnic composition of Canada has been constructed over time. After outlining debates about whether or not ethnicity should be measured in relation to health, we end by considering the practical implications of measuring ethnicity for health service delivery. Our intention is to highlight the complexities inherent to answering the following question: Is access to and use of health services universal and equitable among varied ethnic communities? Our exploration demonstrates how attempts to address this issue contribute to either the reification or redefinition of health and ethnicity statistics, paving the way for potentially powerful and dangerous utilization of statistics.

ETHNICITY AND HEALTH: WHY MEASURING ETHNICITY MATTERS

Previous research concerned with ethnicity and health suggests that there are differences in health status between ethnic[3] groups (Wu et al. 2003, Irby et al. 2006, Lepore et al. 2006), differences in perceptions of health risks and symptoms along cultural lines, and differences in use of health services among varied ethnic groups. Health researchers are often interested in ethnicity because they have observed differences in how members of varied ethnic groups respond to health interventions, or because they have observed that some populations appear to be more susceptible to certain diseases than other populations. For instance, death rates from ischemic heart disease from 1979 to 1993 were higher among Canadians of

South Asian or European origin while markedly lower among Canadians of Chinese origin (Sheth et al. 1999).

Generally, immigrants to Canada are healthier upon arrival than the general Canadian population, but this advantage is lost over time (Chen et al. 1996a, 1996b; Vissandjee et al. 2004). Explanations for this decline over time include adoption of a Canadian lifestyle (and the health risks related to diet, use of alcohol and drugs, etc., more common in Canada than in immigrants' home countries), and that immigrants may suffer the health effects of a disadvantaged position in Canadian society related to lower income and worse living conditions than the Canadian population. Psychological and psychosomatic factors related to difficulties integrating into Canadian society have also been offered as explanations for health declines over time among immigrants.

There are also differences between various ethnic groups in use of health care services. Sharif et al. (2000) report that cultural factors explain low use of preventive health care services among some minority groups. Low health services utilization among nondominant ethnic groups in Canada has been attributed to racism in the health care system, cultural differences in perceptions of health, and language and communication barriers. While ethnicity is often used to describe differences in health outcome and use of health services, no definite standards exist to address which indicators to use for measuring differences.

Once a link is made between ethnicity and a health issue, causality is not necessarily determined. For instance, one can not necessarily tell whether higher rates of death from ischemic heart disease among particular ethnic groups are due to lifestyle, genetics, or other factors, such as social disadvantage; likewise, differences in health service utilization among a specific ethnic group over time can reflect variations in how ethnicity was measured over time, as well as actual changes in health service use. Understanding the causal links associated with the prevalence of a health issue within an ethnic group has important implications for treatment and prevention. For example, if the high prevalence of a disease is linked to diet, interventions aimed at diet change might be most successful. However, if social determinants of health (the economic and social conditions under which people live which determine their health) are found to be causal, stable housing may be a more effective intervention.

Although it may be possible to document increased prevalence of a disease among a particular ethnic group once a working definition of ethnicity is in place, doing so raises many questions. For example, how should a particular ethnic group be defined? How should an ethnically diverse population be approached regarding health services and information? Should various groups and individuals be treated equally or differently in order to serve their best interests? What does it mean to provide ethnic groups with equal (or different) services and information?

ETHNICITY AND RACE

The concepts race and ethnicity are sometimes viewed as synonyms, and sometimes viewed as fundamentally different. Ethnicity is a term often used to explain a sense of cultural belonging and identity, while race is often used to explain genetic characteristics and other biological factors. In his glossary of terms relating to ethnicity and race, Bhopal (2004, 443) defines ethnicity as "(t)he social group a person belongs to, and either identifies with or is identified with by others, as a result of a mix of cultural and other factors including language, diet, religion, ancestry, and physical features traditionally associated with race." Race is characterized thusly: "[b]y historical and common usage the group (subspecies in traditional scientific use) a person belongs to as a result of a mix of physical features such as skin color and hair texture, which reflect ancestry and geographical origins, as identified by others or, increasingly, as self identified" (Bhopal 2004: 444). Bhopal stresses that increasingly these two concepts are used synonymously, especially in the United States, where the combined term race/ethnicity is in common use. Race and ethnicity are often confused in health studies. For example, data may suggest that populations with a common background may have a higher than average occurrence of a particular illness, and the causality may be attributed to genetic or racial characteristics, however data collection may not make it possible to determine if causality flows from what was once commonly thought of as racial factors (e.g., genetics), as opposed to ethnic variables (e.g., diet).

Ethnicity and race are often portrayed as characteristics of "the other," while the "white" ethnic or racial group is then seen as the standard other groups are measured against. This "white" category is itself not easily determined or assigned, as no clear definitions exist (Aspinall 1998). Terms such as "white," "Caucasian," and various terms describing geographic origin such as "European," "Anglo-American," and so on, are often used to describe this category. The predominant terms used in medical research are "white" and "Caucasian," although the latter term has been heavily criticized for its association with what is today seen as scientific racism (Aspinall 1998, Bradby 1996). Still, both terms usually refer to people of light skin color, although no clear definition exists about how light skin must be to be classified as white. Skin color is also relative—a light-skinned Iranian might be viewed as white in Iran and nonwhite in Canada. The term Caucasian may also be problematic for health researchers in that it amalgamates several arguably distinct populations into one category, which may mask significant differences among subgroups of Caucasians.

MEASURING ETHNICITY IN CANADA

Canada is a multiethnic society built on immigration, and is one of the most ethnically diverse countries in the world. In 1996, 44 percent of Canada's total population reported origins other than British, French, or Canadian (Pendakur and Hennebry 1998). In the 2001 Census,[4] 18 percent of the population were born outside of Canada and 13 percent identified themselves as visible minorities (defined as "persons, other than Aboriginal peoples, who are non-Caucasian in race or non-white in colour"[5]). The incidence of people reporting an ancestry of multiple ethnicities has risen in recent years. In 2006, 41.4 percent of the population reported more than one ethnic origin, compared to 38.2 percent in 2001 and 35.8 percent in 1996.

Measurements and Definitions in Use

Variation exists in how data about ethnicity have been collected over time within the same government department, as well as between departments. Such variation serves as a reminder that statistics are socially produced, and the ways that we count and categorize change over time to reflect varying social norms, interests, and worldviews. Complicating matters further, ambiguity of meaning associated with many of the terms used in collecting data about ethnicity in Canada remains.

Statistics Canada is responsible for several population studies and public health surveys,[6] including the Canadian Community Health Survey (CCHS) which contains the most in-depth information about ethnicity,[7] and the population Census, which takes place every five years and is the largest in scope and most influential of Statistics Canada surveys. Canadian Census data about ethnicity focuses primarily on two areas: geographical origin and language. Statistics Canada's health surveys view ethnicity as a constellation of related variables such as place of birth, citizenship, and immigration status (whether a person is a citizen or not, whether they immigrated to Canada or not, and if they immigrated, how long ago they immigrated) and language use (e.g., first language spoken, language spoken at home and work, and what languages one can converse in).

Statistics Canada also makes a distinction between one's ethnic and cultural origins (e.g., the ethnic or cultural groups the respondent's parents belonged to [such as French, Scottish, or Chinese], and one's ethnic/racial status [meaning white, Chinese, South Asian, Black, Latin American, etc.]) and Aboriginal status. In 1996, a new question was added to the Canadian Census about visible minorities, in order to collect information "to support programs that promote equal opportunity for everyone to share in the social, cultural and economic life of Canada,"[8] which reflected the Canadian Multiculturalism Act and the Canadian Charter of Rights and Freedoms. However, the introduction of visible minorities as a variable has provoked controversies because it was seen as a (re)introduction of racial terms in the Census.[9]

Interestingly, while there is overlap in questions asked in surveys administered by different government departments, question wording often varies from survey to survey, making it impossible to compare information gathered through different surveys. For example, the CCHS asks immigrants what year they first came to live in Canada, while the Census asks what year a person first became a landed immigrant.[10] Variation also exists in the extent to which some topics are addressed. For example, the 2006 Census asked respondents which language they speak at home, and although the CCHS asks several questions about language (e.g., which languages the respondent is conversant in, which language the respondent first learned at home and still understands), it omits this question about language, which might prove to be a useful proxy for degree of assimilation in Canadian society, which in turn may help explain differences in utilization of health services or health status. Each different set of questions allows us to view the relationship between the constellation of variables which together tell us something about ethnicity in different ways.

Researchers interested in ethnicity must decide if they will use common and already validated measures (such as Statistics Canada measures) which can support comparison to preexisting data sets, or use new sets of questions that could support development of new insights about ethnicity and its relationship to health. These are important questions in part because, as the editors of this volume have argued, "to count something or someone is to make it/them count in the policy sphere, and by corollary, only that/those which already count tend to get counted" (Sætnan et al., this volume).

Interpreting Ethnic Differences in Health

Comstock et al. (2004) have identified two major concerns associated with studies related to ethnicity. First, considerable potential for a person's race or ethnicity to be inaccurately classified or misrepresented exists, because of challenges associated with operationalizing the notion of ethnicity, outlined earlier. A person's own ethnic identity may differ from the category assigned by others (e.g., those using a study). Second, Comstock et al. (2004) highlight the risk of failing to recognize underlying factors and variables associated with ethnicity. That a study finds marked discrepancies in health between ethnic groups does not by itself explain differences. Factors such as socio-economic status (SES) may play a significant causal role in many cases. Groups with a low SES are likely to come out worse where SES is correlated with health, and specific individuals can score better on a given health variable in relation to socio-economic factors rather than his or her ethnic status. Nevertheless, some research has shown a significant discrepancy in health along ethnic lines even after controlling for SES (e.g., a study about race, ethnicity, and depression in Canada found that socio-economic factors explain some, but not all, differences in mental health between ethnic groups [Wu et al. 2003]). In contrast, another study based on U.S. data found no significant differences in mortality between blacks and whites when mortality was adjusted for per capita income (Sterling et al. 1993). Consideration of the link between SES, health, and ethnicity highlights the need to investigate other factors before concluding that ethnic differences in health have any biological or cultural explanation, and underscores the need to include variables that will allow researchers to assess the impact of SES on health, and disaggregate SES from ethnicity in terms of statistical effect on health.

Some variations in study findings about the role of ethnicity in health differences result partly from how health, ethnicity, and SES are measured, which suggests that results should always be interpreted in light of specific definitions used in studies. If a study finds no difference in self-reported annual income between ethnic groups this does not necessarily mean that no differences in SES exist—it might be the case that one of the groups had greater access to economic goods that were not captured in the study. A possibility always exists that other factors related to the variable of study have not been captured through the means used to measure variables such as SES or ethnicity.

Krieger (1999) outlines four alternative explanations when ethnic/racial differences in health persist after checking for SES. She suggests that SES may have been inadequately measured leaving real differences related to socio-economic factors unaccounted for. Second, differences in health may be due to noneconomic aspects of racial discrimination that are not measured, such as effects of stress or abuse. Third, differences may be due to other unmeasured factors that are neither due to SES nor discrimination (such as cultural differences in nutrition). Fourth, health differences may be due to innate genetic differences.

Differences in SES may in part be explained by discrimination or institutional racism, and hence the relation between racism and social class and ethnic disparities in health can be a question of "both/and" rather than "either/or" (Krieger 2003). This does not imply that the health effects of racism will be fully explained by SES. Rather, it is likely that racial/ethnic disparities in health cannot be reduced to differences in class or SES (Krieger 2000). In addition to material factors such as SES, living conditions, family relations, employee status, and so on, studies (e.g., Williams 1999, Karlsen and Nazroo 2002) have found that racism itself harms health, be it in the form of personal experiences or perceived discrimination of one's ethnic group.[11] Nonetheless, socio-economic differences should be considered when explaining differences in health between ethnic groups (Senior and Bhopal 1994) in order to be able to better distinguish the health modifying effects of SES from ethnicity. Still, correlations between low SES and poorer health outcomes among specific ethnic groups may explain differences, but does not make them go away. This raises questions about when and how ethnicity should be deployed as a statistical measure in health research—and these questions are poltical as much as they are scientific.

ETHNICITY IN HEALTH RESEARCH—
SHOULD ETHNICITY BE MEASURED?

Although there is much interest in assessing health in relation to ethnicity—particularly in countries such as Canada where the Canada Health Act "is designed to ensure that all residents have reasonable access to medically necessary hospital and physician services"[12]—questions remain about whether or not ethnicity should be considered as a variable in health research. Although it may sometimes be beneficial to categorize health data by ethnicity, this can come at a cost and may also have negative consequences. For instance, drawing conclusions based on explanations of ethnicity alone can serve ideological functions or mask underlying social factors that may explain ethnic differences in health (Anand 1999). Linking health outcomes to ethnicity also can contribute to stereotypes based on ethnic factors, for example by portraying specific groups as extra vulnerable or exposed to certain diseases or hazards as a result of behavioral or lifestyle characteristics associated with those groups. McKenzie and Crawford (1996) advise researchers not to draw any conclusions which go beyond the available data, and caution researchers to make sure that there is correspondence between what is measured and interpretation of measurements.

Senior and Bhopal (1994) identified four fundamental problems with ethnicity as a key variable in research about health. It can be difficult to measure; heterogeneity of the populations being studied may not be reflected; there may be a lack of clarity about the purpose of the research; and, ethnocentricity may affect the interpretation and use of data. This implies that ethnicity is never an unproblematic variable to include in a study. Senior and Bhopal recommend adoption of guidelines for using ethnicity as an epidemiological variable, which include differentiating ethnicity from race and acknowledging the complex and fluid nature of ethnicity and limitations of current methods of classifying ethnic groups. They advocate for disclosure and reporting of the basis upon which ethnic classifications are made, and call for investigators to recognize the potential influence of their personal values, including ethnocentricity, on scientific research and policy making. They suggest that caution should be exercised before results about ethnicity are generalized across time, generations, or populations with different migration histories. Results from studies about ethnicity and health should be analyzed and applied to health services planning, and the relative importance of environmental, lifestyle, cultural, and genetic influences should be studied when observing variations in the prevalence of a disease (Senior and Bhopal 1994).

Ethnicity may be used (or abused) as a health variable for political and social purposes. By measuring health discrepancies between ethnic groups, it is possible to detect discrimination and differences in standards and services available for various subpopulations. Still, observing or documenting inequalities alone has little impact on reduction of such inequalities

(Bhopal 2001). Bhopal (1998) observed that close and repeated observation and tracking of inequalities in health between black and white people in the United States has been accompanied by a widening, not narrowing, gap. Data alone will not ameliorate ethnic discrepancies in health, but may be a precondition for developing interventions aimed at improving health of populations that share characteristics (such as diet). It remains unclear to what extent we require even more knowledge about existing differences in health in order to diminish these differences.

In contrast to the view of ethnicity research taken by Bhopal (1998, 2001) and others, Risch et al. (2002) argue that the critique of racial categories in biomedical research is not scientific and objective. Their claim is that racial/ethnic self-categorizations are objectively and scientifically valid, and that from a public policy perspective it is important to know which segments of a population are more susceptible to certain diseases or are more likely to benefit from particular therapeutic interventions. Risch et al. argue that five genetically different human races can be distinguished, according to continental origin: Africans, Caucasians, Pacific Islanders, East Asians, and Native Americans. Risch et al. suggest that such distinctions should not be conflated with politics and value systems, but are rather objective and scientific means to provide better risk assessments and treatments. They argue against a "race-neutral" approach to biomedical research on the basis that ignorance of racial differences will reinforce gaps between groups of different genetic constitution.

This stance is countered by the view that race and ethnicity are social rather than biological categories (e.g., Krieger 2000, 2003). However, Krieger (2000, 2003) also suggests that these categories should still be maintained in order to provide standards against which to measure the effects of racism and discrimination. The rationale here for collecting information about race and ethnicity is that without such categories we have no way to detect discrimination and disparities, and differences will be harder to abolish in the absence of data that demonstrate the existence of health disparities along ethnic lines. Social categories are viewed as no less real than biological categories, and, as such, are valid variables in biomedical and epidemiological research. However, detecting ethnic differences is not the same as saying that these differences reflect inherent ethnic or racial characteristics. Differences in ethnicity and racial characteristics may be correlated with differences in social determinants of health and/or treatment—hence care must be taken to interpret statistical patterns of ethnic inequalities. Fausto-Sterling (2004) shares concerns raised by Krieger (2000 and 2003) about the social determinants of racial and ethnic categories, but is more skeptical about the need for further documentation of such ethnic discrepancies in health, and argues for more action rather than more information and knowledge.

Doyle (2006) argues that genetics does not make race and ethnicity redundant as these concepts also encompass nongenetic factors such as culture, behavior, diet, environment, social status, and so on. This is not an unproblematic argument, however, because making assumptions about such traits as behavior and diet on the basis of racial and ethnic characteristics can reinforce stereotypes and prejudices. Hence it is important to consider whether or not an individual possesses specific risk factors (e.g., smoking), rather than draw conclusions based on general population characteristics. For example, if smoking is more common among Hispanics than Caucasians, knowing this may be critical to the development of successful health promotion programs, but does not accurately yield information about an individual's behavior or susceptibility for smoking related health issues.

It is unlikely that all discrepancies between ethnic groups in health and reactions to treatments can be reduced to specific variables because complex relations and synergies of many factors contribute to making some groups more vulnerable—or protected—than others. Following this reasoning, measuring ethnicity in the context of health research is of questionable use, since drawing conclusions about causality is unlikely. Ethnicity as a (health) variable is arguably a black box where casual relations are at least partially unknown and where the results and connections made will always be open to interpretation and discussion. The concept of ethnicity may only provide insights about possible statistical correlations, and may fall short of explaining causality.

These issues suggest that potential negative consequences of a study, as well as benefits, should be considered in advance of collecting ethnicity data, and the relevance of ethnicity to research questions needs to be scrutinized. Varied means of operationalizing ethnicity should be carefully evaluated, and both studies and policies concerned with ethnicity need to take into consideration how data will be used, and potentially abused, in the future.

ETHNIC CONSIDERATIONS IN HEALTH SERVICES

The question of how to approach ethnic diversity in relation to health issues has practical consequences. Ethnically and culturally diverse populations raise potential challenges for the health care sector. Health provider agencies typically deal with a scarcity of resources and need to decide how to allocate resources among competing needs. For example, service providers must determine to what extent health care can be provided in a country's nonofficial languages, and ideally should consider how varied cultural views and belief systems may influence decisions to seek treatment, as well as treatment compliance.

Numerous approaches can be taken in responding to the challenges that ethnically diverse populations pose in relation to provision of health services. For analytical purposes, we have grouped these strategies into four categories, which we refer to as approaches that stress homogenizing, establishing basic common ground, targeting, and diversity. These categories should not be understood as realistic representations of existing practices, but rather, can be viewed as ideal types, each of which emphasizes certain key characteristics in their pure forms. These categories, which are not mutually exclusive and may appear in overlapping forms, are intended as simplifying tools which existing institutional practices can be measured against. Each is outlined briefly next.

Homogenizing Approaches

Homogenizing approaches imply that ethnic minorities themselves are responsible for adapting to the dominant culture and preexisting systems. Maintaining a stance of homogenization implies that no changes or special considerations should be made to health services to accommodate ethnic minorities because it is the responsibility of these groups to use the established system. The health system's responsibility is one of providing information about available health services, guidelines and standards, which will allow diverse groups to make rational use of the services available according to their own needs. This approach does not account for factors such as language barriers, cultural differences, lifestyle and risk factors, and holds those most likely to require additional or different services responsible. This approach is not culturally sensitive, and may erase or threaten the established cultures and lifestyles of minorities.

Basic Common Ground Approaches

Common ground approaches imply that there are a set of core needs and problems common for all people, regardless of race or ethnicity. Differences are downplayed, and the perception is that everyone shares the same basic needs. These approaches rest on the belief that ethnic differences are just present on the surface, and that underneath the cultural superstructure all humans want and need the same things. This emphasizes what is common rather than what differentiates people. Common ground approaches focus on core services that address what are deemed shared basic needs, rather than on providing services tailored to various groups. Defining common ground and specifying core services required to meet common needs presents challenges, including whether or not services defined as core have been neutrally constructed, or if they have been deemed core services on the basis of specific "naturalized" ethnic groups that are then taken as the standard for others to adhere to. This approach also fails to address what is not included in the notion of common ground. Nondominant groups may be at risk if they have a high prevalence of a health problem that is not widely shared by other groups due to a failure to acknowledge any particular needs or characteristics that may emanate from or be specific to cultural differences. Cultural differences are not accounted for in this strategy.

Targeting

This approach attempts to identify the special needs or characteristics of different ethnic groups, and then target services or information toward them. The emphasis is on the differences between various groups which are treated as separate subpopulations with various characteristics. In this approach, groups characterized with special needs or risk factors warrant special services and information targeted toward them. Examples of this can be campaigns directed toward subpopulations perceived to be of special risk for exposure to HIV infections, or that are characterized with high rates of risk factors such as smoking or substance abuse.

The success of this approach depends upon whether or not the information and perceptions about targeted groups is correct and useful. Lack of flexibility of targeted health service activities (e.g., health promotion materials translated into Cantonese will not be useful to Hindu-speaking populations), and possibly time-consuming reconstruction (if targeted populations change) can pose problems. This approach can further reinforce cultural, ethnic, or racial stereotypes. Targeting particular groups in relation to specific lifestyles and risk factors can contribute to the reproduction of stereotypes that link specific health issues to specific populations (e.g., that link risk of HIV or drug addiction to Aboriginal populations). Any reinforcement of cultural stereotypes may further cultural or racial prejudice and discrimination—be it from governmental authorities, health care personnel, the population at large, or the groups in question themselves. Targeted health service delivery can also inadvertently signal to groups where health issues are less statistically prominent that they are not at risk, which can contribute to increased risk behaviors.

Diversity

This approach emphasizes openness, flexibility, and multiplicity, and seeks to meet each patient or group on their own terms, through provision of services and information according to diverse needs. Populations, patients, or clients are encouraged to express their concerns and needs, which the health system should attempt to meet to the greatest extent possible. Ethnicity should not be regarded as a characteristic of specific individuals or groups, but rather as a characteristic of population diversity, implying that different people may have different needs and should be met and treated accordingly. Rather than drawing conclusions based on an individual's ethnicity about what services that person may need, this approach seeks to approach the population as a whole which is made up of a multitude of different practices and cultures, as well as genetic and biological traits, which constitute sets of challenges that confront health care service providers. This strategy places heavy demands on health care institutions and workers in terms of knowledge, skills, and flexibility, as staff must be able to understand, communicate with, and accommodate a huge diversity of needs. This requires good and reliable information, training of personnel, feedback from users of various ethnicities, flexible technical facilities, sufficient resources, and so on.

DISCUSSION

Each of the approaches to provision of health services just outlined has implications for whether or not ethnicity should be measured in relation to health, and if so, how it should be measured or categorized. Implicit to each approach is the need for somewhat different information, which in turn has implications for health researchers. The homogenizing approach makes research about ethnicity irrelevant to the health care sector as all services available will be the same regardless of the ethnic composition of the population and any special needs or risks associated with various ethnic groups. Although the underlying logic of the common ground approach differs markedly from the homogenizing approach, it suggests ethnicity need not be especially considered or researched—other than to state that there actually exists a common core of needs and problems regardless of ethnic status, and to define this core. Hence, this approach implicitly rests on an assumption that ethnic differences are only of limited medical relevance and do not need to be considered in planning and delivering core health services, but research will still be required to determine what services should be core services.

The targeting approach, on the other hand, is highly dependent upon health research using ethnicity as a variable, as this provides information that makes it possible to assess specific needs and risk factors associated with ethnic groups. Up-to-date and accurate information that considers nuances between, and heterogeneity within, ethnic groups is a precondition for development of successful targeted health services. The risk of stereotyping or misconceptions increases if services and information are based on research using flawed conceptions of ethnicity, or where the categorizations of ethnicity used in the research differ markedly from the target populations. This approach necessitates that health practitioners involved in targeted campaigns or health care work remain up to date about research related to the targeted population(s) they serve.

The diversity approach is less dependent upon accurate information regarding specific ethnic groups. However, knowledge about variation in a population is necessary in order to accommodate diverse needs and prevent increasing discrepancies in health. This approach is highly reliant upon accurate and updated health data, but this information need not necessarily be classified according to ethnicity. What is required is information about the heterogeneities and multiplicities running across the population, and information about correlations of variables that may have implications for service delivery. Rather than fixed ethnic categories, the diversity approach calls for information about the multiplicity of health issues related to a population, as well as the specific correlations between, on the one hand, health factors and risks and, on the other hand, biological and behavioral characteristics. This approach seeks to identify linkages among diverse variables that may influence health, and which may or may not be related to ethnicity. This approach pursues specific and concrete linkages between health and issues such as smoking, diet, language, social networks, genetic factors, and so on, across the entire population, rather than in relation to specific ethnic groups. As such health services and information can be provided according to research that provides insights about specific correlations, rather than characteristics of ethnic groups. Ethnicity as a general category thus becomes redundant in this approach.

CONCLUSION

As the material here suggests, statistics are hardly simple, straightforward descriptions gathered from neutral points of view. The statistics we collect illuminate some populations and some sets of relationships, while obfuscating others. As our discussion about the four approaches to ethnic diversity in provision of health services suggests, our overarching views of the world have implications for what is measured, how it is measured, as well as what underlying perspectives can be supported in policy debates. Each time we make decisions about how ethnicity will (or will not) be operationalized in research, we are often implicitly reifying some views of the world, and undermining others.

In a society that values cultural and ethnic diversity, public policy that specifically identifies universality and equity of service provision as a goal creates an implicit need for measurement to be undertaken in the service of accountability. Yet the act of measuring ethnicity in relation to health remains complicated, in part because of definitional ambiguity of the term. Additionally, our ability to see over time whether or not we are improving the health of various ethnic groups depends partly on maintaining consistency in how ethnicity is measured. Yet maintaining such consistency may constrain our ability to further refine our understanding of ethnicity, and the impact of ethnicity on health. Each of the decisions we make about how we count brings some views of the world into focus while leaving others outside of our visual field.

Although measurement can bring some phenomenon into view, and as researchers we may undertake measurement (for example) in the interest of documenting an inequity, it is important to realize that while measurement can render a phenomenon visible, interpretation of data is still, to a degree, in the eyes of the beholder. Statistics are always subject to interpretation, and pointing out that a particular ethnic group has a disadvantaged health status is no guarantee that corrective actions will be taken—the decision to address health disparities based on ethnicity is more likely to be a values-based decision than a decision based wholly on statistics. Yet at the same time, statistics will likely be required to determine whether or not equity exists. Insuring that health services are equitably delivered in Canada is likely to require a focus on both the values underpinning notions of universal and equitable care, as well as on the statistics required to monitor progress.

NOTES

1. Canadian Health Act, http://www.hc-sc.gc.ca/hcs-sss/medi-assur/overview-apercu/index_e.html (accessed July 14, 2006).
2. "The Canadian Multiculturalism Act reaffirms multiculturalism as a fundamental characteristic of Canadian society. It encourages federal institutions to uphold longstanding values of respect, fairness, and equality with respect to members of diverse groups. The Act helps protect the rights of all Canadians, fosters the full participation of all members of society, celebrates Canada's diverse heritage, and recognizes the vast contributions of all Canadians regardless of their ethnic, cultural, racial, religious, and linguistic backgrounds. It encourages federal institutions to respond to the needs of all Canadians of all backgrounds in their programs, policies, and services" (Statistics Canada 2009, n.p.).
3. In referring to previous research undertaken, we have used whatever terminology authors of previous work have used to describe ethnicity and related concepts such as immigrants, minority groups, and so on.
4. Statistics Canada: 2001 Census: Analysis Series. Canada's Ethnocultural Portrait: The Changing Mosaic, http://www12.statcan.ca/english/census01/products/analytic/companion/etoimm/pdf/96F0030XIE2001008.pdf (accessed July 14, 2006).
5. Canada Department of Justice: The Employment Equity Act, http://laws.justice.gc.ca/en/e-5.401/238505.html#rid-238510 (accessed July 14, 2006).
6. Statistics Canada: Population Health Surveys, http://www.statcan.ca/english/concepts/hs/index.htm (accessed August 23, 2006).
7. For more information, see Statistics Canada: 2006 Census Questions and Reasons Why the Questions Are Asked, http://www12.statcan.ca/english/census06/info/questions/index.cfm (accessed July 14, 2006). The Canadian Community Health Survey (CCHS)—Cycle 3.1, http://www.statcan.ca/english/concepts/health/cycle3_1/index.htm, (accessed August 23, 2006). Concerning other health related surveys, The Joint Canada/United States Survey of Health (JCUSH) uses a shorter version of the ethnicity questions in the CCHS, while the ethnic information in household questions in the National Population Health Survey (NPHS) focuses on languages spoken and first learned.
8. Statistics Canada: 2006 Census Questions and Reasons Why the Questions Are Asked, http://www12.statcan.ca/english/census06/info/questions/index.cfm?S=11 (accessed July 14, 2006).
9. See Bourhis (2003) for a presentation and discussion of these controversies.
10. A landed immigrant is someone who has a legal right to live in Canada and enjoys all of the legal rights of a Canadian citizen other than the right to vote.

Those born outside of Canada normally must become a landed immigrant before they can take out Canadian citizenship, and during the period of time a person is a landed immigrant they must remain in Canada for at least six months of each year and they may be deported if they break the law.
11. See Krieger (2003) for a further discussion on the harmful effects of racism on health.
12. Canada Health act—Main Page, http://www.hc-sc.gc.ca/hcs-sss/medi-assur/index_e.html (accessed July 14, 2006).

REFERENCES

Anand, S.S. (1999) "Using Ethnicity as a Classification Variable in Health Research: Perpetuating the Myth of Biological Determinism, Serving Socio-Political Agendas, or Making Valuable Contribution to Medical Sciences?" *Ethnicity and Health* 4(4): 241–244.

Aspinall, P.J. (1998) "Describing the 'White' Ethnic Group and Its Composition in Medical Research," *Social Science & Medicine* 47(11): 1797–1808.

Bhopal, R. (1998) "Spectre of Racism in Health and Health Care: Lessons from History and the United States," *British Medical Journal* 316: 1970–1973.

———. (2001) "Racism in Medicine: The Spectre Must be Exorcised," *British Medical Journal* 322: 1503–1504.

———. (2004) "Glossary of Terms Relating to Ethnicity and Race: For Reflection and Debate," *Journal of Epidemiology and Community Health* 58: 441–445.

Bourhis, R.Y. (2003) "Measuring Ethnocultural Diversity Using the Canadian Census," *Canadian Ethnic Studies* 35(1): 9–32.

Bradby, H. (1996) "Genetics and Racism," in *The Troubled Helix: Social and Psychological Implications of the New Human Genetics*, eds. T. Marteau and M. Richards, 295–316. Cambridge, MA: Cambridge University Press.

Canada Department of Justice. Employment Equity Act, Canada Department of Justice, http://laws.justice.gc.ca/en/E-5.401/text.html (accessed April 7, 2010).

Chen, J., R. Wilkins, and E. Ng (1996a) "The Health of Canada's Immigrants in 1994," *Health Reports* 7(4): 33–45.

———. (1996b) "Health Expectancy by Immigrant Status, 1986 and 1991," *Health Reports* 8(3): 29–38.

Comstock, R.D., E.M. Castillo, and S.P. Lindsay (2004) "Four-Year Review of the Use of Race and Ethnicity in Epidemiologic and Public Health Research," *American Journal of Epidemiology* 159(6): 611–619.

Doyle, J.M. (2006) "What Race and Ethnicity Measure in Pharmacologic Research," *Journal of Clinical Pharmacology* 46: 401–404.

Fausto-Sterling, A. (2004) "Refashioning Race: DNA and the Politics of Health Care," *A Journal of Feminist Cultural Studies* 15(3): 1–37.

Irby, K., W.F. Anderson, D.E. Henson, and S.S. Devesa (2006) "Emerging and Widening Colorectal Carcinoma Disparities between Blacks and Whites in the United States (1975–2002)," *Cancer Epidemiology Biomarkers and Prevention* 15(4): 792–797.

Karlsen, S., and J.Y. Nazroo (2002) "Relation between Racial Discrimination, Social Class, and Health among Ethnic Minority Groups," *Journal of Public Health* 92(4): 624–631.

Krieger, N. (1999) "Embodying Inequality: A Review of Concepts, Measures, and Methods for Studying Health Consequences of Discrimination," *International Journal of Health Services* 29(2): 295–352.

———. (2000) "Refiguring 'Race': Epidemiology, Racialized Biology, and Biological Expressions of Race Relations," *International Journal of Health Services* 30(1): 211–216.

———. (2003) "Does Racism Harm Health? Did Child Abuse Exist Before 1962? On Explicit Questions, Critical Science, and Current Controversies: An Ecosocial Perspective," *American Journal of Public Health* 93: 194–199.

Lepore, S.J., et al. (2006) "Effects of Social Stressors on Cardiovascular Reactivity in Black and White Women," *Annals of Behavioral Medicine* 31(2): 120–127.

McKenzie, K., and N.S. Crawford (1996) "Describing Race, Ethnicity, and Culture in Medical Research," *British Medical Journal* 312(7038): 1054.

Pendakur, R., and J. Hennebry (1998) "Multicultural Canada: A Demographic Overview 1996." Ottawa: Department of Canadian Heritage.

Risch, N., et al. (2002) "Categorization of Humans in Biomedical Research: Genes, Race, and Disease," *Genome Biology* 3(7): 12.

Senior, P.A., and R. Bhopal (1994) "Ethnicity as a Variable in Epidemiological Research," *British Medical Journal* 309: 327–330.

Sharif, N., A. Dar, and C. Amaratunga (2000) "Ethnicity, Income and Access to Health Care in the Atlantic Region: A Synthesis of the Literature, Prepared for the Population and Public Health Branch," Atlantic Regional Office: Health Canada.

Sheth, T., C. Nair, M. Nargundkar, S. Anand, and S. Yusuf (1999) "Cardiovascular and Cancer Mortality among Canadians of European, South Asian and Chinese Origin from 1979 to 1993: An Analysis of 1.2 Million Deaths," *Canadian Medical Association Journal* 161(2): 132–138.

Statistics Canada (2009) Annual Report on the Operation of the Canadian Multiculturalism Act 2007—2008, Citizenship and Immigration Canada, http://www.cic.gc.ca/english/resources/publications/multi-report2008/part1.asp#diversity (accessed April 7, 2010).

———. 2006 Census Questions and Reasons Why the Questions are Asked, http://www12.statcan.ca/english/census06/info/questions/index.cfm (accessed July 14, 2006).

———. Population Health Surveys, http://www.statcan.ca/english/concepts/hs/index.htm (accessed August 23, 2006).

———. 2001 Census: Analysis Series. Canada's Ethnocultural Portrait: The Changing Mosaic, http://www12.statcan.ca/english/census01/products/analytic/companion/etoimm/pdf/96F0030XIE2001008.pdf (accessed July 14, 2006).

———. The Canadian Community Health Survey (CCHS)—Cycle 3.1, http://www.statcan.ca/english/concepts/health/cycle3_1/index.htm (accessed August 23, 2006).

Sterling, T., W. Rosenbaum, and J. Weinkam (1993) "Income, Race, and Mortality," *Journal of the National medical Association* 85(12): 906–911.

Timmermans, S., and M. Berg (2003) *The Gold Standard: The Challenge of Evidence-Based Medicine and Standardization in Health Care*. Philadelphia, PA: Temple University Press.

Vissandjee, B., M. Desmeules, Z. Cao, S. Abdool, and A. Kazanjian (2004) "Integrating Ethnicity and Migration as Determinants of Canadian Women's Health," *BMC Women's Health* 4: S32.

Wiener, C. (2000) *Elusive Quest: Accountability in Hospitals (Social Problems and Social Issues)*. Somerset, NJ: Aldine Transaction.

Williams, D.R. (1999) "Race, Socioeconomic Status, and Health: The Added Effects of Racism and Discrimination," *Annals of the New York Academy of Sciences* 896: 173–188.

Wu, Z., S. Noh, V. Kaspar, and C.M. Schimmele (2003) "Race, Ethnicity, and Depression in Canadian Society," *Journal of Health and Social Behavior* 44(3): 426–441.

6 Seeing Like Citizens
Unofficial Understandings of Official Racial Categories in a Brazilian University[1]

Luisa Farah Schwartzman

Brazil has a long history of discrimination based on skin color and a well-documented association between people's racial category and their access to resources, patterns of socialization, and family formation (see Telles 2004). However, recently implemented affirmative action policies, designed to address these social injustices, have generated a heated debate over the feasibility or appropriateness of classifying people racially for such policies. Some commentators claim that accurate categorization is impossible in Brazil because Brazilians are a mixed-race people, with no clear racial boundaries. Others suggest that classification is difficult due to "fraud": people can dishonestly declare their racial category to benefit from the policy.

This chapter argues that potential policy beneficiaries often classify themselves differently from what policymakers and advocates intend them to, but not simply for the previously mentioned reasons. An important reason is a mismatch between the worldviews and knowledge that policy beneficiaries and policy designers bring with them when considering the appropriate rules for classifying people for affirmative action purposes.

To understand this disjuncture, I investigate three issues. The first issue is how policy designers learn about the meanings of racial categories and how they transform this knowledge into the categories used in the policy. The second issue is how potential policy beneficiaries translate their previous knowledge about "race" into decisions about how they should classify themselves for the policy. The third issue concerns the nature of the communication between these two sets of actors.

My empirical case is one of the first universities in Brazil to implement racial quotas for admissions, the State University of Rio de Janeiro (UERJ). I describe the context and process of implementation of affirmative action in Brazil and at UERJ, and analyze open-ended interviews I conducted in 2005 with twenty-eight UERJ students. I contrast the perspectives of students who were admitted after the quotas were implemented with those of black movement activists and social scientists who have influenced the policy.

Racial categories in affirmative action forms and their official interpretations emerged in the context of a new "racial project" (Omi and Winant 1994), where black movement activists and some social scientists attempted to diagnose and change the racial order. Policies at UERJ were gradually made consistent with this project. Because of the increasing involvement of the project with state bureaucracy and quantitative social science, policy designers "see like a state" (Scott 1998), that is, through categorical and statistical simplifications of social reality, using criteria that fit the goals of doing policy from a distance (see also Porter 1995).

In contrast to policy designers, students *see like citizens*. They rely on a view "from below," taking into account contextualized and flexible uses of racial categories in everyday life. By being asked to self-classify, they also become active participants in this official categorization process, trying to make sense of the policy and their own relationship to it. This makes them "citizens" and not simply "subjects." This last step requires *reflexivity*: a conscious examination of the workings of society and one's place in it (Giddens 1990).

However, many students have only partial access to the worldview advocated by the new racial project. They combine pieces of this worldview with their own life experiences and with alternative worldviews perpetuated by older racial projects. Some students, who have been directly involved with black movement organizations, make sense of racial categories and affirmative action using a system of beliefs and practices taught by activists. Even then, they may adapt those beliefs to fit their particular life experiences and goals.[2]

Ongoing discrimination and social relationships based on race and color do not automatically translate into self-classification for affirmative action. For students to make this translation, they must acquire more macro-level knowledge about those realities, the policy, and their own relationship with the policy. Efforts to monitor self-classification through surveillance methods such as taking pictures or doing interviews with candidates may work for dealing with "fraud." However, such methods are usually contingent on students' prior self-selection and make this self-selection more risky. They may, therefore, further contribute to denying access from candidates who suffer racial discrimination but do not see themselves as "black enough" for quota purposes.

POLICY DESIGN

A New Racial Project

The implementation of racial quotas in UERJ can be understood as part of a new "racial project" that began to change the interpretation and uses of racial categories since the late 1970s. Winant (1992) defines a racial project as "simultaneously an explanation of racial dynamics and an effort to reorganize the social structure along particular racial lines." He notes that racial projects are "both a discursive or cultural initiative, an attempt at racial signification and identity formation on the one hand; and a political initiative, an attempt at organization and redistribution on the other" (Winant 1992, see also Omi and Winant 1994). Brazil's new racial project results from an alliance between black movement activists and quantitative social scientists who studied racial inequality.

An older racial project, which influenced government policy from the 1930s to the 1970s, preached that Brazil was a "racial democracy" and that Brazilians were the cultural and biological fusion of three "races": whites, blacks, and *índios*. This project supplanted an even older one, with roots in the transition from slavery to freedom in the late nineteenth and early twentieth centuries. This early project portrayed "African blood" as inferior to "European blood," promoting immigration policies to "whiten" and thereby "improve" the population. Though many Brazilians still use ideas from older racial projects, discourse promoted by advocates for the new racial project has gradually reached mainstream debates and government agencies, providing an alternative understanding of Brazilian society (Skidmore 1995, Nobles 2000, Daniel 2006, Silva 2006).

The roots of the new racial project can be traced to the beginnings of the black movement in the early twentieth century and to the social scientific studies in the 1950s (Fry 2001, Guimarães 2003, Telles 2004). Until the 1970s, the project primarily focused on forging Afro-Brazilian identities and debunking the racial democracy myth. Afterward, the project became gradually involved in measuring racial disparities statistically and designing policies to address them. In the late twentieth century the project became increasingly mainstream, influencing debates in the media and government policies. This meant an intensifying collaboration between black movement activists, the state apparatus, and an increasingly quantitative social science.

In the late 1970s, Carlos Hasenbalg and Nelson do Valle Silva finished their PhD dissertations in the United States (Silva 1978, Hasenbalg 1979) and returned to Brazil, inaugurating a new line of race research based on statistical analysis of racial disparities (see Fry 2001). They found that people classified as *pretos* (blacks) and *pardos* (browns) had similar socioeconomic outcomes, which were in turn different for those of *brancos* (whites). This conclusion was used as a justification to group *pretos* and *pardos* together in statistical analyses and description of data.

Though Hasenbalg and Silva called the sum of these categories *não-brancos* (nonwhites), it later became common among social scientists to call this group *negros*. As we will see later, in everyday language the word *negro* (another word for "black") is often restricted to people at the darker end of the color spectrum (Bailey 2008). However, black movement activists have adopted a broader definition of *negro*—including Afro-Brazilians with lighter skin tones—to promote black consciousness among a larger constituency (Guimarães 2003).

This move from *não-branco* to *negro* reflected a desire by some social scientists to be aligned with black movement discourse, and to influence social policy. For example, Ricardo Henriques—an economist whose work influenced the media's framing racial inequality and the writing of Brazil's report in the Durban conference in 2001[3]—told me in an interview that he decided to use the *negro* label instead of *não-branco* when, while dialoguing with the black movement in the preparations for the conference, he found that the label *negro* resonated better with them.

The new interpretation of the label *negro* as a statistical category allowed black movement activists to attach a "scientific" legitimacy to both the broader definition of the term "negro" and to claims about the existence of race-specific disadvantages in Brazilian society.

During the democratization period of the 1980s and 1990s, an increasing number of black movement activists were elected to public office. Since 1995, pressured by black movement activists, but also under a president who had started his career as a sociologist of race in Brazil, the federal government started working with those activists to decide on the country's official stand at the Durban conference. The Brazilian delegation presented an official report at the conference that had many proposals to combat racial inequality, including race-based affirmative action in universities (Peria 2004, Telles 2004, Htun 2004, Machado 2004).

The conference and its media coverage influenced the design and passing of the state legislation that introduced race-based affirmative action at the two universities run by the State of Rio de Janeiro—the Universidade Estadual do Rio de Janeiro (UERJ) and Universidade Estadual do Norte Fluminense (UENF)—followed by similar initiatives in other public universities across the country (Peria 2004, Ramos 2005, Silva 2006).[4] Since then, race-conscious policies have expanded to other branches of the government, to private universities and a few private companies, and congress is discussing legislation that will make racial categorization in social policy even more pervasive.

Racial Quotas in UERJ

In 2001, the Rio de Janeiro State Legislature approved a law that would institute a 40 percent quota for *negros* and *pardos* in the two state universities, UERJ and UENF. Though the law was conceived and approved without much debate, subsequent changes in the law were influenced by several organized sectors of society, notably university administrators, black movement organizations, and the courts (Peria 2004, Ramos 2005, Machado 2004). This debate was also indirectly informed by official statistics and their social scientific interpretations. The racial quota law was implemented together with a 50 percent quota for students from public high schools, which overlapped partially with the racial quota.

After many lawsuits against UERJ challenging the eligibility criteria and the policy as a whole, the university administrators proposed a change in the quota law, suggesting 20 percent for *negros* and *pardos*,

20 percent for public school students, and 5 percent for disabled people and members of "other ethnic minorities." The proposal was accepted, but with a modification: in a meeting with the Secretariat of Science and Technology that involved black movement activists and other organizations from UERJ, the category *pardo* was eliminated (Peria 2004). Later, the Rio State Legislature added a family income limit to the eligibility requirements. Students who entered UERJ during the second year of quotas followed the new rules.

Influential in this process of changing the law—and later in pressuring other universities around the country to adopt affirmative action—was black movement activist and priest Friar David dos Santos (known as Frei David) and his NGO Educafro, an organization that prepares *negro* and *carente* (needy) students to take university entrance exams, while also giving them lessons about citizenship and black consciousness (*consciência negra*). Educafro was a split from a similar organization called PVNC (Pré-vestibular para Negros e Carentes). Educafro and PVNC were the black movement organizations that most influenced students I interviewed, a few of whom had been students in these organizations while preparing for the entrance exam.

MISUNDERSTANDINGS AND "MIS-CLASSIFICATION"

Much debate on the practical feasibility of racial quotas in Brazilian universities focuses on the incongruence between how policy designers intend the quotas to be used and how students implement them. Two kinds of arguments have been made: (1) that students would mark their "race" for admissions based on an *identity* that is inconsistent with the policy's categories or intentions, or (2) that they would mark their "race" based on a rational calculation of the material benefits of qualifying for the quota.

The first kind of argument generally focuses on the idea that Brazilians are (or see themselves as) racially mixed, and that racial boundaries are imprecise, and therefore classifying between *negro* and *branco* is impossible (e.g., Benjamin 2007). However, some defenders of racial quotas claim that in Brazil most people know who is *negro* and who is *branco*, but are concerned with the possibility of "fraud." According to this perspective (e.g., Medeiros 2004), classification in racial quota systems can be challenged by cynical people, who are usually identified as *brancos*, but who strategically display a deliberately long-forgotten *negro* ancestor when convenient for justifying quota eligibility.

Table 6.1 suggests the limits of these accounts. The rows in the table represent how students at UERJ declared their "race" in a socio-economic questionnaire that would *not* be used to make decisions on admissions. The columns represent what they chose in the "self-declaration" that would allow them to enter the quota system. Both questions were asked before admissions, but there is only data for students who were admitted. In the "self-declaration," students had to declare "under the penalties of the law" if they were "negro" or "pardo." If they were neither, they were told to write an "N" or leave it blank.

A few students switched categories between the two questionnaires. Some of this instability can be explained by self-interested, rational calculation consistent with the "fraud" explanation. Greater incentives seem to cause a greater adherence to nonwhite labels: a larger percentage of private school students than public school students who had chosen *branco* in the socio-economic questionnaire (8 percent vs. 2 percent) chose *pardo* or *negro* in the "self-declaration" form, where admissions were at stake. Because there was a quota for all public school students, those students had a smaller incentive to claim *negro* or *pardo*.

However, "fraud" cannot explain all inconsistencies in classification. Most *brancos* consistently classified themselves. In contrast, a larger percentage of students who classified as *pardo* in the socio-economic questionnaire did not take advantage of the quotas, reacting *against* their self-interest: about 30 percent of pardos from the private school entrance

Table 6.1 Classification for Quota Purposes as % of Classification in Socio-Economic Questionnaire

		Self-Declared as *negra/parda* for admission through the racial quota (%)		Total (Absolute Numbers)	
		Regular Exam	Public School Quota Exam	Regular Exam	Public School
Classification in socio-economic questionnaire	Branca	8	2	1,430	873
	Negra	93	93	232	317
	Parda	69	84	749	727
	Amarela	26	5	39	56
	Indigena	61	21	33	19
	Did not answer	35	43	282	152
	Total	35	47	2,765	2,144

Source: Departamento de Seleção Acadêmica (DSEA), Subreitoria de Graduação, UERJ
Note: Only valid answers from students who were admitted in the exam are represented in this table.

exam chose not to qualify for racial quotas, compared to about 15 percent of pardos from public schools.

The inconsistencies between how policy designers expect students to classify themselves for quota purposes and how students actually self-classify are not simply the result of students' "identity" or "fraud," but also of misunderstandings between students and policy designers regarding appropriate criteria for classification. To investigate these misunderstandings, we need to examine more closely how these two sets of actors perceive the relationship between social reality and the policy. From the previous section, we already know how policy designers look at these issues. Next I discuss how and why designers' logic clashes with students' perceptions.[5]

Antonio's Story

> *Antonio*: They had there: "declaration", what are you, you are . . . *negro, pardo, branco* [. . .]. What happened is that I didn't consider myself *branco*. I still don't consider myself [*branco*] until today. Even though today I have a different view of what *pardo* means. Then I thought that it meant something that was not simply *branco* [. . .]. Why? Because they classified the thing in such a simplistic way, *negro, pardo, branco*, if you are not in the extremes you are in the middle, which is wrong [. . .]. That was not really the intention of the legislator, nor of the people who fought for the quota. [. . .] So, what happened, I put there that I was *pardo*.

"Antonio"[6] understood *pardo* as something that did not fit either the *branco* or the *negro* labels. Later, he read a newspaper article by Frei David criticizing a person who looked *branco* but who had opted for the quotas because his great-grandmother was *negra*. That made Antonio think that he should not have marked *pardo*.

> *Antonio*: For him [Frei David] the issue was the following: it was whether your appearance would lead to discrimination [. . .]. If in a situation, for example, such as employment, would someone avoid hiring you because you were . . . do you understand? Then I thought that in this criterion of his I should not have taken the spot.

I asked Antonio whether his situation was similar to the person with the great-grandmother that Frei David had described.

> *Antonio*: Not in my case, my mother is almost *negra*. Because my mother is the daughter . . . what we call *mulato*, she is the daughter of a *branco* with a *negra*, but she is closer to *negra*. My father is very *branco*, because he is the son of Portuguese [. . .] So what happened? Of course, me and

my brother came out lighter than my mother [...] So, what happened ... I felt closer to this reality than that guy. And he declared himself [*pardo*], so, I mean, this generated a big polemic. I was already feeling guilty for having declared myself, but then I noticed that there were lots of people who had done this and were much further away from this situation. But the guy understood that *pardo* is someone who is not *branco*. And many people understood this, the truth is that it was not ... No wonder this term disappeared. Now the quotas are for *negros*.

Though Antonio does not believe that he suffers discrimination, his situation cannot be easily described as "fraud." He does not just have a distant *negra* great-grandmother; he has close relatives that he defines as *negro*, *branco*, and *mulato*. This makes it difficult for him to claim either the *branco* or the *negro* labels.

Later in the interview, he said he did not understand the argument that quotas should benefit those who suffered from discrimination: for him poverty, not discrimination, was the main source of *negro* disadvantage. When I asked him why *negros* were poor, he talked about how *negros* had been exploited since the formation of the country, by coffee growers, then by industrialists who had been part of the agrarian elite. The disadvantage was inherited. It was inherited by some *brancos* too but mostly by *negros*. Then he connected this story to this own family history:

> *Antonio*: My mother, for example [...]. Her mother was a maid in a family's home, and her boss had a relationship with her, you see ... and she had her [my mother] and afterwards she ... obviously the boss did not recognize [the child] [...] and she did not even have the means to raise my mother. My mother was raised by her godmother, I mean, exactly in this process of discrimination of the *negro* [...] What happened was that the social process here [in Brazil] created such a huge economic abyss between the two classes that there was no need to mark a separate place for the *branco* [...]. So there was always racism in Brazil, the *negro* always came out worse off in this story. So what happened, I didn't really create a very perfect identity in this story because of the whole mixing that happened, and because my mother is almost *negra* and my father is *branco*.

For Antonio, being *negro* or *pardo* has to do not only with facing discrimination, but also with a long history of economic and sexual exploitation that has resulted in poverty.

Antonio's connection to the label *pardo* is neither opportunistic nor due to a natural "fact" of his identity. Rather, he uses his family history and his ties to his relatives to decide his racial status, which he sees as "something that is neither *negro* nor *branco*" which, he has learned, is what the category *pardo* is supposed to mean. Because Antonio's classification is based on a conscious reflection upon the meaning of this category, he can adapt this classification based on new considerations. As he learns about affirmative action, he tries to understand and incorporate policy designers' rationales. In his view, the elimination of the *pardo* option for quotas provides a potential solution to Frei David's account of the great-grandmother problem.

> *Antonio*: Now they took it out, the word now in the law is *negro*. So [. . .] if you feel like you are almost *negro* you declare yourself . . . *negro*. You are not really *negro*, but a very dark *pardo*, you have all the phenotype of a *negro* you declare yourself [. . .] But a person that is much closer to *branca*, they will not [. . .] only if they are very cynical.

Indeed, the category *pardo* was eliminated as a criterion for entering through the quota system. However, it did not happen for the reason Antonio imagined. According to Peria (2004), in February 2003, the State government held meetings with representatives of the universities (UERJ and UNENF), professors, university staff, and representatives of black movement organizations to change the legislation in time for the next entrance exam. Peria (2004) says that the decision to exclude the category *pardo* was suggested by a representative of the black movement, who argued:

> A segment of the black movement and a segment of academia, based on statistical data, consider that it is legitimate to join *pretos* and *pardos* in another category—that of *negros*. Why? Because the distance between *pardos* and *brancos* is a large distance and the distance between *pardos* and *pretos* is always small, measured by indicators such as infant mortality, wages, education, etc., etc. [. . .]. If you talk about *negro* you talk about *pretos* and *pardos* together, forming *negros*, afro-descendants. Carlos Hasenbalg prefers *não-brancos*, in sum, the name you want to give.[7]

According to Peria, after this activist spoke, the university representatives and the government secretary accepted the modification without problems and sent it to be approved by the Governor. The black movement, informed by social scientists, has defined *negro* to mean the sum of *pretos* and *pardos*. This, in turn, was the definition that prevailed when new criteria for the quota were introduced.

Antonio's story shows a transition from a more naïve decision-making process, based on the question *What am I?*, which stems from his experience with race in everyday life, his physical appearance, his family history, and his place in Brazilian society, to a more sophisticated one, where he asks, *Are quotas for me?*, assessing the goals of the policy and the legitimacy of his claims given those goals. Antonio's story shows that neither of these questions have straightforward answers. Furthermore, it suggests that Antonio has only partial exposure to the ideas that have guided the design of the policy, and therefore he fails in his effort to adjust to policy designers' expectations. Next, I discuss how other students at UERJ answer the questions *What am I?* and *Are quotas for me?*

WHAT AM I?

When answering the question *What am I?*, students combine experiences with racial classification in everyday life with their interpretation of official categories. Students usually understand the label *negro* based on their previous everyday experiences with racial categorization or on contact with black movement organizations such as Educafro. They often ignore policy designers' understanding of the label *negro*, which includes *pretos* and *pardos*. Although some students use the label *negro* more broadly, including Afro-Brazilians of lighter skin tones, they often do so to denote origin, ethnic identity or class exploitation rather than to demarcate potential victims of discrimination.

Students rarely report using the label *pardo* outside interactions with bureaucracies. However, through these interactions, students have learned what this category is "supposed to mean." Sometimes they understand *pardo* as "something that is neither *branco* nor *negro*," or as synonymous with the more commonly used categories that indicate racial mixture, such as *mulato* or *moreno*.

Although clear-cut, uncontested identification as *branco* is common among UERJ students (some of whom I interviewed), especially in the more prestigious departments, this chapter does not analyze their experience. I focus here on individuals who can make some claim to the labels *negro* or *pardo*. Though many students have flexibility in their self-definition, family ties and physical appearance do constrain the legitimate options for self-classification (see Teixeira 2003).

Translating from Everyday Life

Research on racial categories in Brazil has found that racial labeling of the same person in everyday life varies according to the nature and purpose of interaction and the physical and social distance between the classifier and the classified (Sheriff 2001, Sansone 2003, Maggie 1991). My data confirms these findings. Contextual variation in the labels students apply to themselves can be explained by two factors. First, as some students come from "multi-racial" families and/or have an ambiguous physical appearance, they can legitimately claim association with different racial categories when talking about different realms of life. Second, students may want to use *hard* or *soft* terms depending on the context. Hard terms stress difference, and are often used to offend or to joke. Soft terms downplay difference and are often deemed less offensive (see also Sansone 2003).

Because of relatively close interaction among people of different skin colors (Telles 2004), Brazilians who discriminate have an interest in keeping their classification system flexible enough to spare their friends and relatives. When asked how she came to think of herself as *negra,* Andreia told me this story:

> I was friends with this girl, [. . .] both from her paternal and her maternal grandparents, they came from Italy. So they were very *brancos* and everything. So I remember that we were talking and all, and then I told her: oh, my father is *negro.* Then she turned to me and [said]: "No, [Andreia], your father is not *negro,* what are you talking about, he is *moreno!*" As if I were devaluing [*desqualificando*] my father by calling him *negro.* But look, *negro,* when you call a person *negro,* you are not devaluing anyone. On the contrary, you are affirming the race of this person.

As Andreia points out, there would be no need for her friend to reject the label *negro* for Andreia's father if that label did not have a negative meaning. By calling Andreia's father *moreno,* Andreia's friend is making an exception for one person while keeping the status of the label "negro" unquestioned. Andreia, on the contrary, is tying her father back to other *negros* and thus advocating for the "group" as a whole, friends or strangers.

The messiness of everyday racial categorization, however, is the result not only of a flexible *morality,* but also of the different *uses* of categories. My interview with Julia illustrates the multiple uses of "race," even though she uses the terms *negra* or *preta* to describe herself in all contexts. Having dark skin and a middle-class background, and having entered UERJ through racial quotas, Julia favors racial quotas because of her experience of being the only *negra* person in several environments. She does not like when people who "are *negra* like myself" call themselves *escurinha* (diminutive of "dark") or *moreninha* (diminutive of *"morena"*). "I think that it is a lack of identity, of . . . I don't know, of cultural identity," she explains. For Julia, then, calling oneself *negra* means recognizing one's cultural heritage. Being a victim of discrimination has influenced her "self-recognition" as *negra.* She said that she began to think of her color when, as a child, an older student told her to get out of the school bus because *"preto* didn't sit in the bus."

Even though Julia often stresses the need for people to take on their *negro* identities, she also invokes the idea of the Brazilian nation as resulting from a mixture of *brancos, negros,* and *índios.* She described fondly a class project organized by her former school teacher:

> *Julia*: Each history topic, especially when the topic was Colonial Brazil, she would do a project around this. And once in school she did . . . uh

... peoples that helped to form the Brazilian matrix, etc., etc., and then there had to be *índios*, there had to be *negros*, and so on, and the *negros* in the school did not want to take part. Then I participated and was the only one.

Here Julia recounts being taught two seemingly contradictory topics: that Brazilians are mixed and that she was *negra* herself, representing one pure element of this mixture. This is a familiar narrative from the old "racial democracy" project, being propagated by the Brazilian educational system. She, however, integrates it into a narrative of black pride, which is consistent with the discourse of the new racial project.

Another student, Joana, was admitted after the quotas law changed, so she was not allowed to apply to both the public school quota and the racial quota. She chose the public school quota. When asked if this was because she didn't consider herself *negra*, she said:

> *Joana*: It's like this [...]. I think that, if a *negra* person wants some policy [...], it is because they feel discriminated against. Even though I don't have good hair, [...] I never felt discriminated against *because of my color*. At some moments, [...] because of my financial condition, but not because of my color, because of my appearance. I know that it exists. Doesn't it? So I think like this: if I felt disadvantaged by society, I think it's an option. [...] But I didn't feel [disadvantaged], even though ... To say that I'm not *negra*? My grandmother is *negra*. Certainly, I could, I think that ... they should accept it, they could not refuse it.

Joana believes she has the right to call herself *negra*. However, she does not think that she suffers discrimination, nor that she should take advantage of racial quotas. She chose the class-based criterion (the public school quota), where she feels disadvantaged. When talking about her family, she says that some people are *brancos*, some are *negros* and she is somewhere in-between.

Joana's depiction of her hair as "bad," however, implies that her hair is still subject to negative judgments, which she seems to have internalized. This suggests that she may be overestimating her freedom to "choose" her racial status, at least in certain situations. The possibility that Joana is not aware of racial discrimination against her is therefore a challenge for this kind of affirmative action policy.

Learning from Official Categories

While some students use their everyday experiences to answer the *What am I?* question, others, especially those that classify themselves as *pardos*, learn how they are "supposed to be classified" from previous classificatory encounters with bureaucratic institutions.

"I don't consider myself *negro*, I only consider myself *pardo*," says André, only to later question the *pardo* category itself, saying that "biologically, regarding genotype, *pardo* doesn't exist. It's *branco*, three kinds of *mulato* and *negro*." Though he does not see the category *pardo* as real, he has learned that *pardo* applies to different kinds of *mulato*, which he sees as biological categories. When asked if he had ever marked his color before the quotas, he said that he had been classified as *pardo* in the army (the Brazilian army does not use self-classification).

Similarly, Carla learned what *pardo* meant from her brother, whose job as a doctor included classifying people. When asked if she had ever discussed the issue of color with her family, she said:

> *Carla*: When I did the enrolment for UERJ I went to talk to my mother, my father and my brother. To discover what color I was ... I said "[Brother's name], what color am I?" Then he said "When a patient goes there and I have to", you know, he does anamnesis [medical history], he said he had to put the color of the patient. He said: "when someone like you shows up I put *pardo*."

When Carla took the university entrance exam, her family already had experiences with official racial classification in other instances, where the term *parda* was institutionalized. After learning the "right" answer, Carla classified herself as *parda* in the quota form.

ARE QUOTAS FOR ME?

Who Are Quotas For?

Students do not always agree among themselves or with policy advocates about the goals of quotas. Julia chose the quotas for *negros* because she believes that it is important to have more *negros* in the university, since she was the only *negra* in many situations in her life. Joana thinks that racial quotas should target those who suffer discrimination. Antonio thinks that the disadvantage of *negros* could be addressed by helping those of lower socio-economic status.

This lack of consensus also shaped the design of the policy, which resulted from an interaction between actors with different goals. The public school quota and the change in the law that added an income limit to the quota reflects a common interpretation at the Rio State Legislature that the quotas were addressing class disadvantage. (Ramos 2005; Fry 2005). Some black movement activists also see quotas as tools for raising black consciousness (Silva 2006).

Not all students have deep moral dilemmas about applying for quotas. Though Ana supports public school quotas as "an emergency measure," she opposes quotas for *negros* because she thinks it "segregates" *negros* and contradicts efforts of *negros* to be equal. Nevertheless, she took advantage of the racial quotas "just to be sure" because she wanted to be admitted "whichever way." She classified herself as *parda*, because classifying herself as *negra*, she said, would be too hypocritical. When asked if she is *negra*, she says that her family is "all mixed" and that in Brazil it is hard to say who is *negro*. Though ambivalent about having applied for racial quotas, she noticed during enrollment that there were people "much lighter than me" that were doing the same, probably reassuring her of her *relative* eligibility. Although her approach was pragmatic, it was not "fraud": she still claimed to be within the bounds the law, given its imperfect design.

Am I Discriminated Against?

If one agrees that affirmative action should focus on discriminated people, one is left with the question of who suffers discrimination. Joana claims not to suffer discrimination despite her "bad" hair, but can we rely on her perception? Antonio thinks that *pardos* do not suffer discrimination, while the quota law was based on the assumption that they do.

The disjuncture between Antonio's (and Joana's) understanding of discrimination and those of policy designers results from two ways of learning about discrimination: observation of discriminatory *processes* vs. deduction from observation of *outcomes*. Gabriel, a politically engaged, older student who used to work as a teacher, describes his experience with *institutional* discrimination:

> *Gabriel*: I was only able to do well professionally when I went through exams to work in the public sector. Because then there is no color, you do the exams, get the highest grade, get classified in the first place, and nobody can kick you out. But in private schools, I never worked in private schools more than two years. Always by chance, it happens by chance, [. . .] the class always gets smaller, someone always takes my place.

Gabriel cannot observe the process of discrimination directly, but has to deduce it through its effects. In what he calls his "civil life," though, discrimination becomes more visible: "people look sideways, if I am in a place where there are only *negros* people get more [. . .] with a certain fear."

Deducing discrimination in employment requires observing patterns and theorizing about unobserved processes. This method is also used in statistical research, which shows discrimination by controlling for confounding variables, and deducing discrimination from the residual. Using this method, social scientists concluded that *pardos* and *pretos* suffer similar impersonal discrimination—in the job market and the educational system—but suffer less interpersonal discrimination—in family, neighborhood, and friendship relations (Telles 2004). Since interpersonal discrimination is easier to notice than impersonal discrimination, it may be harder to notice discrimination against *pardos*. Both survey research (Bailey 2008) and interviews with students suggest that most Brazilians do not believe that *pardos* suffer discrimination or deserve race-based affirmative action.

Students who learn about this more "hidden" discrimination often do so through black movement organizations such as Educafro or through the media. Both sources use statistical research to prove discrimination but, for political reasons, frame it as discrimination of *brancos* against *negros*, including *pardos* within the *negro* category. This reinforces some students' beliefs that *pardos* do not suffer discrimination.

DIRECT ENGAGEMENT WITH THE NEW RACIAL PROJECT

The new racial projects' original interpretation of racial categories is not always understood by students, who often only gain access to the project's discourse through exposure to the media, through the institutional design embedded in the law, and through contact with acquaintances involved with the black movement. Often students get pieces of the projects' message, together with ideas from the older projects, and assemble this information in their own ways to make sense of particular situations. Some students, however, are exposed to more complete "packages" of the new project's ideology.[8] They explicitly reject intermediary categories like *pardo*, *mulato*, or *moreno* and embrace the broader meaning of *negro*, so as to debunk the "myth of racial democracy."

"Gustavo" took classes at Educafro to prepare for the entrance exam. Besides the normal curriculum, Educafro also offers "citizenship classes" where Gustavo says, he learned about "black consciousness." Asked if he had never thought of those things before, he said:

> *Gustavo*: No, I *had* thought about it. But I didn't have a systematization of how you can put this in practice in your life. [. . .]. So Educafro helped me a lot, even in this search for self-esteem, for you to accept yourself as *negro*. [. . .] So, this question, even when I fill in a socioeconomic questionnaire, or something like this, I always put, well, what is your color? *Preto*. Because *pardo* was a legal invention that was created to detach this mentality from the *negro*. [. . .] in Brazil, you are seen as *negro* or not by the color of your skin. If you are a little bit *branco*, a little bit *claro*, your ancestors don't matter, you are considered *branco*. So if you are not *preto*, in the material sense of the color, you are considered, "oh, no, he is *pardo*, he is *moreno*, he is *mulato*." Precisely to try [. . .] to take this focus out of race.

Similarly to less engaged students I interviewed, Gustavo adopts *both* a broad and a narrow concept of blackness, distinguishing between *negro* and *"preto in the material sense of the color."* However, Educafro has taught him to view the broader category of *negros* as people who share a common struggle and a common burden of discrimination, and to explicitly reject mixed-race labels.

Nonetheless, Gustavo's understanding of *negros* as participating in a common struggle for social justice is not only about racial discrimination, but also about class exploitation. When discussing the importance of black consciousness in his own experience, he says:

> *Gustavo*: I accept myself, I even teach that the *negro* should accept himself as *negro*, should have an attitude toward his peer of self-recognition, of supporting each other and to take this message forward that you are capable [...] Because [...] there is a tension in the [...] poor neighborhoods, you live your whole life preparing to obey, to be the servant of your father's boss. [...] Then the poor youngster, the *negro* youngster, not just the *negro* youngster, but the poor youngster, he does not have this idea that he has the intellectual capacity to seek something bigger.

Similar to other students I interviewed that went through PVNC or Educafro, Gustavo talks about *race* consciousness that is tied (and more or less equivalent) to a *class* consciousness. It is therefore natural for him to apply the lesson that he learned at Educafro to poor people *in general*.[9] To be incorporated by students, black movement ideology often has to compete (or "cooperate") with other ideologies, worldviews and common sense, being interpreted according to students' particular life experiences.

CONCLUSION

Students make decisions on how to classify themselves for affirmative action based on what they have learned from previous everyday interactions, from prior contact with bureaucratic institutions, and from the media. Students' understandings of "race" can be traced to different racial projects, but they usually do not receive those ideas as coherent systems of meaning. Rather, they often receive them as partial stories and symbols that they use and combine selectively to make sense of particular situations (see Swidler 1986). Only some students, most of whom have been involved with black movement organizations, incorporate the ideology of the new racial project so as to displace ideas from other projects. In contrast, policy designers' views of the legitimate criteria for classification for affirmative action have been influenced by the new racial project, and shaped to fit with a bureaucratic rationality of state policy. This means that these two kinds of actors often do not agree on the criteria for selecting people into quotas.

Paradoxically, the concern with "fraud" and the emphasis on discrimination as a criterion for affirmative action may be reinforcing some students' view of *negro* as a category that is restricted to darker-skinned Afro-Brazilians. This may lead to the self-exclusion of some people who suffer discrimination or who inherited racial disadvantages from their ancestors.

Although official racial classification in Brazil is inherently imprecise, better communication between students and policymakers regarding the meaning of the categories in use and a greater awareness of discrimination patterns could improve the extent to which quotas target their intended beneficiaries. Perhaps, however, "imprecision" should be viewed not as a problem but as an inherent part of a (somewhat) participatory process.

NOTES

1. A longer version of this chapter has been published as in the Journal of Latin American Studies (Farah Schwartzman 2009).
2. My account of how students use racial categories is consistent with Swidler's (1986) theory of culture. See footnotes in Farah Schwartzman (2009).
3. The World Conference Against Racial Discrimination, Xenophobia and Related Intolerance.
4. Public universities in Brazil have more competitive admissions than private universities. People who can afford to go to private schools have a better chance of getting into public universities.
5. I recruited students by broadcasting on e-mail lists and lectures, by approaching them in UERJ corridors, and through limited snowballing. Most students interviewed entered the university after the implementation of racial quotas. Of those, fourteen studied law, four studied education (*pedagogia*), four studied social work, four studied nursing, and two studied medicine. The students' ages varied from eighteen to forty-four years. Before the interview I gave all but one a questionnaire, where they had to fill out their "color or race" using the same options as those in the census. Nine classified themselves as branco, seven as preto, nine as pardo and two students refused to answer. Fifteen were male and thirteen were female. Most lived in lower-middle-class neighborhoods of Rio (in Zona Norte and Zona Oeste).
6. I use pseudonyms for all students to protect their confidentiality.
7. Cited in Peria (2004), my translation.
8. I use the term "ideology" to refer to a coherent and highly rationalized system of ideas, not distinguishing here between "false consciousness" and accurate depictions of reality.
9. According to Maggie (2001), PVNC was already a compromise between race-centered and class-centered views of its members and students, reflected in the name "*negros e carentes*." Educafro probably inherited some of this perspective.

REFERENCES

Bailey, S. (2008) "Unmixing for Race-Making in Brazil," *American Journal of Sociology*, 114 (3): 577–614.

Benjamin, C. (2007) "Tortuosos Caminhos," in *Divisões Perigosas: Políticas Raciais no Brasil Contemporâneo*, eds. P. Fry et al., 27–34. Rio de Janeiro: Civilização Brasileira.
Daniel, R. (2006) *Race and Multiraciality in Brazil and the United States: Converging Paths?* University Park, PA: Penn State Press.
Farah Schwartzman, L. (2009) "Seeing Like Citizens: Unofficial Understandings of Official Racial Categories in a Brazilian University," *Journal of Latin American Studies* 41(2): 221–250.
Fry, P. (2001) "Politics, Nationality and the meanings of 'Race' in Brazil," *Daedalus*, 129 (2): 83–118.
———. (2005) *A Persistência da Raça. Ensaios Etnográficos sobre o Brasil e a África Austral*. Rio de Janeiro: Civilização Brasileira.
Giddens, A. (1990) *The Consequences of Modernity*. Stanford, CA: Stanford University Press.
Guimarães, A.S.A. (2003) "Intelectuais negros e modernidade no Brasil," *Working Paper* CBS-52–04. Centre for Brazilian Studies, Oxford University. Oxford, UK.
Hasenbalg, C. (1979) "Race Relations in Post-Abolition Brazil: The Smooth Preservation of Racial Inequalities," PhD dissertation, University of California, Berkeley.
Henriques, R. (2001) "Desigualdade Racial no Brasil: Evolução das Condições na Década de 90," *Texto para Discussão* 807. Rio de Janeiro: IPEA.
Htun, M. (2004) "From 'Racial Democracy' to Affirmative Action: Changing State Policy on Race in Brazil," *Latin American Research Review* 39: 60–89.
Machado, E.A. (2004) "Desigualdades 'Raciais' e Ensino Superior: Um Estudo sobre a Introdução das 'Leis de Reserva de Vagas para Egressos de Escola Públicas e Cotas para Negros, Pardos e Carentes' na Universidade do Estado do Rio de Janeiro (2000–2004)." PhD Dissertation, Rio de Janeiro: PPGA/IFCS, UFRJ.
Maggie, Y. (1991) "A Ilusão do Concreto: Análise do Sistema de Classificação Racial no Brasil," Tese de Professor Titular. Rio de Janeiro: IFCS/UFRJ.
Medeiros, C.A. (2004) *Na Lei e na Raça: Legislação e Relações Raciais, Brasil-Estados Unidos*. Rio de Janeiro: DP&A.
Nobles, M. (2000) *Shades of Citizenship: Race and the Census in Modern Politics*. Stanford, CA: Stanford University Press.
Omi, M., and H. Winant (1994) *Racial Formation in the United States: From the 1960s to the 1980s*. New York: Routledge.
Peria, M. (2004) "Ação Afirmativa: Um Estudo Sobre Reserva de Vagas Para os Negros nas Universidades Públicas Brasileiras: o Caso do Estado do Rio de Janeiro." Masters Thesis. Rio de Janeiro: Museu Nacional, UFRJ.
Porter, T.M. (1995) *Trust in Numbers: The Pursuit of Objectivity in Science and Public Life*. Princeton, NJ: Princeton University Press.
Ramos, C. (2005) "'Nem tão pobres, nem tão negros.' Um estudo de caso sobre os alunos indeferidos no vestibular/2004 da UERJ" Masters Thesis. Rio de Janeiro: PPGA/IFCS, UFRJ.
Sansone, L. (2003) *Blackness Without Ethnicity: Constructing Race in Brazil*. New York: Palgrave MacMillan.
Scott, J.C. (1998) *Seeing Like a State: How Certain Schemes to Improve the Human Condition Have Failed*. New Haven, CT: Yale University Press.
Sheriff, R.E. (2001) *Dreaming Equality: Color, Race and Racism in Urban Brazil*. New Brunswick, NJ: Rutgers University Press.
Silva, G.M.D. (2006) "Ações afirmativas no Brasil e na África do Sul." *Tempo Social, Revista de Sociologia da USP* 18(2): 131–165
Silva, N.V. (1978) "White-Nonwhite Income Differentials: Brazil." PhD dissertation. University of Michigan.

Skidmore, T. (1995) *Black into White: Race and Nationality in Brazilian Thought*. New York: Oxford University Press.

Swidler, A. (1986) "Culture in Action: Symbols and Strategies," *American Sociological Review* 20: 305–309.

Teixeira, M.P. (2003) *Negros na Universidade: Identidade e Trajetórias de Ascensão Social no Rio de Janeiro*. Rio de Janeiro: Pallas.

Telles, E.E. (2004) *Race in Another America: The Significance of Skin Color in Brazil*. Princeton, NJ: Princeton University Press.

Winant, H. (1992) "Rethinking Race in Brazil," *Journal of Latin American Studies* 24 (1): 173–192.

7 Ideas in Action
"Human Development" and "Capability" as Intellectual Boundary Objects

Asunción Lera St. Clair

INTRODUCTION

The notions of human development and capability (HD-CA[1]) as espoused by the United Nations Development Programme (UNDP) have challenged the dominance of purely economic conceptions of development and poverty based on individualistic and market-related approaches. Initially elaborated by an independent think tank, the Human Development Report Office (HDRO), HD-CA seems to have steadily influenced the organizational and ideological goals of UNDP, and arguably has been crucial in revamping and strengthening the role of this institution in the UN System, especially in relation to more influential development agencies, notably the World Bank. The UNDP today explicitly endorses a view of development based on concerns for the well-being and dignity of people and evaluates the role of development in terms of securing people's freedoms to pursue the lives they have reasons to value; arguably placing concerns for ethical norms above the cognitive norms of the discipline of economics.

In this chapter I assess the strategy by which HD-CA has successfully challenged dominant notions of development. It focuses on how ideas travel across the multilateral system and the extent to which values (both moral and nonmoral) influence such processes. I argue that ideas in development straddle a shifting divide between science and global politics in the making; ideas are subjected to what social studies of science refer to as "boundary work": management of the ongoing tensions that arise between the realms of knowledge and politics; between science and policymaking; the boundaries among diverse disciplines, worldviews, and interests; processes that include and often lead to a periodic redefinition of what is knowledge and what is nonknowledge. I depict HD-CA as intellectual "boundary objects," and the HDRO as a "boundary organization"—two types of tools for inter-realm boundary management. I conclude by proposing that a third way to stabilize these tensions is to imbue ideas with three explicitly ethical norms (accountability, equity, and deliberative participation) in such a way that the production of knowledge for development is framed and regulated by concerns for global justice. If development policymaking is to be informed by ethical principles rather than driven only by cognitive or ideological concerns, these ethical values cannot be simply added on to development policy, but rather ought to be essential elements of knowledge production that underpins this policy.

IDEAS AND THE MULTILATERAL SYSTEM: TRAVELING THE BOUNDARY BETWEEN SCIENCE AND POLITICS

Ideas are a driving force in the current political debate as to what appropriate development policy is and how to conceptualize poverty. Ideas in development are concepts which powerfully influence development policy, yet they are often created by the UN System itself.[2] Following insights from sociology of science and science and technology studies (STS), I propose that a fruitful way to analyze knowledge about development and global poverty in the UN System is to view ideas and theories neither as pure scientific truths nor as knowledge separated from the realm of politics. UN System institutions are hybrid sites of coproduction of both knowledge and politics, ideas and social order (St. Clair 2006a). In addition, UN institutions that perform or are responsible for research are frequently subject to problems of delegation, given that those who provide funds for research and those who perform it are embraced by the same organizational structures or somehow dependent on the funding and support given by the UN System (St. Clair 2006b). These institutions are best seen as regulatory agencies similar to national regulatory institutions such as the U.S. Food and Drug Administration (FDA) or the Environmental Protection Agency (EPA). Within these regulatory institutions, science and politics are linked in complex and often conflictive ways. They produce knowledge-based policymaking which does not fit the old linear model that first knowledge is elaborated, then politics responds and acts upon such knowledge. It is more accurate to view knowledge production as a process which social studies of science refer to as "boundary work."

As Sheila Jasanoff (1990, 2004) and Thomas Gieryn (1995, 1999) have argued, what demarcates science from nonscience is not simply the presence of a particular set of necessary disciplinary, cognitive, or epistemic characteristics, but is also related to contingencies and strategic behavior. The separation of knowledge from nonknowledge is thus a messy combination of all these factors, as Jasanoff has elaborated in her analysis of U.S. advisory committees, the interactions these have between politicians and scientists, and the increasing participation of private sector actors and citizens groups (2004). The more complex and ill-structured a problem is, as is the case with development issues and global poverty, the more important is the role of strategic behavior and contingencies, the role of other competing ideas and theories, and the broader political and ideological background against which these ideas are viewed.

This perspective aims to make visible the normative, social, and political aspects embedded in scientific knowledge. Boundary work aims to explain the linkages and negotiations that are part of what appears objective and value-free codified knowledge. It reminds us that the demarcation of what is or is not knowledge is often traced not only by the analytical qualities of ideas, but according to sets of strategies and contingencies in which political interests, patrons funding research, other established ideas, and policy and social goals play their own substantial roles. This perspective is fruitful as it helps understanding the complicated aspects of knowledge-based policymaking processes (St. Clair 2006b).[3]

In traveling through this messy space between knowledge and politics the role of ideas is fundamental. Ideas in the UN System need to respond to, adapt to, negotiate between, and act as interfaces and trading zones between a whole variety of users with their own values, interests, and social, economic, and political goals. In particular, ideas need to speak to both scientific and political audiences. I suggest that it is therefore fruitful to view ideas in development policy as "intellectual boundary objects," and specifically to look at boundary concepts.

The term boundary object was first elaborated by Star and Griesemer (1989) in their study of the formation processes of a museum of natural history in Berkeley, California, at the beginning of the twentieth century. Star and Griesemer wanted to account for the collective work and cooperation among a host of partners with different backgrounds, disciplines, levels of expertise, missions, and expectations. These included natural scientists, amateurs, professionals, functionaries, visionaries, and patrons, among others. "In conducting collective work, people coming together from different social worlds frequently have the experience of addressing an object that has different meaning for each of them" (Star and Griesemer 1989: 412). The divergent points of view of all these actors required an analytical tool to account for the possibility of common representations; to ask what were the translation mechanisms at play, how can we talk about reconciliation of goals and expectations, and how can we account for the flow of objects and ideas through networks of different kinds. Star and Griesemer argue that these problems of cooperation and coherence are in fact common to all science.

Boundary objects, these authors argue, are objects which are both plastic enough to adapt to local needs and to the constraints of several parties employing them, yet robust enough to maintain a common identity across sites. They have different meanings in different social worlds but their structure is common enough to more than one world to make them recognizable, a means of translation (Star and Griesemer 1989: 393).

Boundary objects may be all sorts of things, ideas as well as things, or even people or processes: "the requirement is that they be able to span boundaries separating social worlds, so that those on either side can get behind the boundary object and work together toward some goal" (Gieryn 1995: 414–415).

Boundary objects allow for the possibility of forming some sort of shared understanding among conflicting views, missions, and interests. They are a social construction that serves as a *trading zone* where diverse groups share information and may be able to arrive at shared meanings; and ideally, they should be plastic enough, or contain an open-ended component to allow the possibility to connect different value choices and world views. These concepts may also function as *interfaces* between different stages of a process and thus make it possible to translate certain types of expertise into implementable notions. Even though boundary objects are often ill-structured, often inconsistent and ambiguous, they serve to *get the job done* as trading zones among competing views. Institutions that enable such roles could be categorized themselves as boundary organizations (Guston 2000).

Recent applications of the boundary view are used to propose more legitimate, salient, and credible knowledge. Two significant research initiatives, the Global Environmental Assessment Project (GEA) and the Network for Science and Technology for Sustainability (NSTS), have recently examined the efficiency of global institutions where knowledge and policy about environmental change is elaborated in terms of boundary work and boundary organizations.[4] This analytical perspective from the field of social studies of science is also relevant to understand and perhaps improve the role of ideas in the UN System to achieve this set of institutions' policy goals.

Ideas in development are best defined as concepts more than simple slogans or "buzzwords," given that they often have a substantial intellectual basis. Two of the main characteristics of ideas in the multilaterals are their bridging capacity between policy and research as well as their need to become operationalized. Ideas arise and are developed in the interplay between the two domains of academia and policymaking, often deriving their credibility from the former (Bøås and McNeill 2004). Yet once within the multilateral system, ideas are subject to a process of consensus-building among many partners and interests within the UN System, to adaptations and competition with other more established ideas and goals of these institutions and their organizational cultures and functions. As globalization processes become more relevant and governance actors expand in number, ideas are increasingly tested and reframed in concepts that can include, or at least address, concerns and interests from nonstate actors (NGOs, civil society groups, and the private sector). In addition, as development and poverty have become widely accepted as multidimensional problems, ideas have had to adapt to interdisciplinary assessments, and thus be able to share meanings across disciplines. In particular they have to encounter—and, sometimes, counter—understandings and terms from the dominant social science discipline of economics, and in particular, neoliberal economics.

In many multilateral agencies, these processes involve a *depoliticization* of ideas and a tendency of economics to become the main cognitive channel to construct definitions, methodologies, and policy proposals in addressing development and poverty reduction (Bøås and McNeill 2003). In addition, given that many institutions in the UN System offer knowledge-based policymaking, ideas must be policy oriented and are often subsumed to preestablished policy goals. Like scientific theories and disciplines, policies have not only moral or ethical but also nonmoral normative constraints. Policies must fulfil certain normative criteria in order to be considered successful—such as testability, quantification, efficiency, or simplicity, norms that are best seen as cognitive values. The fact that the dominant cognitive values of policymaking fit particularly well with the cognitive values of neoliberal economics is a key source of power for such ideas. For example, whether a concept is suitable to quantification and thereby easily measurable has been an important reason for the differing pace of institutionalization within the UN System (St. Clair 2004).

According to McNeill (2004), the analytical qualities of ideas that challenge mainstream ideology, for example social capital, are distorted as a result of the pressures to depoliticize and economize them. I argue that similar processes occur with regard to the ethical content of concepts and ideas. Value conflicts and trade-offs are commonplace in the battlefield of development agencies, and ethical values that challenge hegemonic ideas and policy goals tend to be distorted or supplanted by thinner versions of such values, and often simply substituted by economics' cognitive values. In particular, cognitive values such as measurability, simplicity, and efficiency tend to prevail over concerns for global justice or any other ethical principle that challenges the dominance of a market ethic. The widely shared assumption that dominant ideas are value-free entails that "hidden norms" move freely and shape knowledge and policy for development and poverty in particular ways (see Gasper 2004). The role of these hidden values is of enormous importance, as they not only assume certain hierarchies of ethical values (for example, consumerism and accumulation above equity and solidarity) but they also define the scope of what are considered relevant facts. For example, the particular cognitive values become decisive in factual claims such as how many poor people are there in the world, as well as for formulating predictions about future global poverty trends (Vandermoortele 2002, Wade 2004). The history of the notion of basic needs from its initial formulation in the International Labor Organization (ILO) Report *Employment, Growth and Basic Needs: A One World Problem* (ILO 1976), the work of Mahbub ul Haq (1976) and Paul Streeten (1981) to the simplistic and distorted use of the approach by the World Bank under Robert McNamara, and its eventual demise, is a history that can be rephrased as a conflict between moral and nonmoral values that led to decisive distortions of basic needs as the idea traveled within the multilateral system.

Ideas that travel the boundary successfully are adopted easily, often in a short time but also perhaps because they are easily malleable (a case at hand is the idea of social capital). Ideas that challenge the dominant players, especially their cognitive and ethical values, become distorted or rejected.

Thus what makes ideas successful and quickly adopted are not necessarily their analytical qualities, but their capacity to either adapt to the boundary (even if distorted) or to travel the boundary while maintaining their basic shape. This latter case, I argue, applies to the twin notions of HD-CA as a value laden intellectual boundary object elaborated by the HDRO and today espoused, partly, by UNDP.

HD-CA AS BOUNDARY OBJECTS

Most ideas proposed or endorsed by UN System institutions follow similar processes as described earlier. For example, Gasper applies fruitfully the boundary view to situate the notion of "human security" as a concept and as a discourse (Gasper 2005). But boundary work is, as stated previously, a battlefield. Ideas become "intellectual boundary objects" as they are forced to satisfy the often contradictory demands of diverse knowledges and diverse political demands. Ideas matter and regulate the processes of knowledge formation, the negotiations among different disciplines, users, actors, partners, and the diverse scales of reference about complex and highly politicized problems such as global poverty and development within these institutions. Ideas in development are forced to become interfaces linking various interests and ideologies, actors and their values and concerns; they must link and provide a space where shared understandings (even if superficial) may enable dialogue among diverse intellectual communities.

In the UN System, boundary work is unavoidable. After the 1990s the boundary character of ideas in development is more clear and explicit because the number of partners in the process of policymaking has increased considerably—with the participation of NGOs, the private sector, and increasingly citizens themselves. Also, as anthropologists, sociologists, and political scientists join development agencies' staff and operational teams, there is a greater need for ideas able to bridge these groups' differences with the more established disciplines of economics and development economics. In regard to specific views on poverty, boundary work has become more relevant because of the now widely accepted notion that poverty is multidimensional. Bridging between and reaching all dimensions is now a necessity for any poverty definition.

Nevertheless, the use of boundary objects does not necessarily entail consensus on issues or goals. Neither do they determine by themselves how to implement the policy recommendations that may follow from their use. These notions may simply represent partial agreements or a shared vocabulary, which eventually may come to mean what the most powerful actors think it means. Thus successful boundary objects about poverty and development are no guarantee that knowledge formation about these social issues is either fair or representative of all stakeholders.

Ideas traveling the boundary of multilateral institutions are subject to a variety of pressures. Boundary work often includes cognitive and moral value conflicts. Discussions about the limitations and advantages of the Human Development Index, for example, are evidence of such conflicts, leading some authors to seek to "rescue the notion of human development from the HDI" (Fukuda-Parr 2003). In an organizational culture where technocratic and quantitative views predominate, this may have two relevant consequences. First of all, ethical concerns are often left out structurally in boundary processes, which means that even though we may see the use of certain ethically loaded ideas, their ethical content ends up being overridden by nonmoral values and by normative claims about the status of expert knowledge. This is the case, arguably, of the World Bank (St. Clair 2006a). The second substantial consequence is that the tacit moral content of what is deemed rational reigns without a competing view of what ought to be done about poverty and leads to overlapping the notion of development effectiveness with cognitive values of dominant ideologies and politics. Alternative scientific understandings that take as their point of departure not only ethically explicit issues imbued in the boundary objects themselves but also a different view of what "effectiveness" means, tend to be consistently relegated as allegedly nonscientific. Ideas with explicit ethical value content that challenge dominant orthodox ones are deemed unable to be implemented precisely because they usually do not share the cognitive values of orthodox policy (the priority to quantification and simplicity). Even if those ideas manage to become part of the boundary work, because unsuccessful past policies or pressures from outside criticism lead to the search for new concepts, their ethical content tends to be emptied out during the boundary work performed. And ideas with proper analytical qualities which may have raised substantial new issues and questions are often transformed into different types of boundary objects. An instance of these processes diverting toward one partial meaning of ideas is found in the history of the "basic needs." HD-CA, on the other hand, seem to have been explicitly conceptualized by veterans of that history with an eye to withstanding such boundary pressures. Their ethical meanings are strategic, designed to remain stable as they become boundary objects with the UNDP and other development agencies and even strategically designed to reach the wider public.

The HDRO and UNDP are both hybrid institutions providing knowledge-based policy decisions for their own development projects, or policy suggestions for client countries. And the intellectual drives of these institutions have been decided by substantial and decisive interactions between action-oriented charismatic leaders, such as Mahbub ul Haq, and academics of worldwide prestige such as Amartya Sen, Gustav Ranis, and Paul Streeten. Building new knowledge for alternative policymaking entailed making cognitive as well as political choices, as Inge Kaul argues, that have shaped the identity of UNDP (Kaul 2003). Given that processes of knowledge creation, expert institutionalization, and community formation are in themselves political exercises, UNDP and the HDRO can be seen as sites of coproduction of both global knowledge and global politics (St. Clair 2006b).

The HDRO has acted as a boundary organization, attempting to distance itself from the possible politicization of the knowledge it proposes, yet able to produce policy-oriented research. As a semi-independent think tank, the HDRO seems to have the freedom to pursue research as in an academic milieu (free from principal-agent constraints and much of the boundary work proper of the UN System), while at the same time having the imprimatur of the UNDP.[5] Although I do not address the issue here, it is worth assessing the extent to which the HDRO could become a well-functioning boundary organization, able to maintain an optimal flexibility between the two sides of the boundary. But as in the case of the application to environmental change, highly politicized, ill-structured, and complex global problems require strategically and analytically well-thought-out boundary objects. The strength and rationale of the institution relies on the boundary objects it has put forward and this, I argue, is the case of HD-CA and the HDRO.

The first (written) explicit mention of "human development" was in 1986, during the Islamabad North South Roundtable (Haq and Kirdar 1987), partly as a reaction to the increasingly obvious negative consequences of structural adjustment policies on vulnerable segments of societies. But it was with the creation of the HDRO that Mahbub ul Haq, who had defended human-centered views of development since the late 1960s (Haq 1976, 1995), was able to *put the idea into action*. The collaboration between Haq, Paul Streeten, and Amartya Sen and their respective intellectual and policy insights led to what this chapter refers to as the twin boundary objects HD-CA. The *Human Development Report Series* (HDRS) can also be seen as boundary objects in themselves, or at least, as sets of boundary objects. As is common in the UN intellectual work, HDRS is the result of consultations with academia and other external think tanks, commissioned papers, and often external committees, while the final editorial work is an exclusively in-house task. The work of elaborating HDRS, however, is closer to academic research than standard UN boundary work because the final editorial work is still "independent" from the political principals. Arguably, this is characteristic of an independent think tank, not reflective of UNDP official policy, and has facilitated the elaboration of ethically informed ideas that challenge mainstream orthodoxy. Haq has stated that this independence was explicitly aimed at avoiding possible distortions of the idea brought about by the usual "consensus-building" process that UN agencies go through (Haq 1995). He realistically envisioned the boundary forces that shape and distort ideas.

The rapid acceptance of the human development perspective within the UNDP is due to the boundary potential of the ideas themselves, and to the vision of elaborating a statistical index accompanying those ideas, thereby accommodating the dominant cognitive values of simplicity and quantification (St. Clair 2004). Referring to the wisdom and vision of Mahbub ul Haq after his sudden death in 1998, Sen claims that Haq "took on the leadership of large armies of discontent that were gunning, somewhat sporadically, at the single-minded concentration on the GNP" (Sen 2000). Coordinating discontent was helped by having a boundary object and a boundary organization institutional framework.

The success of the human development perspective is also due to the boundary potential of the combination of the concepts of capability and human development, which in my opinion form together a unique boundary object. The HD-CA boundary object has been able to win over, at least partly, the economic and political forces of UNDP, although it took several years until the idea became officially accepted.[6] Since then, strategies developed by the HDRO's publications and their ideas have influenced the choice of many other goals in UNDP. In the words of its architect, Haq, human development aims to be a "judicious mix of market efficiency and social compassion . . . [at the time it offered] a candid, uninhibited development dialogue that would serve the interests of the global community" (Haq 1995: 28). Yet that "judicious" mix was possible because Sen's notion of capability was integrated within human development, providing the philosophical underpinnings of the idea while endowing it with the authority of the cognitive values of Sen's approach which, elaborated over many years, has aimed at offering an alternative conception of development economics and poverty that expands their dominant informational basis to include concerns for the quality of life, social justice, entitlements, freedoms, and rights. Sen's approach to development originates from a critique of the reductionism and hegemony of utilitarianism in moral philosophy. It represents a social and political ethics based on pluralism that places its main normative force in the ideas of public deliberation and freedom.[7] This multidimensional view of well-being and poverty places the emphasis not on goods and consumption, nor on preference satisfaction or happiness, but rather on opening people's choices to live productive and creative lives according to their needs and interests, or, as Sen often formulates it, those lives that people have reason to value.

In addition, one of the core analytical concepts of capability, the notion of choice, is predisposed to perform well at the boundary of hegemonic development aid approaches dominated, as stated earlier, by economism and market values. "Choice" speaks the language of economists, and allows for different interpretations by those at the boundary. But at the same time it does not easily allow wholesale appropriation and reductionism. Clearly, one can answer questions about *choices of what?* or *choices for whom?* in many different ways, but if development is about opening choices, in principle the choices refer to everyone's. Choice and capability speak very well to grassroots groups concerned with the poor themselves.

Ideas in Action 145

To make the approach more suitable for boundary work, Sen has always refused to provide a list of capabilities (something philosopher Martha Nussbaum has done but so far with limited success). As Alkire has rightly argued in her analysis of the dimensions of human development, part of the challenges lying ahead for any multidimensional view of development is that it must be "user-friendly" in a way that does not distort the basic analytical and value content of the idea. Basic needs, Alkire claims, were user-friendly in the World Bank because needs were very easy to reframe in terms of commodities (Alkire 2002). Sen's reluctance to specify priority capabilities is related to the possible distortions that come with the necessity to be not only operationalized and used by practitioners, but also to speak to many actors. The debate on why and how to specify capabilities, I suggest, may be more fruitfully advanced as has happened in the many discussions about competing scientific ideas in the literature of social studies of science.

I suggest to differentiate the capability approach as a boundary object of the UN System from the capability approach as a mere academic, analytical, and moral category. Such a distinction may also help to move forward debates about the appropriateness of capability as the main fundament of human development, or even particular interpretations of human development, as the shifting divide of the boundary between knowledge and global politics changes and transforms itself as the conditions of global politics and global knowledge change and evolve over time. For example, Des Gasper points out that even though the concept of capability offers a fruitful balance among disciplines and between formal theorizing and policy relevance, Sen's approach would benefit from further engagement with disciplines like psychology and sociology or with alternative political views, such as those developed by communitarians, to enrich its analysis of human agency, the role of culture, or the weight of the community in people's individual lives (2000). Gasper rightly points out how Sen's approach evolves from his involvement with both economics and Anglo-American philosophy, and how his formulation of the capability approach still involves a very thin conception of personhood, one that fails to be adequately self-reflective. Gasper characterizes Sen's approach as *cautious boldness*, and claims that Sen's desire to influence mainstream economics and major policymakers leads him to a style of *gentle persuasion* that can be confused with political timidity. A major shortcoming, Gasper states, is that "Sen does not engage with the behavioural realities of commodity fetishism, commodity superfluity and addiction, and other unfreedoms and side-effects generated by wealth and commodities" (Gasper 2000: 996). A puzzle arises, which Gasper elaborates in a later paper assessing the relations between capability and human development:

The capability approach has been fruitful and the HDRs are in many respects an important advance on earlier mainstream treatments of development. Yet viewed from outside economics CA seems primitive in some ways, insufficient as a theory of well-being, and hardly a theory of the "human" in human development (HD). Amongst possible weaknesses are: its extreme emphasis on choice; obscurities in key concepts; and its emergence from a dialogue between economics and philosophy without much involvement from psychology, sociology and anthropology (Gasper 2002: 14).

Thus "human" development may just not be that human. All the wisdom in the way the concept had performed boundary work has also led to a weakening of the idea itself and even perhaps to some sort of stagnation. Human development remains economistic, and dominated by economists who consider themselves not only the experts, but "omnicompetent, because now 'human' too" (Gasper 2002:14). Focus on such limitations can however divert our attention, I argue, from the needed supraterritorial social and political economy analyses of the problems of development, and especially in regards to global poverty.

If we make a distinction between capability as scholarly analytical concept and as the underpinning of the HDRO's boundary object HD-CA, Gasper's points should be assessed in relation to possible changes in the shape of the boundary itself, rather than as mere analytical shortcomings of HD-CA. I suggest that the success of HD-CA itself has helped transform the boundary on which ideas of development are in play. HD-CA has even been instrumental in the eventual acceptance of the Millennium Development Goals (MDGs), and for example, in increasing awareness about multidisciplinarity and the limitations of economism not only within UNDP but many other UN System institutions, including the World Bank. In fact, as Gasper himself notes, we can define the boundary in policy work as moving from economism in the direction of human development (2004).

Similarly, the important critique offered by Thomas Pogge (2003) of the capability approach may be sharpened if one distinguishes between capability as a moral category and capability as it functions within the boundary object HD-CA. For instance, one of Pogge's points is that capability theorists claim that social justice requires special compensatory benefits for the naturally disabled or disfavored, yet according to UNDP's applications of the capability perspective, social justice does *not* require special compensatory burdens on those naturally advantaged or favored. The Human Development Index (HDI), Pogge argues, is inconsistent with the official intent of the capability approach. For example, Pogge claims that "the HDI encourages policymakers to withhold resources from those who have special needs that make their life expectancy more expensive to extend. Contrary to the advertised intent of the capability approach, this component of the HDI constitutes discriminations *against* the naturally disfavoured" (Pogge 2003: 64). Regarding the other two components of the HDI—primary, secondary, and tertiary school enrolment and GDP per capita—Pogge claims that these are plainly resource-based and not capability-based. Furthermore, and perhaps most important, Pogge argues that the way the HDI aggregates data conceals the extent to which inequalities mitigate or aggravate one another. "An institutional order is surely more unjust if these inequalities aggravate one another, that is, if those with the lowest life expectancy, those who are illiterate, those who lack schooling, and those with the lowest incomes are always the same people" (Pogge 2003: 66). In addition, the HDI scores, Pogge claims, "convey no sense of what life is like for the majority of people in a poor country. And they encourage the thought that each country is solely responsible for its own" (Pogge 2003: 69). That is, Pogge claims that HD-CA, as operationalized in most UNHDR work, tends to reinforce methodological territorialism and diverts our attention from needed supraterritorial perspectives.

Pogge establishes, moreover, an *unusual* relationship between the appalling economic history of the 1990s, which has seen both substantial increases in poverty in many regions of the world as well as in inequality, and capability in its boundary concept form. "I am suggesting," Pogge claims,

> that the large increase in global inequality and the consequent persistence of massive severe poverty in the developing world *have contributed* to the stunning success of the capability approach in international organizations as well as in popular and academic discourse. Capability metric tends to conceal the enormous and still rising economic inequalities which resource metrics make quite blatant. And they may also exaggerate the relative aspects of poverty, thereby lending new respectability to the old nationalist exhortation that protecting our own poor (in the rich countries, where our normative reflections are produced and consumed) must take precedence (Pogge 2003: 71, emphasis added).

Pogge's critique, I argue, not only invites a debate as to whether resource-based or capability-based measurements are best. As suggested earlier, in order to engage on a fruitful debate it is important to disentangle Pogge's argument by distinguishing capability as a moral category, as an analytical tool of welfare economics, and as the fundament of the HDRO's boundary object HD-CA. I propose, rather, to direct our attention toward an analysis of the changes in the shape and elasticity of the boundary work today on both sides, the knowledge side and the global politics side, and not only of the analytical qualities of capability-human development in its diverse forms outside the field of action.

Pogge raises an objection to the role of ethically grounded ideas as they perform boundary work that could be extended to all boundary objects acting within the UN System. Ethical ideas *per se* do not necessarily lead to placing concerns for global justice at the boundary work of multilateral institutions or donors. Increasing inequities and unfairness may even contribute in perverse ways to the success of some ethically grounded boundary objects. The problem is then not only distortions and misinterpretations that affect the analytical qualities of ideas and their value content, but that ethically grounded ideas *per se* are not enough.

What is important is to formulate knowledge that uses, for purpose of framing the discussion and redirecting our attention, counterfactuals based on the possibility of a future ruled by global justice. That HD-CA is a fairer notion than development as economic growth does not in itself make it a model for fair development, and it may not lead either scientists or politicians to focus our attention on the structural causes of maldevelopment, of persistent and increasing poverty and suffering. Haq was visionary 15 years ago. It may be time for different boundary strategies that put forward reformulated, bolder versions of HD-CA. This may be perhaps in the minds of some, as new notions, such as those put forward by Human Security Now, may add on to HD-CA greater strength, broader meanings, less caution and more political force. It is, I suggest, work both for human development theorists themselves as well as for followers of the capabilities approach to respond to challenges of the contemporary boundary between knowledge and politics. As Jasanoff and others have demonstrated in empirical case studies, making transparent the boundary between knowledge and politics may lead to improved ideas, to improved policymaking. In the next section of the chapter, I propose one such strategy for furthering the debate on human development.

IDEAS IN ACTION: ETHICALLY IMBUED BOUNDARY OBJECTS

HD-CA shows us that the elaboration of boundary objects based on specific ethical concerns, from the early stages of knowledge formation and with a certain awareness of the processes of boundary work proper of the UN System, may lead to a better management of value conflicts and to more effectively regulated boundary work. Nevertheless, further work on HD-CA would benefit from addressing explicitly the necessary negotiations, conflicts, and pressures that such ideas may face during open UN boundary work, as well as addressing changes in the boundary itself. An ideal would be the transformation of institutions coproducing knowledge and social orders to become well-managed boundary organizations. A more modest proposal entails preparing the ideas for the tensions and value conflicts they would have to endure as they are translated into tools for action at the same time as they include concerns for global justice.

The proposal entails regulating the boundary work performed by these expertised institutions by using ideas imbued with three specific ethical principles so as to guarantee the presence of considerations about global justice and act as incentives for leading toward the stabilization of the boundaries between knowledge and politics. Explicit use of ethical values by development agencies does not itself entail an appropriate treatment of the tensions and conflicts leading toward increased global justice. Something more is needed.

A more fair system of transnational relations requires attention to processes of knowledge formation. I propose the inclusion of three norms for the formation of boundary concepts: accountability, equity, and deliberative participation.

Accountability mechanisms are needed in order to control the dominance of technocratic views, and to limit the power of experts hired by plutocrats and oligarchs who sideline knowledge claims simply because they may not suit their goals. There is a pressing need to devise ways to hold responsible those making policy decisions. Development institutions ought to be accountable for their ideas, planning and operationalization of their projects, and proposals. An important way to improve the knowledge produced by experts who work for these organizations is to put on their shoulders the burden of proof, to make them accountable for avoidances or mistakes that could have been foreseen.

The principle of *equity* aims to break the pervasive avoidance of discussing the relations of inequality to poverty. Equity here refers to the instrisic equal value of all human beings. It refers to both material and procedural equity. Equity in the distribution of material conditions is a prerequisite for equality of opportunity to increase people's standard of living according to their talents and effort, but also a prerequisite for procedural equity. Without the equitable participation of the views, values, interests, social worlds, and approaches of all those at the boundary table (including experts and nonexperts), it is not possible to talk about accountability, because it is in reference to equity that the question *accountability for what* must be answered. For example, it is difficult to envision institutional accountability mechanisms without a previous equitable participation of all those affected by the policies at hand (or at least by their empowered representatives) and by the knowledge informing those policies. Both material and procedural equity are needed to build shared understandings among in principle opposing views and interests from all stakeholders. Only if both developing and developed countries perceive transnational polices as equitable would it be possible to have feasible consensus and arrangements leading toward global justice.[8]

The third principle, *deliberative participation*, aims to regulate the formation and operationalization of boundary objects and boundary work.[9] It may lead to the formation of partnerships among a variety of very different actors, groups, and academic disciplines with diverse political and research capacities, with diverse interests and normative social, cultural, and political views, and with different and often contradictory cognitive and ethical values. Deliberation and participation are crucial from the early stages of knowledge formation, including the empirical foundations of policymaking, especially given recent research addressing possible flaws in empirical accounts for poverty trends (e.g., Pogge and Reddy 2003, UNDP 2003, Wade 2004, St. Clair 2006b). Deliberative participation may lead to a more open and fair debate as to what ought to be the conceptual foundations of a more adequate methodology for assessing trends in income poverty.

These three principles could guide the formation of knowledge systems and could be applied to all the stages and levels of knowledge formation: that is, to the choice of data deemed relevant in the conceptualization of the problems and the policy proposals, to the choice of partners and the characterizations of who is or is not an expert, to the choice of disciplines elaborating boundary objects, to the diplomatic and scientific processes of boundary work, and to the organizational culture and dynamics of the boundary organizations themselves. Altogether and viewed as working holistically, these principles may help in building a needed nonhegemonic consensus, the shared understandings required for reaching favorable thresholds of knowledge and policy recommendations with enough salience, credibility, and legitimacy among partners with diverse criteria for these three attributes. The proposal aims too to improve the levels of fairness and social justice achieved by these same knowledge systems. HD-CA already shares some of these imbued values. But more can be done.

CONCLUSION

I have argued that the boundary perspective, in particular the assessment of ideas within the UN System as boundary objects, helps us to rethink the strategies, achievements, and challenges ahead for human development. HD-CA are best seen as unique boundary objects of the HDRO which have influenced UNDP and much of the rest of the UN System. Global development institutions, in addition, are best seen as sites of coproduction of both knowledge and politics. Given that ideas still get distorted and transformed as they become objects acting along the boundary between knowledge and politics, and given that explicit ethical principles are not sufficient, by themselves, to change debates about appropriate choices for action, I have suggested to imbue boundary objects with three main ethical principles. Accountability, equity, and deliberative participation as principles regulating knowledge formation processes may help to bring concerns for global justice into the UN System at a fundamental level: by changing the framework of the debate, by shifting boundaries, and thus enabling new forms of boundary work.

In his recent writings, Amartya Sen is becoming more outspoken—less cautious—in addressing the fairness or unfairness of current global dynamics, the outmost boundary. Discussing whether globalization is benefiting the poor or not, Sen argues that it is wrong to claim that globalization is not unfair to the poor because some comparative analysis shows they also benefit from it. What matters is not whether the poor are getting richer or not, but what types of choices one contemplates in analyzing a possible fair distribution of the benefits of globalization. That is, Sen claims, the real choice exercise (Sen 2002). In other words, Sen reminds us that what is important are the types of counterfactuals we take as possible when framing ideas and choices, the same idea proposed by Thomas Pogge. HD-CA has operated within a limited range of choices, constrained by its need to move well while operating at a policymaking interface that was alien to its content, defensive to the types of alternatives proposed. Its architect, Mahbub ul Haq, knew such a boundary very well, as an insider to the UN System institutions, and as an insider to an important idea with substantial analytical and ethical shape, basic needs, that was distorted and eventually negated because of poor boundary performance. But the boundary has changed, and it may be time to expand, to reformulate, and to reframe ideas in development with counterfactuals that take as a goal, not the overcoming of economistic views of social and individual life (past states of the boundary), but rather furthering the goal of global justice. Work in that direction may require bolder, sharper boundary objects that not only help raise needed moral awareness for global poverty, but that also shift the boundary itself.

In short, given that global ideas are often proposed and endorsed by global institutions, UN System organizations are somehow sites of coproduction, coproducing knowledge and social order. The chapter suggests that the policies and discourses of some global organizations are the product of relations within and between them (as well as other external influences), but also of the motivations and beliefs of individuals within the organizations as social beings, i.e. relating to the organization as a whole and to their colleagues within it.

Furthermore, the values that ideas entail, their ethical and cognitive content, and their political content are all crucial elements in determining the fate of these ideas. Language matters, cognitive values matter, ethical language matters. It structures or frames the ways in which actors envision a particular issue, the descriptions and methodologies developed to conceptualize and map their extent. Language structures the way we think, the way organizations may define themselves, and thus the way we act and the way organizations may act. Some terms direct toward a particular set of actions, while others are open to multiple interpretations and may lead to no particular commitments.

This chapter elaborates further this theoretical framework in relation to ethical ideas (in particular capability and human development) and the increasing concerns with human rights in many global organizations. Different multilateral organizations often have ambivalent attitudes to fundamental ethical issues. Regarding the increased role of global normative frameworks such as "human rights" the tendency is to narrow those so as to make them consistent with ongoing ideas and policy frameworks inside the organizations, to technocratize the language of rights while maintaining the moral legitimacy and prestige that follows from the use of human rights language. Ideas may also have political content if they present an issue (poverty, for example) as a matter of power relations. In fact, seeking to decide how to define poverty, or to prevent poverty from being defined in particular ways, is in itself a political act, evidence that language and ideas do indeed matter. The characterization of global ideas such as "capability" and "human development" as intellectual boundary objects, permits us to disentangle the role of language and ideas in forging global politics.

NOTES

1. In this chapter I use the acronym HD-CA in singular to refer to the general approach adopted by the UNDP. I use it as plural when referring to the constituent elements or boundary objects of the approach.
2. For an analysis of ideas elaborated by the UN System along the years see the United Nations Intellectual History Project (www.unhistory.org), in particular Emmerij, Jolly, and Weiss (2001 and 2005). See also Bøås and McNeill (2004).
3. For a more detailed analysis of the notions of boundary work and the coproduction of knowledge and politics among multilateral actors see St. Clair (2006a and 2006b).
4. The Global Environmental Assessment (GEA) is a "collaborative, interdisciplinary effort to explore how assessment activities can better link scientific understandings with effective action on issues arising in the context of global environmental change. The Project seeks to understand the special problems, challenges and opportunities that arise in efforts to develop common scientific assessments that are relevant and credible across multiple national circumstances and political cultures (Guston et al 2000)."
5. For an extended discussion of the problems of delegation in multilateral agencies which also perform research see St. Clair 2006a.
6. According to Richard Jolly this can be dated to the mid-1990s (personal communication).
7. Sen's approach shares many of the fundamental theses put forward much earlier by John Dewey.
8. A similar requirement for equity is needed in the case of another highly politicized and complex global problem, climate change. See Brown (2002).
9. I take the notion of deliberative participation from David Crocker (2003). The goal is to expand the use Crocker makes of the term beyond a discussion on the capabilities approach.

REFERENCES

Alkire, S. (2002) "Dimensions of Human Development," *World Development* 30(2):181–205.
Brown, D. (2002) *American Heat: Ethical Problems with the United States' Response to Global Warming.* Lanham: Rowman & Littlefield.
Bøås, M., and D. McNeill (2003) *Multilateral Institutions: A Critical Introduction.* London: Pluto Press.
———. eds. (2004) *Global Institutions and Development: Framing the World?* London: Routledge.
Crocker, D. (2003) "Participatory Development: The Capabilities Approach and Deliberative Democracy," Working Paper, Institute of Philosophy and Public Policy, School of Public Affairs, University of Maryland.
Emmerij, L., R. Jolly, and T. Weiss (2001) *Ahead of the Curve? UN Ideas and Global Challenges.* Bloomington, IN: Indiana University Press.
———. (2005) "Economic and Social Thinking at the UN in Historical Perspective," *Development and Change* 36 (2): 211–235.

Fukuda-Parr, S. (2003) "The Human-Development Paradigm: Operationalising Sen's Ideas on Capabilities." *Feminist Economics* 9(2–3): 301–318.

Gasper, D. (2000) "Development as Freedom: Moving Economics Beyond Commodities—The Cautious Boldness of Amartya Sen," *Journal of International Development*, 12(7): 989–1001.

———. (2002) "Is Sen's Capability Approach an Adequate Basis for Considering Human Development?" *Review of Political Economy*, 14(4): 435–461.

———. (2004) *The Ethics of Development: From Economism to Human Development*. Edinburgh, UK: Edinburgh University Press.

———. (2005) "Security Humanity: Situating Human Security as a Concept and Discourse." *Journal of Human Development* 7(2): 221–245.

Gieryn, T.F. (1995) "Boundaries of Science," in *Handbook of Science and Technology Studies*, ed. S. Jasanoff, 339–443. Thousand Oaks, CA: Sage.

———. (1999) *Cultural Boundaries of Science; Credibility on the Line*. Chicago, IL: University of Chicago Press.

Guston, D. (2000) *Between Politics and Science: Assuring the Integrity and Productivity of Research*. Cambridge, UK: Cambridge University Press.

Guston, D., et al. (2000) "Report of the Workshop on Boundary Organizations in Environmental Policy and Science," in *Global Environmental Assessment Project*. Cambridge, MA: John F. Kennedy School of Government, Harvard University.

Haq, M. ul. (1976) *The Poverty Curtain: Choices for the Third World*. New York: Columbia University Press.

———. (1995) *Reflections on Human Development*. New York: Oxford University Press.

Haq, K., and U. Kirdar (1987) *Human Development, Adjustment, and Growth*, Islamabad: North South Roundtable.

International Labor Organization (1976) *Employment, Growth and Basic Needs; a One World Problem*. Geneva: International Labor Organization.

Jasanoff, S. (1990) *The Fifth Branch: Science Advisers as Policy Makers*, Cambridge, MA: Harvard University Press.

———. ed. (2004) *States of Knowledge; The Co-production of Science and Social Order*. London: Routledge.

Kaul, I. et al. (Eds.) (2003) *Providing Global Public Goods: Managing Globalization*. Oxford: Oxford University Press.

McNeill, D. (2004) "Social Capital and the World Bank," in *Global Institutions and Development: Framing the World*, eds. M. Bøås and D. McNeill, 108–123. London: Routledge.

Pogge, T. (2003) "Can the Capability Approach be Justified?" in Special issue of *Philosophical Topics; Global Inequalities*, eds. M. Nussbaum and C. Flanders, 30(2). Extended quoted version in www.etikk.no/globaljustice/

Pogge, T., and S. Reddy (2003), "Unknown: the Extent, Distribution, and Trend of Global Income Poverty," www.etikk.no/globaljustice/ (accessed April 10, 2003).

Sen, A.K. (2000) "A Decade of Human Development," *Journal of Human Development* 1(1): 17–23.

———. (2002) "Global Inequality and Human Security" Lecture 2, Ishizaka Lectures, Tokyo, February 18, www.ksg.harvard.edu/gei.

Star, S.L. and J.R. Griesemer (1989) "Institutional Ecology, Translations and Boundary Objects: Amateurs and Professional in Berkeley's Museum of Vertebrate Zoology 1907–39," *Social Studies of Science*, 19(3): 389–420.

St. Clair, A.L. (2004) "The Role of Ideas in the United Nations Development Programme," in *Global Institutions and Development; Framing the World*, eds. M. Bøås and D. McNeill, 178–192. London: Routledge.

―――. (2006a) "The World Bank as a Transnational Expertised Institution," *Journal of Global Governance* 12 (1): 77–95.

―――. (2006b) "Global Poverty: The Co-Production of Knowledge and Politics," *Journal of Global Social Policy* 6 (1): 55–77.

Streeten, P. (1981) *First Things First: Meeting Basic Human Needs in the Developing Countries*. New York: Oxford University Press for the World Bank.

United Nations Development Programme (2003) *Human Development Report 2003*. New York: Oxford University Press.

Wade, R. (2004) *Governing the Market. Theory and the Role of Government in East Asian Industrialization*. Princeton, NJ: Princeton University Press.

Vandermoortele, J. (2002) "Are We Really Reducing Global Poverty," New York: Bureau for Development Policy, United Nations Development Programme.

Part III
Accountability and Manageability

8 Labeling and Tracking the Criminal in Mid-Nineteenth Century England and Wales

The Relationship between Governmental Structures and Creating Official Numbers

Chris Williams

INTRODUCTION

In the United Kingdom in the nineteenth century, large-scale surveillance of perceived criminal populations necessarily involved a number of innovative governmental technologies. This chapter will concentrate on one of the more significant outcomes: the use of counting to assess the effectiveness of government policy. It will consider the ways that statistical thinking, in the form of the auditing of a large number of transactions, affected policy and governmental statistical practice, in the process providing a case study in the relationship between micro- and macro-level statistical practices. First, I will note two contrasting views of how British government structures worked, describe how they generated statistics of criminals "at large," and how these figures were co-constructed by central government and the reporting bodies. Next, I will show how the resulting figures were deployed politically to justify further government action based around the registration of all criminals. Following a description of the activity of this Habitual Criminals Register, I will show the ways that its performance was measured within government and the statistical techniques used to back up the competing explanations for its perceived failure.

"Habitual criminals" and released convicts of various classes were one of the first groups of people that the British state decided to keep track of in the nineteenth century. Hence they have received a significant amount of attention from historians: this has included Wiener's work (1990) on the relationship between surveillance and penal policy more generally, and that of Stevenson (1986) and Stanford (2007) on the politics of the adoption of surveillance and its effectiveness. Others have examined the relationship between calls for increased surveillance of the criminal population (and condemnations of the consequences of perceived lax surveillance) and moral panics surrounding street crime (Bartrip 1981, Davis 1980, Sindall 1990). Sekula (1986) and Ireland (2002) have shown how photography became a key technology to "fix" the criminal, though, as Cole (2002) has shown, this generated a large amount of information related to identity which was hard to search: one focus here will be on the state's quest to find a way of keeping track of this information.

Traditionally, historians have explained the government of the United Kingdom in terms of the action of a small clique of generalists, whose main task was to manage the appearance of government (Moran 2003: 34). A different position has been advanced by Agar, who argues that by the early twentieth century Whitehall (the traditional shorthand for the British executive government and its civil servants) was concerned to make use of information to best increase the effectiveness of government (Agar 2000: 10). These two competing views of government can be equated to an extent with the two technologies of governing—"uncertainty" and "risk"—noted by O'Malley (2004: 13). Moran's "club government" appears to be better articulated toward the management of uncertainty, given its propensity to both act on and generate a series of pictures of political possibilities, figurations, and narratives, each to an extent dealing with uncertainty via "generalized foresight." Conversely, Agar's technocrats would deal with risk, seeing their activity as a response to a series of calculable dangers inherent both within the world at large, and within the governmental structures designed to deal with it. A style of management which emphasizes personal authority and discretion is more readily associated with "club government"; that which relies more on actuarial calculation, with technocracy.

NINETEENTH-CENTURY BUREAUCRACY AND CRIMINAL STATISTICS

Here, I would like to open up the process of government as a particularly relevant object of scrutiny, paying as much attention to the language, forms, and structures of the file and the register—the technologies of government—as to the information contained within them. The answer to the statistical question *How many?* is the product of a long chain of events, which rests on structures of bureaucracy, and on the ways that particular bureaucracies process information and make decisions. The reduction of people and activities to countable units involves a number of standardized steps, usually following sets of rules for manipulating information. Most of this paper is about these pre-statistical steps, rather than about statistics *per se*.

That which is acted upon by government is simplified. The institution which analyzes also employs a series of codified practices. In the government context, this is almost always done via the practices of bureaucracy, the hierarchical form of which has still been best expressed by Weber (1986: 956). British civil service government is about traditions of simplification and modification, which reconstruct information so that it can be processed. These advances in governmental techniques were associated with government servants who controlled semiautonomous agencies—Edward Henderson in the Metropolitan Police and Edward du Cane in the Department of Convict Prisons. Both men had a background in the army, and Henderson had occupied du Cane's role before promotion to the police.

Their experience of running large disciplined organizations appears to have been useful to them, thus bearing out the observation by Dandeker (1990: 101) that bureaucratic surveillance was often pioneered in military organizations. Both agencies reported to the Home Office, the government department with responsibility for implementing all aspects of police and penal policy. The Home Office's default governmental responsibility for a wide variety of internal affairs had threatened to overwhelm it administratively since the early nineteenth century, thus it had developed a long tradition of spawning external agencies to deal with routine business, while the civil servants in the department proper concerned themselves with policy (Chester 1981: 258–259). Each agency had an incentive to represent itself as bureaucratically competent, predictable, and reliable. The surviving files relating to this issue indicate that as well as this, they also tended to be inclined to cooperate with one another rather than to compete.

Statistical thinking in relation to crime began with the production of the first series of criminal statistics in 1810. These related only to the numbers and outcomes of indictable (serious) crimes prosecuted, and were published on the initiative of MPs (members of Parliament) who wanted to use them as ammunition against the death penalty (Gatrell and Hadden 1972: 341). These, and some series of prison returns published subsequently, served as the raw material for many arguments from the 1820s onwards about the causes of crime, which were often subordinated to other debates, such as the utility of universal education (Rawson 1841). These discussions, though, were conducted "out of doors" by members of the public rather than by members of the government. They concerned a definite and known population of individuals who could be located in a variety of legal, personal, and geographical contexts.

THE STATISTICS CONCERNING "CRIMINALS" AT LARGE

The 1857 County and Borough Police Act was a result of a compromise between local governments which wished to retain the responsibility for controlling their own police, and central government, which tended to want more uniformity in policing practice, even if that entailed more central control. The Act obliged police authorities for the first time to transmit a variety of measurements to the central government, in the shape of the Home Office. This accompanied the introduction of central funding to the police, although making the returns was never a criterion for receiving the government grant: the only figure of merit which was used for this process was the proportion of police to population, determined by an annual inspection (Emsley 1996: 34).

The annual police returns included a number of runs of data related to crime, including indictable offenses reported under various headings, numbers arrested and numbers prosecuted. The returns also mandated the police forces to return annual estimates of individuals and premises falling under the following headings:

- known thieves and depredators;
- receivers of stolen goods;
- prostitutes;
- suspected persons;
- vagrants and tramps

Each was then divided by gender and into those over and under sixteen years of age. The premises that had to be counted were as follows:

- houses of bad character;
- resorts of thieves and prostitutes, themselves divided into
    - public houses;
    - beer shops;
    - coffee shops;
- brothels and houses of ill fame;
- tramps' lodging houses

None of these numbers derived from prosecutions: the Home Office was clear that police forces should record individuals and addresses that fell into this category whether or not there was actionable evidence against them, thus creating a class of information about people and proprietors who were "known" to be breaking the law despite the lack of proof which would convince a court (Home Office 1869a).

These returns were managed and framed in a number of ways. In the initial period of operation of these statistical returns, all communication with the local authorities was about the interpretation of the instructions. Critiques of the numbers received did not begin until 1861, once central government had a few years' figures to work on. So, for example, in 1861, the Home Office wrote to the Chief Constable of the County of Berkshire, noting that the "decrease of numbers in Table 3 [suspicious characters in various categories] is remarkable, amounting in some instances to more than half" and demanded "information, or observations" in explanation (Home Office 1861). Similar letters flowed elsewhere that year and in subsequent ones (Home Office 1862a, 1863). Too-spectacular movement was worthy of a follow-up, but so was a suspicious regularity: in 1863 the returns of suspicious characters and locations in Hertfordshire was returned with a note pointing out that this had remained exactly the same as in previous years, and considering that "as the criminal population must to a certain extent be of a variable character these same numbers cannot correctly represent the depredators offenders and suspected persons at large" (Home Office 1863).

These figures were massaged and to an extent created from the center. This was possible because unlike the other returns they did not reflect a legal activity such as court action, an arrest, or even a report of a crime (though the crime reports were generally subject to extreme distortion by the forces that originated them). The Home Office indeed was keen to point out that they should not reflect convictions, merely suspicion, reprimanding local authorities who sought to base their counts of vagrants on the numbers actually arrested for this crime, or their returns of receivers of stolen goods, and their houses, on those convicted (Home Office 1862b). In the case of most of the forces, the Home Office could chide and plead, and remind that they could withhold the police grant (which amounted to a quarter of the costs before 1873, half thereafter) if returns of a sort were not sent off, although in practice, this was never done. The grant was only ever withheld when forces failed to appoint an adequate number of men in proportion to the population. In the case of the Metropolitan Police, largely funded by and ultimately responsible to central government, they could, and did, respond to the inconsistent tendency to label all brothels as places of resort for thieves, by returning the forms along with instructions to fill them in "correctly" (Home Office 1873a).

This list of "known" thieves often generated extreme variations from year to year, on which the Home Office sometimes, but not always, remarked. The attention paid to the fluctuations appears to have fluctuated itself (Home Office 1869b). Concern seems to have been the product of the arrival of new Home Secretaries, just as a rise or fall in the figures of crime correlates pretty well with the appointment of a new chief constable (Tobias 1967: 258, Wiener 1990: 146). The figures for criminals at large and their places of resort were, like those for the number of crimes reported, a "discretionary statistic."

THE REIFICATION OF ESTIMATES

This particular run of numbers—the bad characters at large and their haunts—was an example of statistics as the performance of governance, rather than as an indispensable link in the operation of government. Nevertheless, it was also an important source of the authority of the Home Office over the provincial police (Home Office 1875). Before 1914, the central government machinery was happy to maintain a hands-off attitude to the local police forces on most matters (Williams 2007), but they were keen to assert their authority to appear to standardize their statistical returns. Partly because of this, although the constituent figures were the product of inconsistent counting practices around the country, and may have exhibited wide fluctuation from year to year, in the annually printed criminal statistics they emerged as definite numbers which implied an all-seeing government gaze. The knowledge of the "criminals at large," seemingly an exact understanding of a criminal community and its haunts, served as a springboard for measures designed to deal with this community as a body.

Official recognition of a criminal population supposedly at large created more political support for moves to contain this population, in an example of the way that counting creates the subject of government as much as it merely illuminates it.

In the 1860s, concern over the extent of crime, stoked by the final ending of convict transportation to Australia, led to a series of "moral panics" about street robbery, which culminated in legislation aimed at making the criminal law more punitive toward the persistent offender (Sindall 1990: 130–147). One of these pieces of legislation—the Habitual Criminals Act (HCA)—was introduced in 1869 as a measure to "suppress crime." The government minister introducing it in the House of Lords in 1869 linked it to a number of factors, but one of the most powerful was the size of the "criminal class" (Times 1869a). After detailing the numbers convicted of serious offenses (14,207 in 1867) in prison and released on licence, he conceded that "the number of the criminal class is far larger." His evidence for this statement was drawn from the averaged statistical returns of the years 1863–1865, and he stated the numbers of known thieves (22,959), receivers (3,095), prostitutes (27,186), suspected persons (29,468), and vagrants (32,938)—"making a total of 115,646." The figures, which the Home Office's civil servants must have known to be vague in the extreme, were here presented as unalloyed facts in the service of penal policy, and marshaled alongside the very much more reliable numbers which were generated by the prison system keeping track of its inmates, and the courts tallying their activity.

The HCA mandated the creation of a register of serious criminals, which would record all those convicted of more serious crimes, in order to make it possible for the courts to punish repeat offenders more severely. It would also make it easier for police who had arrested suspects and were unsure of their identities to make sure, through reference to physical description and distinguishing marks. An influential body of opinion considered that the proper management of information was one crucial way in which the problem of the criminal population—especially that part of it which was taken to be mobile, durable, and protean—could be controlled. In 1878, a Parliamentary Committee of Inquiry into the organization and operation of the Metropolitan Police's detective forces heard Inspector Shore reply, in response to a leading question on the uses of a general list of criminals:

> We should then have a general register of all cases, and it would be kept in one place and as the inspectors were brought into the office to be consulted with reference to certain classes of crime, they would see that certain people were in existence who committed the particular kind of offences, and in my opinion the offenders would be detected in a very short time, because those men would be put under observation (Home Office 1878: 27).

Shore was aware of some of the difficulties of such projects for total information awareness—he responded to the vision of "one great network" with

"I certainly think it would be very beneficial, but at the same time I do not see exactly how it is to be done" (Home Office 1878: 25)—but he was also part of an era which looked to forms of total centralized understanding to solve social problems. This era advertised its ability to cope with social problems in often remorselessly numerical terms. Throughout the country, the effectiveness of local police forces was also portrayed largely in terms of numerical returns (Williams 2003). The pattern was repeated in London where the annual report of the Metropolitan Police Commissioner listed, among other measurements relating to the force's capacity, challenges, and achievements, the fact that during the previous year London's patrolled area had grown by "226 new streets and two squares, a total length of 38 miles and 722 yards" (Times 1872).

The introduction of habitual criminal legislation led to a new form of thinking about statistics. It was different in that it involved definite numbers which were linked to administrative practice and outcomes. It gave the Home Office statutory responsibility for a management task which they had to monitor. The area that concerned the department's civil servants was two-fold: on the one hand, the actual group of criminals at large, and their potential danger to the public, and on the other, the efficiency and effectiveness of those subordinate agencies of the department which were entrusted with monitoring this population. Both problems were susceptible to numerical analysis. This reflexive analysis was an example of a wider commitment to governmental statistical thinking which underpinned bureaucracy by holding out the prospect of a "uniformising, apparently impartial and modernising mould, to which bureaucrats at all levels were expected to conform" (Woolf 1984: 165). Measurable outputs were expected, and the process whereby they were arrived at was also, increasingly, measured.

THE OPERATION OF THE REGISTER

The following account of the way that the register evolved will pay particular attention to the modes of understanding which were used to diagnose (and represent) relative success or failure of the way that the register was administered. It was launched with fanfare as an attempt to map and track a specific class of mobile criminal. As it was being implemented, *The Times* wrote that:

> Legislation in the nature of this Habitual Criminals Act can only be justified on the hypothesis that the population is divisible into two orders—those who habitually obey and those who habitually break the law. Criminal statistics amply vindicate such a division, and testify to its consequences in the shape of a multiplication of crime (Times 1869b).

It concluded that while in the past, the State had been unable "to supervise the criminal population," now there was "machinery available to govern it." In a widely reported speech in 1870, Lord Derby, the ex-Cabinet Minister who was active in the local Discharged Prisoners Aid Society, said of it:

> We have taken a great step within the last year or two in the Habitual Criminals Act. We have begun to recognise the fact, that though isolated offences cannot be prevented, yet it is possible, and it is our duty, to protect society against the existence of a class who live by making war on property (*Glasgow Herald*, 1870).

The classification and "fixing" of a class of career criminals, therefore, was seen as reasonable, in tune with the spirit of the age, and attainable.

The register was implemented according to the best estimates of rational administration. By the 1860s, the labor roles in the civil service had been defined and codified for the first time. Following the Trevelyan review of the civil service, which was largely designed to increase efficiency by eliminating patronage in favor of a meritocracy, the various roles of the clerk and the (inferior) "writer" had been codified (Chester 1981: 307–311). Thus it was possible to define a role in terms of a set of competencies which it would require—for example, writing, basic arithmetic, and dictation. This role would be assigned a salary, and the candidate had to be examined in each competency before his appointment could be confirmed (Home Office 1870). The data collection efforts, and the collation effort of the Habitual Criminals Register, followed standard structures of bureaucratic governance (Keymer 1922: 132–135). Nodes were identified and defined: in this case, 220 police forces and 150 prisons. Communications were sent *ex officio* to chief officers, and requested from them, whatever the precise administrative arrangements of the node (Home Office 1869c). One point which this process illustrates to the twenty-first century observer is that the standard bureaucratic language remained in the process of formation. The description of the Habitual Criminals registration process in 1873 reads "All Descriptive forms after Registration, are Docketed, Numbered, and put away for easy reference" (Home Office 1873b). "File" was not yet a verb.

Technical definitions were created for categorizing words in general usage. From 1857 the Home Office included the results of inquests within the Criminal Statistics, and correspondence with Coroners reveals an instance of the often arbitrary assignment of numbers to concepts, and vice versa: the categories of "infant," "child," and "adult" which are used for the returns are clarified to mean children under seven years of age, seven- to sixteen-year-olds, and people over sixteen years of age (Home Office 1857). Codes, usually but not always in the form of abbreviations, were used on registers to describe physical features of the criminal in the register (Metropolitan Police 1900).

The initial register, begun in 1869 and kept by a special department in the Metropolitan Police, did not last long in its initial form. The Prevention of Crimes Act 1871, introduced in an attempt to improve identification, made the photographing of criminals compulsory. Henderson, the Metropolitan Police Commissioner, hoped that this would lead to an increase in the overall efficiency of the register (Times 1872) but this did not come to pass. Following a review in 1874 of its operation in England and Wales (a single unit for criminal justice purposes) and Scotland and Ireland, it was substantially shrunk. The Registry was amended in 1877, to contain the details only of those people who were guilty of serious offenses and about to be released from prison, which were forwarded to London as standard forms with a photograph (Home Office 1877). This led to a reduction in the number of names added to it every year from around 15,000 to around 3,000. Its operation was also taken over by the Convict Prisons Department in 1878.

REFLEXIVE THINKING ABOUT THE REGISTER

The key point in the evolution of the Habitual Criminals Register was in 1874, when it was examined, and found to be a failure. This followed a number of critical reports from within the Home Office and the Metropolitan Police. At the time, two main explanations were advanced for the failure of the first iteration of the register. Both relied on statistical thinking; one is more convincing than the other. The first was that it was too big to be searched effectively. Eventual overload had been predicted by a Home Office civil servant at the time that the register was set up (Home Office 1871). It was also given as the official reason when the register was restructured in 1877:

> Up to the present time all persons convicted of "crime" as defined by the Statute applying to this subject (now the Prevention of Crimes Act, but formerly the Habitual Criminals Act, 1869) have been registered and photographed, and a large number of names of criminals has thus been accumulated, respecting the majority of whom the information which the registry is intended to furnish is not likely to be required, and the usefulness of the registry as a means of identification is much impeded by the difficulty and loss of time involved in searching among too large a mass of names and descriptions (Metropolitan Police 1877).

The second, more convincing explanation was that it was underused because other methods were better. This explanation, advanced from within the Metropolitan Police, relied on an assessment of the rate at which the register was used by various police forces in the rest of England.

Edmund May, the Met's Chief Clerk, (who had up until the year before been arguing for more staff for the Habitual Criminals Office), submitted a detailed numerical analysis of the actual business which suggested that its business was too slight to justify paying a chief clerk and two assistants (Home Office 1873c). The Met also canvassed its divisional superintendents as to the effectiveness of the Office. They replied that although they were obliged to contact it in cases of disputed or unclear identity, there were few or no cases where its intervention had proved significant: in almost all cases, they had identified the suspect through other channels (Home Office 1874a).

The staffing of the Office was constant over this period, and in 1874, May was convinced that it was, if anything, overmanned: this opinion was based on a count of the letters (around 600 per year) received by the office (Home Office 1874e). Given that, the figures in Table 8.1 do not paint a picture of a register suffering from information overload: the hit rate of the searches is increasing, despite the increase in the size of the register. Furthermore, a further memo revealed that of the 3,957, 3,006 were made by the Met, and 571 of the identifications were of those; for 90 percent of the country, four years of operation had produced just about 1,000 requests, of which 300 were successful (Home Office 1874b).

The practice in the 1870s, and up to the 1890s at least was for police forces to rely not on the central register in London, but on "routes"—requests for information sent to neighboring and other forces—dealing with identification of suspects (Home Office 1895, Home Office 1874c). A "route" was a form containing the relevant personal details of the suspect, often with a photograph included, it also contained a list—headed "route" in a position that must explain how these forms got their name—of the places to which it was to be sent. If any of the prisons or police forces which received the form could identify the suspect, they contacted the requesting body; if not, they merely passed the "route" on to the next institution on the list; either way, they recorded their activity on the back of the form. The selection and

Table 8.1 Identification from the Register, 1870–1873

	Requests	Identified	Percentage	Register Size
1870	262	39	15	31,764
1871	1,007	168	16	59,754
1872	1,385	331	24	88,452
1873	1,303	352	27	117,568
Total	3,957	890	22½	

Source: Home Office (1874e). N.B., This table is a copy of a table in May's memo, *except* for the final column, "Register Size" which is added here as an aid to analysis.

ordering of places that it was sent to—the route's route—was a matter for the local knowledge of the detective or prison officer concerned. It was a craft skill which had to be learned on the job, rather than a routinized skill of the kind employed in the construction of the centralized register. It also differed in timing, since it was a task that was only carried out when it was needed, unlike the regular reporting of information to the central register.

The Met Commissioner himself recognized the strength of this practice in 1874, writing of the provincial forces that they:

> have no disposition to avail themselves of a Register kept by a local body like the Metropolitan Police, and I believe that they get more information by enquiring of the police and governors of gaols in places to which the person may be surmised to have resorted (Home Office 1874).

The central government's analysis of the register's effectiveness revealed a propensity for using statistics. Here we see numbers used to govern: rates, means, and totals are calculated and put to use, but the topic in question is not the criminal, or society, but the best way to run the register. Chief Clerk May also used sampling to establish his case—taking 350 entries "promiscuously" from the register (like "filed," "random" had yet to be used in this context), he pointed out that 273 of them were convicted for minor offenses, and extrapolated from this to the register as a whole (Home Office 1874b). He was then able to calculate that if those committing minor offenses were not recorded, the resulting workload would be suitable for a police inspector and a sergeant, at an annual cost of less than half its current level (Home Office 1874b).

Henderson, summing up May's analysis of the register for the Home Secretary, was also careful to rely on his numerical evidence to back up his claims. He noted that the proposed staffing by police officers was possible because the work of the office was "mechanical" (Home Office 1874d). From within the Home Office's secretariat, Godfrey Lushington also marshaled numbers in his summing up of the scheme's effectiveness, including comparative identification rates through the registers in England, Scotland, and Ireland (1875). Finally, Du Cane, the head of the Convict Prisons office, which ended up in charge of the truncated register, also relied on an audit of the practice of the office when making his case. He too repeated the charge of incipient information overload, while admitting that in action, the register was largely bypassed by the "route" system. He also used a sampling technique to prove his point, noting that "of 3,073 men, indexed under the letter A, during the six years the Registry has been in operation" only 358 have more than one entry, and only 89 have more than two thus "about the only names on this Register which serve any useful purpose are these 89, out of the whole 3073" (Home Office 1876).

The actual solution to the problem of the register's failure was two-fold. Within the Home Office, a list of legal criteria for inclusion on the register, deriving from the type of sentence rather than the fact of conviction, was constructed, which was arrived at in order to create one of manageable length, and this formed the basis for amending legislation in 1876 (Metropolitan Police 1877). The register itself was taken from the Met, and given to the convict office to manage. The dream of a national register of all criminals continued, but everyday police detective practice remained based on local records, accessed on a bilateral and networked basis. The model of a centralized register which could comprehend the whole of a defined criminal class, and in doing so allow them to be "fixed," did not work.

CONCLUSION

The tale of the Habitual Criminals Register shows how, in Moran's words:

> The synoptic gaze of a regulatory state could hardly function unless it had something to gaze on . . . The Victorian revolution in inspection and data gathering was indeed impressive; but it was still puny compared with the social range demanding surveillance (Moran 2003: 46).

We must remember the historical experience of the state's weakness and ignorance, even as we acknowledge the desire of states to be strong and knowledgeable. Nevertheless, this case has shown how, in the top echelons of the Home Office's professional agencies at least, Moran's idea of "club government" did not apply. Instead, policy and practice were debated within Whitehall in a technocractic style, in a manner consistent with the patterns suggested by Dandeker and Agar. What we have seen is another example of one of the new technologies which emerged for governing the "dangerous classes" as "'actuarial entities,' statistically knowable and calculable risks" (O'Malley 2004: 17). Before the individual could be knowable, the governmental structures themselves had to be. In writing of the "criminal justice state," Zedner endorses Harcourt's assertion that "actuarialism chose us" (Zedner 2009: 259, Harcourt 2007: 32). But it did not: to some extent, we chose it.

The concept of "information overload" was understood in the 1870s, but it was not this that led to the failure of the first iteration of the Habitual Criminals Register, and it had yet to materialize when the register was drastically scaled down in 1877. Instead, it was used as an excuse for other failings. Although the all-knowing centralized system had a powerful rhetorical force, users preferred networked information exchange over centralized systems, tacit knowledge over centralized institutionalization. And government was sufficiently self-regarding to diagnose this problem and react to it. An examination of the micro-practices of how the "back office" governed statistical surveillance reveals that in this period, numbers were as likely to be used to govern within the state apparatus as outside it.

ACKNOWLEDGMENT

As well as the help and encouragement from the editors of this volume, I need to acknowledge a number of welcome suggestions by Robert M. Morris, who has corrected many of my misapprehensions about the nineteenth-century Home Office and its practice. My attendance at the conference at which an earlier version of this chapter was presented was funded by the Open University's Arts Faculty.

REFERENCES

Agar, J. (2000) *The Government Machine: A Revolutionary History of the Computer.* London: MIT Press.
Bartrip, P.W.J. (1981) "Public Opinion and Law Enforcement: The Ticket-of-Leave Scares in Mid-Victorian Britain," in *Policing and Punishment in Nineteenth-Century Britain,* ed. V. Bailey, 150–183. London: Croom Helm.
Chester, N. (1981) *The English Administrative System, 1780–1870,* Oxford, UK: Clarendon Press.
Cole, S. (2002) *Suspect Identities: A History of Fingerprinting and Criminal Identification.* Cambridge, MA: Harvard University Press.
Dandeker, C. (1990) *Surveillance, Power and Modernity: Bureaucracy and Discipline from 1700 to the Present Day.* Cambridge, UK: Polity Press.
Davis, J. (1980) "The London Garotting Panic of 1862: A Moral Panic and the Creation of a Criminal Class in Mid-Victorian England" in *Crime and the law: the social history of crime in Western Europe since 1500,* eds. V.A.C. Gatrell, et al., 190–213. London: Europa.
Emsley, C. (1996) *The English Police: A Political and Social History.* Harlow: Longman.
Gatrell, V.A.C., and T. Hadden (1972) "Criminal Statistics and Their Interpretation," in *Nineteenth Century Social History: Essays in the Use of Quantitative Methods for the Study of Social Data,* ed. E.A. Wrigley, 336–96. Cambridge, UK: Cambridge University Press.
Glasgow Herald (1870) "Lord Derby on Crime and Pauperism" Tues January 18th 1870.
Harcourt, B. (2007) *Against Prediction: Profiling, Policing and Punishing in an Actuarial Age.* Chicago, IL: University of Chicago Press.
Home Office (1857) TNA [National Archives] HO 155/1A "Statistics entry books," Circular, p. 22, 13 Jan 1857.
———. (1861) TNA HO 155/1A "Statistics entry books," Letter of 21st Oct 1861 to Chief Constable Berkshire: 255.
———. (1862a) TNA HO 155/1A "Statistics entry books," Letter of 13th Nov 1862 to Chief Constable Dorset: 303.
———. (1862b) TNA HO 155/1A "Statistics entry books," Letter of 13th November 1862 to Head Constable Leicester: 304.

170   *Chris Williams*

———. (1863) TNA HO 155/1A "Statistics entry books," Letter of 13th Oct 1863 to Head Constable Hertford Borough Police: 334.

———. (1869a) TNA HO 155/1A "Statistics entry books," Letter to Staffordshire Police: 155, Dec 1869.

———. (1869b) TNA HO 155/1A "Statistics entry books," Letter to Head Constable of Bath, 10 Dec 1869: 452, also similar letters, all from the Home Secretary.

———. (1869c) TNA HO 45/9509/16260 "Habitual Criminals Registry. Organisation and staffing," Memo, Sept 21st 1869.

———. (1870) TNA HO 45/9509/16260 "Habitual Criminals Registry. Organisation and staffing," Letter of 5th Jan 1870, Henderson to Home Office.

———. (1871) TNA HO 45/9509/16260 "Habitual Criminals Registry. Organisation and staffing," Memo of 21 Jan 1871.

———. (1873a) TNA HO 155/1A "Statistics entry books," Letters of 26 Dec 1873, 20 Jan 1874 to Metropolitan Police Commissioner: 523, 4.

———. (1873b) TNA HO 45/9509/16260 "Habitual Criminals Registry. Organisation and staffing," Memorandum of duties performed in the Office of Registrar of Habitual Criminals, 2nd April 1873.

———. (1873c) TNA HO 45/9509/16260, "Habitual Criminals Registry. Organisation and staffing," Memorandum "Habitual Criminals Office" by E.G. May, April 2 1873.

———. (1874a) TNA HO 45/9509/16260, "Habitual Criminals Registry. Organisation and staffing," Report from Metropolitan Police, April 1874.

———. (1874b) TNA HO 45/9518/22208, "Register of Habitual Criminals. Working of and transfer to Prison Department," Memo from EG May, June 17 1874.

———. (1874c) TNA HO 45/9518/22208, "Register of Habitual Criminals. Working of and transfer to Prison Department," Prisoner record, Ipswich County Gaol, March 24, 1874.

———. (1874d) TNA HO 45/9518/22208, "Register of Habitual Criminals. Working of and transfer to Prison Department," Memo from Henderson (Met Commissioner) to Cross (Home Secretary) 22 June 1874.

———. (1874e) TNA HO 45/9518/22208, "Register of Habitual Criminals. Working of and transfer to Prison Department," Memo from May, Met Chief Clerk, May 16th 1874.

———. (1875) TNA HO 45/9518/22208, "Register of Habitual Criminals. Working of and transfer to Prison Department," Printed memo by G. Lushington, Jan 1 1875.

———. (1876) TNA HO 45/9518/22208 "Register of Habitual Criminals. Working of and transfer to Prison Department," Memo by E. Du Cane, 10th Feb 1876.

———. (1877) TNA HO 45/9518/22208 "Register of Habitual Criminals. Working of and transfer to Prison Department," Prevention of Crimes amendment Act, 1876, Memo from R A Cross 15th March 1877.

———. (1878) TNA HO/45 9442/66692 "Report of the departmental commission on Detectives," Evidence of Inspector Shore, p. 27.

———. (1895) TNA HO 45/10421/R99, "Statistics," Letter from Chief Constable of Wolverhampton 28 Oct 1895.

Ireland, R.W. (2002) "The Felon and the Angel Copier: Criminal Identity and the Promise of Photography in Victorian England and Wales" in *Policing and War in Europe*, ed. L. Knafla, 53–86. Westport: Greenwood Press.

Keymer, E.S. (1922) *The Government Clerk's Companion: A Complete and Comprehensive Exposition of the System of Office Procedure Employed in Government Secretariats*. Chunar: Sanctuary Press.

Metropolitan Police (1877) TNA MEPO 3/88, "Habitual Criminals Act, 1870 and Prevention of Crimes Act, 1871: Correspondence" Letter from Liddell, 15th March 1877.
———. (1900) TNA MEPO 6/12a, "Register of Habitual Criminals," 1900.
Moran, M. (2003) *The British Regulatory State: High Modernism and Hyper-Innovation*. Oxford, UK: Oxford University Press.
O'Malley, P. (2004) *Risk, Uncertainty and Government*. London: Glasshouse.
Rawson, R.W. (1841) "An Enquiry into the Condition of Criminal Offenders in England and Wales, with Respect to Education," *Journal of the Statistical Society of London* 3(4): 331–352.
Sekula, A. (1986) "The Body and the Archive," *October* No. 39: 3–64.
Sindall, R. (1990) *Street Violence in the Nineteenth Century: Media Panic or Real Danger?* Leicester, UK: Leicester University Press.
Stanford, T.G. (2007) "The Metropolitan Police 1850–1914: Targeting, Harassment and the Creation of a Criminal Class." Unpublished PhD thesis, University of Huddersfield.
Stevenson, S. (1986) "'The Habitual Criminal' in Nineteenth-Century England: Some Observations on the Figures," *Urban History Yearbook*: 37–60.
Times (1869a) "Parliamentary Intelligence" Sat Feb 27th 1869.
———. (1869b) "Our Police Report" Tues Aug 25th 186.
———. (1872) "The Metropolitan Police" Mon Aug 19th, 1872.
Tobias, J.J. (1967) *Crime and Industrial Society in the Nineteenth Century*, Harmondsworth, UK: Penguin.
Weber, M. (1968) *Economy and Society: An Outline of Interpretative Sociology* [Ed by G. Roth and C. Wittrich], New York: Bedminster Press.
Wiener, M.J. (1990) *Reconstructing the Criminal: Culture, Law and Policy in England 1830–1914*. Cambridge, UK: Cambridge University Press.
Williams, C.A. (2003) "Catégorisation et stigmatisation policières á Sheffield au milieu du XIXe siècle," *Revue d'Histoire Moderne et Contemporaine* 50(1): 104–125.
———. (2007) "Rotten boroughs? How the Towns of England and Wales Lost Their Police Forces in 1964," in *Corruption in Urban Politics and Society: Britain 1780–1950*, eds. J. Moore and J.B. Smith, 155–175. Aldershot, UK: Ashgate.
Woolf, S.J. (1984) "Towards the History of the Origins of Statistics: France, 1789–1815" in *State and Statistics in France 1789–1815*, eds. J.-C. Perrot and S.J. Woolf, 81–94. London: Harwood Academic Publishers.
Zedner, L. (2009) "Epilogue: The Inescapable Insecurity of Security Technologies," in *Technologies of Insecurity: the surveillance of everyday life*, eds. K. Aas, H. Gundhus, and H. Lomell, 257–270. London, Routledge.

9 From Categorization to Public Policy
The Multiple Roles of Electronic Triage

Ellen Balka

INTRODUCTION

In a brilliant advertising campaign,[1] a pill bottle lies empty on its side, surrounded by the pills that have fallen out of it. The caption reads "a lawyer sees patent infringement, an economist sees a bullish market, a psychologist sees addiction—question every angle." This graphic and text—one of several used in an advertising campaign that was developed to promote a university as "the interdisciplinary university"—is based on the theme that a single artifact can be seen from multiple points of view. Such is the case with medical records, the data that constitute them, and the activities which are coordinated by those records.

In this chapter, I follow an intellectual trajectory similar to that implied by the advertisement just described in relation to the triage process that occurs in hospital emergency rooms. Triaging in an emergency room is the activity that staff perform to determine the acuity of a patient, in order to ensure that patients with the most severe ailments are seen the most quickly. Drawing on ethnographic studies of triaging systems in two Canadian hospitals—a study of the introduction of a patient registration system into an adult emergency room (Sharman 2003; Schuurman and Balka 2009), and a study of a pediatric emergency room's triage system that was undertaken in the wake of a problematic electronic triaging system (Balka and Whitehouse 2007; Bjorn and Balka 2007)—I consider the multiple ways that data collected during the process of triaging and treating patients are used in a number of decision-making processes. In this chapter, my focus is not on the triaging activity per se. Rather, I am concerned with the electronic trace(s) of the triaging activity—the data produced during the triaging process, the places those data go, the other data with which triage data are combined, and so on. Ultimately, I am concerned with how triaging is viewed by varied actors, all of whom have a stake in both triaging activities, and the data collected in relation to triaging, which are used in subsequent decision-making activities.

Elsewhere (Balka 2003) I have written about how local work practices may be tied to seemingly disconnected and far away activities of remote, or nonlocal actors; I have tried to show how what happens in local work environments is often connected to activities that occur in a sense "off stage," at a distance from—but not disconnected from—use. More recently, I have also been concerned with how interactions between the multiplicity of actors who come together to configure systems for local use—some actors part of the organization that is implementing the technology, and some actors who are external to that organization—influence what Suchman and Jordan (1989) referred to as technology design-in-use (Balka and Wagner 2006).

My exploration here continues this line of inquiry which attempts to situate local work practices within broader contexts, in an effort to explicate the multiplicity of interests that come to bear on specific technology designs and use contexts. While it may seem obvious that in looking at a pill bottle and pills that "a lawyer sees patent infringement, an economist sees a bullish market, a psychologist sees addiction," the multiple and divergent ways that a software program and the data it is enlisted to produce are seen by various groups often remain either unexplored in relation to health information technology or are "lost in translation" between the requirements stage and the prototype stage of system development, with costly and potentially disastrous results.

Here I continue to explore the way that technology in general and health information systems in particular are situated within broad networks of actors, and how the data which health information systems produce assume agency in broader debates. These topics are addressed by focusing on what different groups of users see when they see a triaging system. I draw on Star and Griesemer's (1989) notion of boundary objects in discussing the varied ways that triaging data are viewed, and I situate triaging activities in a broader array of complex systems that are both organized by, and which help organize work practices in emergency rooms. Through my treatment of triaging activities, I show how medical records, the data that constitute them, and the activities which are both constitutive of those records and which those records coordinate can be seen in many different—and at times divergent—ways. Medical records are sites where "multiple stories about multiple patients and organizations are at stake" (Berg and Bowker 1997), and here I hope to bring some of those stories to life through the entry point of triaging. I begin by introducing the concept of boundary object (Star and Griesemer 1989). After an overview of triaging activities, I go on to discuss triaging scales as classification systems, and the work that data collected during triaging activities accomplishes in broader policy debates about health systems. Material covered here demonstrates how data collected for one activity—in this case the triaging of patients—can enter into policy debates, and may be pressed into service in managerial and other debates, where the limitations inherent to data may be obfuscated in the interest of broader controversies and policy debates.

BOUNDARY OBJECTS

In their 1989 article that introduced the term boundary object, Star and Griesemer explain that scientific actors face many problems in trying to ensure information integrity amidst diverse activities. They suggest that actors trying to solve problems come from diverse social worlds, and must establish a common operating mode in order to succeed. When the worlds of diverse actors meet, challenges arise because the objects and methods required to generate new knowledge mean different things to different actors. Cooperation between groups of actors requires reconciliation of these diverse meanings. Star and Griesemer (1989: 389) suggest that scientists and other actors contributing to science "translate, negotiate, debate, triangulate and simplify in order to work together."

Objects of scientific inquiry inhabit multiple social worlds, because conducting science requires intersectional work (ibid.). Scientific work is heterogeneous and yet requires cooperation; the diversity must be managed, and Star and Griesemer (ibid.) suggest that such diversity is managed in part through boundary objects. An analytic concept, the term boundary objects is used to describe "those scientific objects which both inhabit several intersecting social worlds [. . .] and satisfy the informational requirements of each of them. Boundary objects are objects which are both plastic enough to adapt to local needs and the constraints of the several parties employing them, yet robust enough to maintain common identity across sites. They are weakly structured in common use, and become strongly structured in individual site use" (ibid.: 393).

Boundary objects can be abstract or concrete and have different meanings in different social worlds. Yet their structure is common enough to more than one world to make them recognizable. Creating and managing boundary objects is a process that is critical in developing and maintaining coherence across intersecting social worlds (ibid.), and in ensuring the successful use of information systems in complex organizations.

TRIAGE AS BOUNDARY OBJECT

Triage has many meanings. Derived from a French word meaning "sorting, selection, choice," triage has been used in English since 1727 to indicate the action of assorting according to quality. For example, coffee beans have long been triaged.[2] In the context of medical care, triage refers to "the process of sorting people based on their need for immediate medical treatment as compared to their chance of benefiting from such care."[3] Triage provides a means through which limited medical resources must be allocated to maximize the number of survivors (in cases of extreme illness). Although triage is performed in varied settings including disasters and wars, the discussion here revolves around triaging in the context of hospital emergency rooms.

The triage area of a hospital serves as a physical point of entry into the acute care medical system, as well as a boundary between the waiting area and treatment areas of an emergency room. Triaging is a process which determines who receives quick service and who must wait. It is characterized by specific practices (for example, a structured interaction between triage nurse and patient—if the patient is able to speak), which revolve around a classification system that produces a score which begins a patient's process of being seen within an acute care setting, where the patient must be tracked and classified in a number of different ways, related to varied domains which come to bear on patient care. Triage can be understood as both a literal place-based boundary, and, as I'll suggest later, data resulting from the triage process can arguably be viewed as a boundary object—arguably, because while triage data possess many of the attributes Star and Griesemer (ibid.) have identified as characteristics of boundary objects (e.g., actors using triage data come form diverse social worlds, and most work together to obtain more useful triage data), it isn't clear that a reconciliation of the diverse meanings of triage data has occurred between the varied actors, all of whom have a stake in data resulting from triage.

TRIAGE AS A CLASSIFICATION SYSTEM

Triage is a critical component of emergency care (Durojaye and O'Meara 2002). Usually performed by a registered nurse, triage is performed when a patient arrives at an emergency room. Triage nurses assess patients and allocate an appropriate category indicating the urgency of care required. Triaging is based on patient signs and symptoms rather than diagnosis. Gouin et al. (2005) suggest that the main role of triage is to assign priority to patients who need urgent care, and to predict the nature and scope of care which patients will likely require. The triage process results in the assignment of a score—the triaging process is one which has as its end goal the classification of a patient's acuity. Implicit in this definition is the notion that triaging systems are classification systems, and the process of triaging is a process through which classification system(s) are applied.

Although the triage function seems relatively straightforward (nursing staff ask patients questions and assign a score based on patient responses), as Gouin et al. (ibid.) have pointed out, historically there has been considerable variation in triage scales used throughout the world, which (among other things) has made it difficult to conduct research about the effectiveness of interventions, or to compare one emergency room to another. At the same time, there have been extensive efforts throughout the world to standardize triage scales. A national five-level triage acuity scale was developed in Australia in 1997, and a collaboration between the Canadian Association of Emergency Physicians and the National Emergency Nurses Affiliation developed a five-level Canadian Triage Acuity Scale (CTAS), which has become mandatory in most of Canada (ibid.). However, dissatisfaction with the CTAS is visible. For example, rural emergency room physicians have indicated that since the implementation of CTAS, "a number of adverse effects from its implementation have been encountered in rural EDs" (SRPC-ER Working Group 2002: 271).

Among the challenges rural physicians reported related to the use of CTAS were inappropriate categorization of patients, the use of a different triage scale by ambulances resulting in an ER unprepared for seriously ill patients, a lack of reference to obstetrical patients in the CTAS, and inappropriate triage of pediatric cases (SRPC-ER Working Group 2002).[4] Limitations related to the use of CTAS by pediatric emergency room staff have contributed to the development of a Pediatric CTAS, which was derived from the adult CTAS (Gouin et al. 2005). Prior to the development of the Pediatric CTAS which was undertaken as a vehicle for measuring case mix and to ensure timely access to intervention, there was no widely accepted tool for triaging of pediatric patients. Although the Pediatric CTAS is increasingly used in Canada's Pediatric ERs, many regions of Canada do not have Children's Hospitals, and even in those areas that do have children's hospitals, children are frequently seen (and triaged) in adult ERs, where CTAS (rather than Pediatric CTAS) is used, which contributes to problems with data standardization.

Bowker and Star (1999) suggest that classification systems are segmentations of the world. Classification systems are sets of boxes (either metaphorical or literal) into which things can be placed, and which do some kind of work, such as knowledge production work (e.g., they serve as one component of an indicator system that allows us to understand how well ERs function), or bureaucratic work (e.g., in the case of triage scores, they serve as an ordering system which determines who is seen when, and may also be used to measure case mix in an emergency room, which in turn may be used to set staffing levels, or ensure that comparisons in service between ERs are fair).

The tension between knowledge production work and bureaucratic work is evident in documents describing how to implement the CTAS, which "attempts to more accurately define patients [sic] needs for timely care and to allow EDs [emergency departments] to evaluate their acuity level, resource needs and performance against certain operating objectives" (Beveridge et al. 1998: 2). Use of triaging scores is both oriented toward caring for patients (efficient management of an ED requires a team of providers capable of correctly identifying patients' needs, setting priorities, and implementing appropriate treatment, investigation, and disposition (ibid.: 2), and at the same time has been developed to "allow ED's to evaluate their acuity level, resource needs and performance against certain operating objectives" (ibid.).[5] Hence triage not only serves as a guide for patient treatment, but also plays a role in the production of knowledge used for management purposes. Often data elements (such as patient acuity data) also play an important role in the production of research knowledge, for example, about the impact of ER wait times on patient outcomes.

CLASSIFICATION SYSTEMS AND INFRASTRUCTURE

Star and Ruhleder (1996) suggest that infrastructure is relational. Something "becomes infrastructure in relation to organized practices" (ibid.: 113), and "infrastructure occurs when the tension between the local and the global is resolved" (ibid.: 114), and when local practices are accommodated by a larger scale technology, which can be used in a natural, ready-at-hand fashion. The recent interest in development of triage acuity scales (whether in Australia, or for use in triaging the adult or pediatric population in Canada) reflects what Wiener (2000) has called a quest for accountability, a quest that would remain unimaginable at the scale currently being pursued, were it not for the existence of current computer and communication systems, which offer the possibility of extending both the reach and uniformity (in theory) of data collection activities over multiple locations.

Bowker and Star (1999: 16) point out that "classification systems are integral to any working infrastructure," and Star and Ruhleder (1996) have argued that infrastructure emerges in relation to organized practices. Successful infrastructure can be used in a natural, ready-at-hand fashion, and technological systems that achieve infrastructural status succeed when local practices are accommodated by a larger scale technology, which is used in a natural, ready-at-hand fashion (ibid.). Information infrastructures can be difficult to analyze (Bowker and Star 1999), because they disappear—or fade into the background—when they work well. Additionally, the complexities of computerized information infrastructures often lack easily recognizable material expressions (one can't see data as it moves through a computer, and the underlying information system architecture is often invisible to all but information system departments and system designers). These properties of infrastructures—that they often lack a material expression and that the classification systems embedded in them are often not obviously visible in information systems—may contribute to the obfuscation of the multiple domains or arenas in which triaging activities are carried out.

THE MULTIPLE DOMAINS OF TRIAGE

Bowker and Star (ibid.: 33) have pointed out that "infrastructures are never transparent for everyone, and their workability as they scale up becomes increasingly complex." In our work in emergency rooms, we have noted many instances where infrastructure becomes visible, and in its visibility, signals a sort of infrastructural failure. This occurs when information systems that are implemented with the intention that they will become boundary objects (malleable enough to work for varied communities of practice and to meet the needs of multiple domains) fail to satisfy the informational requirements of all of the social worlds or domains which they inhabit. This type of failure preceded our work with a pediatric emergency room that suspended use of an electronic triaging system. When objects that are introduced with the intention that they serve as boundary objects fail to effectively bridge between social worlds, communities of practice, and domains of use, they instead assume the role of borders, which render the different communities that exist around them visible. In the case of our research partner, the differences between adult and pediatric triaging processes were amplified as staff attempted to use an electronic triage tool that had been designed for an adult population, in a pediatric setting.

Notwithstanding the difficulties just identified concerning the suitability of various triaging systems for use with particular populations (and these are, of course, significant), on another level, on the face of things, triage on its own seems relatively simple: one follows conventions and rules to determine how ill a patient is likely to be, and on the basis of the triage interview a triage score is assigned to a patient, which in turn determines how quickly the patient is likely to be seen.

THE DOMAIN OF CARE

One way to think about triage is as an entry point into what I call the domain of care—that is, it marks the commencement of a process of obtaining care from a medical establishment. Within the domain of care, the focus is on caring for the patient, and making the patient better. The primary actors in the domain of care are health processionals and practitioners who perform the activities of care, such as taking patient histories, delivering treatments, and ordering tests and diagnostic procedures in an effort to improve a patient's health state.

Berg and Bowker (1997) suggest that medical records represent different practices of reading and writing that are intertwined to produce different patient's bodies, different body politics, and different bodies of knowledge. They also suggest that "as organizational infrastructure, the medical record affords the interplay and coordination of divergent worlds" (ibid.: 513). Such is the case of triage, which represents only one way of "reading and writing" a patient's body. The triage score attached to a patient is a description only of the urgency with which a patient should receive care: it tells nothing of who the patient is, where they live, whether they have been in the hospital before, what tests they have had or what medications they are currently taking, or even how sick they were when they entered the emergency room. Yet these ways of knowing a patient are all activities that are undertaken once a patient passes through triage, and which help care providers (physicians, nurses, social workers, etc.) assess the patient and provide care.

Bowker and Star suggest that "systems of classification... form a juncture of social organization, moral order, and layers of technical integration. Each subsystem inherits, increasingly as it scales up, the inertia of the installed base of systems that have come before" (Bowker and Star 1999: 33). Although a triage system can work on its own and, as a stand-alone system, need not be integrated with other systems, pragmatic concerns (such as registering a patient in the system so they can be located as they move through the various hospital departments) often give rise to a desire to connect triage systems to patient registration systems. In some hospitals patient registration systems have been developed primarily for keeping track of patients[6] (as is the case in one of the ERs we have worked in), while in other settings the patient record may (to varied degrees) be integrated in with some form of electronic medical record (as is the case in the other ER we have worked in).

Although triaging is fundamentally about determining how quickly a patient requires care, it is also a point of entry (one of many which also include ambulatory care clinics, day surgery centers and preadmission clinics for planned hospital visits) into the hospital, and, consequently, in both hospitals we have studied, the triaging system is closely intertwined with the patient registration system. Although often in ERs there is not a direct feed of triage data into a patient registration system, any data that are collected in the ER must maintain continuity with admissions data that are collected at other points of entry, such as day clinics and preadmission clinics.

CREATING INFORMATIONAL CONTINUITY IN THE DOMAIN OF CARE

In all instances where computer-based medical records are linked (for example, if a stand-alone laboratory information system feeds data into an electronic medical record, or triage data collected from a stand-alone system is integrated with data from a patient registration system), a mechanism must be developed to ensure that the correct records are integrated—a problem that can balloon to extreme proportions in the face of a poorly designed system (see Balka 2005). The quality of informational continuity required when merging electronic patient records requires a consistent means of identifying patients across computer systems, and, at times, across locations. Integrating systems that were designed as stand-alone systems can present many challenges. For example, consolidation of patient information from multiple sources requires that records with slight variation but representing the same person must in some way be identifiable as parts of a single patient's record. In the case of one system developed in part to meet this need in one of the ERs we have worked in, this proved to be quite a complicated and costly exercise.

Although only four million residents live in the province (a portion of whom are served by the health agency), a system upgrade implemented at one hospital in order for the local computer system to communicate with an integrated medical record required the conversion of over sixty-five million patient and data records from the old system to the new system.[7] Once this process had begun, it was anticipated that reconciliation of these duplicate records would require the labor of twenty full-time staff members over a six-month period, at a cost of in excess of $1 million dollars.[8] This effort was being supported provincially by an initiative called the Enterprise Master Person Index (EMPI)[9] designed to address issues arising because patients are identified differently from system to system. Although this may seem to be an extreme case, it highlights the challenges associated with the electronic traces of a triaging system, as a patient moves throughout the domain of care, and in the process, interacts with multiple information systems, each of which may have sprung from the needs of different communities of practice (symbolized in Figure 9.1), and which incorporate varied classification systems and their attendant information architectures.

As Figure 9.1 suggests, the electronic trace of triage data crosses into several domains: the access domain, the federal accountability domain, the local accounting and quality domain, and the research domain, all of which interact with the care domain (Balka and Whitehouse 2007). Each of these domains may be comprised of multiple professional groups, many or all of which may have their own legacy computer system, or at the very least,

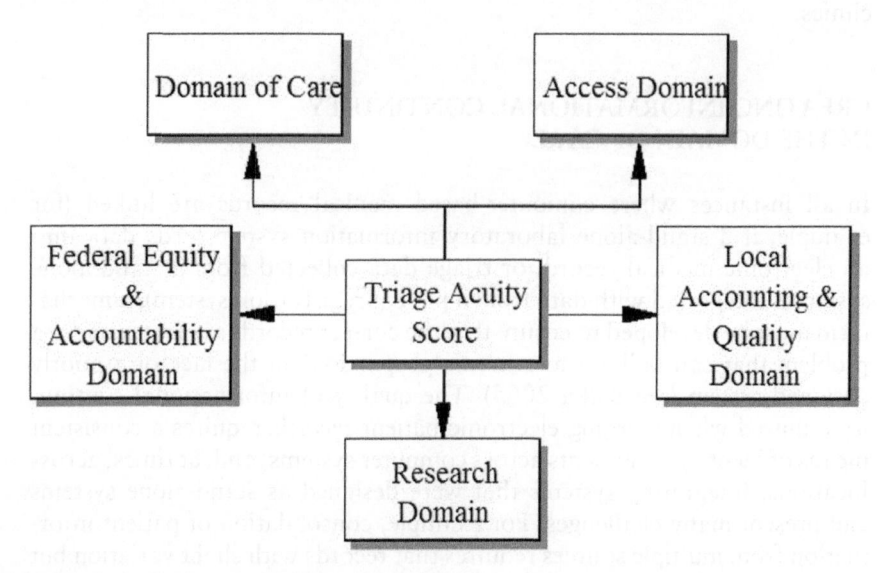

Figure 9.1 The multiple domains of electronic triage.

preferences about the classification system which gives coherence to their portion of the care domain. The patient registration system is included here because the ability to correctly identify patients (as well as past tests and test results along with past medical history) is central to safe care of patients.

Figure 9.1 represents examples of phenomena that exist in our research context, rather than presenting a definitive picture of the interaction of triage data with other data sources. Because the computational environment varies so much from setting to setting, it is impossible to create a graphic that captures all of the variability that exists in the way that triage data may be called upon to enter into varied domains within the health sector, and also captures the multiplicity of computing systems, social environments, and classification systems that knit them together and to one another. The value of Figure 9.1 lies in its articulation of multiple domains (into which most health computing systems would likely fall), and its ability to show how triage data, like the pill bottle that can be seen from many different perspectives, can be viewed (and used) within the context of multiple decisions, debates, and arguments that occur within the domains. Each of the other domains is described briefly next.

THE ACCESS DOMAIN

The accessibility of emergency rooms in Canada's hospitals has become a "hot button" issue, with the public, the medical community, health policy makers, and the press all engaged in public debates about how long people must wait prior to being seen in an emergency room. The length of time a patient waits prior to being seen in an emergency room is of great concern to Canadians. This measure is being used as an indicator of the health of the health care system (Chan, Schull, and Schultz 2001). Reports of long wait times cause panic in communities (Canada Newswire 2004, Guardian 2005, Kirkey 2004, Steyn 2004), and calls for action from health system managers. ER wait times play a significant role in resource allocation decisions in hospitals (Calgary Health Region 2005, McIntosh 2005, Sin 2005, Young 2005). Hospitals with long ER wait times come under scrutiny, and senior management of hospitals issue directives to middle management to shorten patient waiting times in the ER. Emergency wait times are also invoked in discussions about poor service. Publicity of long wait times often sets off a series of programs to rectify long wait times. Recent examples of effects of such publicity include opening additional beds (Giuffre 2005), providing improved care outside of hospitals, quicker access to specialists within emergency rooms as well as access to liaison nurses charged with improving communication related to care within emergency rooms, setting up discharge lounges, purchasing new equipment, improving information systems within emergency departments (Capital Health 2004) and allowing ER triage nurses to order diagnostic tests (Sin 2005). Another response is for managers to engage in data gaming (e.g., making a change to how data are reported or which data are reported in order to show more favorable outcomes). While there are a range of possible responses to ER waiting time data, it is easy to demonstrate that these data are the linchpin for a number of decisions. The domain in which these decisions take place is the access domain. The access domain is populated by a wide range of health practitioners (both inside and outside of ERs, policy makers, and human resource departments, as well as the computer systems used by each of those groups to develop an understanding of issues related to the accessibility of Canada's ERs.

Within the context of the access domain, data from the triage process (the level of a patient's acuity) is combined with patient registration data to produce information about how long patients wait for care, in relation to the acuity score they are assigned through the triage process. Data about the level of patient acuity and wait times are also used to generate a picture of an ER's case mix and volumes, which in turn is used to determine staffing levels. These data may be used to determine appropriate staffing levels for ERs, or to develop programs aimed at diverting less acute patients away from emergency rooms, as a means of lessening the demand for emergency services. Data resulting from the triage and registration process may also be used as a source of information by a clerk who is responsible for locating a bed for a patient who must be admitted from the ER. In these contexts, triage data ideally would be compatible with data from other systems such as the patient registration system, as well as data from the bed database, as debates within the access domain—the domain where local health systems attempt to address issues of ER and broader health system access—may depend upon triage data as an important input into decision making that occurs in the access domain. For those operating in the access domain, triage data is essential in determining whether ER wait times expose patients to increased health risks, and are essential in gaining insight about upstream interventions that could reduce ER demand.

THE FEDERAL EQUITY AND ACCOUNTABILITY DOMAIN

The Federal Equity and Accountability Domain is the domain in which the distribution of health resources is monitored by the federal government, as an element of ensuring accountability to Canada's federal health act. Canada's health act sets out several principles for the Canadian health system, which include equitable, accessible, and universal care. Although the underlying principles upon which the health act (and Canada's health system) is built have been specified at a federal level, responsibility for delivery of care in a manner consistent with the federal health act falls to the provinces in Canada, with costs for health care delivery shared by the federal government and provincial governments. In practical terms this means that in order for the federal government to ensure that health services are being delivered in a manner that is consistent with the federal health act, some form of monitoring must take place. Provincial governments are responsible for the day to day management of health within provinces, but must demonstrate to the federal government that they are complying with the Canada Health Act in delivery of care. Within this context, provincial governments may collect data about health service delivery within their province, and the federal government may require the provinces to report data about health service delivery, in order to determine whether or not the provinces are meeting their obligations. Because provinces are responsible for reporting, reporting requirements at hospital level vary among provinces and territories.

In addition, there is historical variation in wait time data elements that have been collected, and considerable variation exists in how data are collected locally, and how those data are reported by individual hospitals and health regions in Canada (regional health provider institutions) in their annual reports (e.g., Ottawa Hospital 2004, University Health Network 2005, Capital Health 2004). For example, in one hospital's emergency room, six different times are recorded on the computerized record: a patient's arrival time at emergency, the time they are triaged, the time they are taken to an emergency department area other than the waiting room (either the treatment area or acute area), the time a patient is seen by a doctor (an optional data field which doctors enter inconsistently), and the time a request to admit a patient is made. Other hospitals using different information systems (which often play a significant role in determining which data points are collected and which are not) capture times at different intervals in the patient's ER waiting period.

When data are collected via computer, semantic complexity may be introduced. Computer systems often require users to enter values into fields (such as the time a patient arrives in an emergency room, or the time the patient is seen by a physician). To the extent that each manufacturers' emergency room computer system varies, the specifics of the data collected may vary (e.g., one system may request time of arrival in an ER, another may request time seen by a triage nurse, etc.). However, both fields may be called arrival time. In each instance the semantic content of the database is superficially comparable but a closer look reveals critical differences. Although data about emergency room wait times has become an actor in health policy debates in Canada and elsewhere, "there is no established standard or definition for measuring ER wait times."[10] Although the limitations of the emergency room wait time data set are significant, the data are used and do play a significant role in health care debates. Often unbeknownst to staff who enter data that are used to derive wait times, the data they generate as they interact with triage, patient registration, and medical record systems trickle from the care domain into both the access domain and the federal and accountability domain, where they play a role in significant policy debates.

LOCAL ACCOUNTING AND QUALITY DOMAIN

Hospitals are typically large institutions that are characterized by a high degree of hierarchy, and, due to their size, a certain amount of decentralized control. In many hospitals, the first areas that were computerized related to finances. One consequence is that in many cases, data related to finances are the most complete data that exist, and often data about costs of providing services are used as a proxy for service delivery and/or health outcome. Another consequence of the early computerization of finance departments is that the processes used by finance departments are often well entrenched in established practices, which in turn are often reified by legacy computer systems. Although many of the administrative functions at a hospital (such as costs) may be monitored through a central finance office, budgets are typically decentralized, with each area (e.g., the emergency room) or groups of areas (e.g., out patient services) sharing a single budget.

Finance department practices may collect data according to some metrics (e.g., a cost center that is related to who manages a unit), but may not allow the manager of that unit (or anyone else) to disaggregate costs, for example, according to type of worker (e.g., doctors vs. nurses vs. clerks). Triage data that are collected may be combined with cost data at some point in order to determine (for example) what costs are associated with patients of various acuity levels, once those patients enter the system. The ability to combine ER data with existing financial data as well as data about what additional services a patient may have used while in a hospital may play a pivotal role in whether or not an ER is allocated additional resources for staffing. Although Figure 9.1 may imply that the local accounting and quality domain is a domain that is perhaps secondary to the broad landscape in which triage data may be used, in reality this may be in some sense the most important domain, to the extent that it may have the greatest direct impact on the resources available to an ER.

THE RESEARCH DOMAIN

Each of the domains outlined here are a context in which triage data may be used, and each of the contexts may, to varying degrees, exert influence over what data look like (e.g., because data collected in one domain must be compatible with data collected in another domain), and which data are collected. Many factors come to bear ultimately on which data are collected, including the relative strength of parties involved in decision making about information systems, the data architecture built into existing and available computer systems, local, provincial, and federal reporting requirements, and so on. In spite of the existence of ample data, the data that exist may or may not be helpful to researchers in understanding, for example, whether or not the interventions carried out in an emergency room have a positive impact on patients. Typically, researchers have collected data at local levels, and at times through multisite studies, but data collected for research purposes were not necessarily integrated with data collected by hospitals, or, in cases where it may be, compatibility may extend only as far as a single computer system, rather than the numerous systems a single system may interconnect with.

Recent advances in networked computers, and the ability to combine data from multiple sources have left open the possibility for engaging in more ambitious—and theoretically more robust—research that can (for example) link treatment options to health outcomes. Researchers may face several challenges. For example, the data that are collected may be missing information that is central to the question researchers are seeking to answer (e.g., it may be difficult to determine what the staff mix was who treated a patient), or may exist in a format that is not ideal. Researchers typically want to collect much more extensive data than what is required for treating a patient, and must typically make do with far less detail than what is perceived as desirable by researchers. The research domain is another domain in which information systems are discussed, and where output from information systems is used. At times, research endeavours can have a significant influence on the architecture of information systems used in health care settings, however, more often than not researchers' needs are poorly met with in-house data collection systems, and they set up parallel data collection systems in order to ensure they have the information they require.

CONCLUSION

Emergency room wait time data that fill newspaper articles and policy debates are the product of patients who need emergency care, staff who may not be aware of their role in producing data used in decision making and policy debates, information systems in use in hospitals (each with its own embedded classification systems and categories), and institutional work practices which result in the extraction of these numbers for use in multiple domains, where the numbers influence change in numerous and varied health policy and resource allocation debates. ER wait times might be used by patients to push for more services, by staff to pursue improved staffing, or by management to argue for improved nonemergency services.

As Smith (1990) argues, texts—including statistics—are not benign, but rather are active, and data collected as part of the triage process is no exception. Smith suggests that text (and I argue statistics) both reflect social relations, and play an active role in constituting social relations.

Each of the domains outlined previously represents perspectives from which ER information systems can be seen, and broadly signal communities of interest—each of which may include multiple communities of practice—which information infrastructures must bridge. If the information system introduced into such a complex environment fails to be taken up as infrastructural technology—technology which is plastic and malleable enough to be used successfully by varied social groups—or, put another way, if it fails to serve as a boundary object, breakdowns occur. In one ER we have worked in, an electronic triaging system was introduced and abandoned within one year of its initial implementation, amidst numerous complaints from staff about work practice issues (which could constitute an additional domain). Prior to the point that the electronic triage system was abandoned, staff and management reported problems related to classifications used to generate the triage score, which was based on the CTAS, rather than the Pediatric CTAS. A poor fit between the classification system upon which the electronic triaging software was based and the classification needs of users prohibited the system from assuming infrastructural status. It can be argued that although the system was introduced with the hope that it would become a boundary object, ultimately the system failed to achieve infrastructural status, and while the apparent use of triage data in multiple domains might suggest that triage data are a boundary object, the failure of the data to satisfy the informational requirements of all of the domains where it is used signals its failure as a boundary object.

Berg and Bowker (1997) suggest that medical records represent different practices of reading and writing that are intertwined to produce different patient's bodies, different body politics, and different bodies of knowledge. They also suggest that "as organizational infrastructure, the medical record affords the interplay and coordination of divergent worlds" (ibid.: 513). Within each of the domains outlined earlier, different bodies of knowledge are required and produced in relation to different patient's bodies, and the domains offer an analytic framework for thinking through the myriad and divergent worlds that an electronic triaging system must enter into. The specific domains outlined earlier are derived from and reflect emergency health service delivery in a particular Canadian province. While the domains may prove analytically useful in a Canadian context, some domains will also be useful in thinking about the many ways that health data are seen in other national contexts as well. For example, the care context will exist in all health settings, as will an accounting and a quality domain. Other domains (e.g., the access domain or federal equity and accountability domains) reflect debates that may be particular to a country's health system at a particular moment in time. Hence the work that statistics do in general and that emergency room wait time data does in particular is contextually dependent upon broader debates, often national in scope.

From Categorization to Public Policy   187

Bowker and Star (1999) suggest that "systems of classification [...] form a juncture of social organization, moral order, and layers of technical integration." (ibid.: 33). Our work in emergency rooms lends support to this way of thinking about systems of classification. Elsewhere (Balka in press, Schuurman and Balka 2009) I have documented various sociotechnical challenges associated with some of the information systems discussed earlier, and the classification systems upon which they are built. My hope in introducing the concept of domains and a graphic representation of the interconnectedness of health information systems is that I can help practitioners—who are typically anchored in one domain or another—to develop a broader perspective of the information systems they are both using and making decisions about, and whose use produces numbers which figure prominently in the policy debates that health decision makers oversee.

Debates about the appropriateness of CTAS raised by rural and pediatric physicians and nurses reflect tensions in systems of categorization, which, as the case of triaging illustrates well, have real-world implications. On the other hand, the failure to regularly include mention of triage scores in most reporting about emergency room wait times serves to oversimplify the data—and hence the nature of the problem—in the public's mind. Such oversimplifications, which neglect to link ER wait times to patient acuity, fuels debates about whether or not healthcare should be privatized, and make it virtually impossible for members of the public to see to what extent patients waiting long periods of time prior to being seen in emergency rooms have low acuity scores, and hence need not be treated in an emergency room. The failure to link low patient acuity to ER wait times effectively squelches debates about inadequate healthcare resources elsewhere in the system, inadequacies which result in increased use of emergency departments. Hence debates about emergency room wait time data demonstrate three important points about statistics and the State. First, tensions around categorization of triage scores demonstrate that categories have politics. Second, the aggregated use of ER data in varied domains, often for management purposes, demonstrates how data come to be used for purposes other than what they were originally intended for, often with real-world consequences. Third, the failure to report ER wait time data in relation to triage scores demonstrates how oversimplification of numeric concepts can influence public policy debates. Taken together, issues raised in this chapter point to the need for a more statistically literate public.

NOTES

1. The advertisement is for Canada's York University, which is marketing itself as "the interdisciplinary university." The advertising campaign can be viewed at http://www.yorku.ca/yfile/special/banners.pdf. This link rotates the subject of the advertisement, and the example cited here is one many ads that

follow the theme of demonstrating that a single artefact can be seen from multiple points of view.
2. http://www.medterms.com/script/main/art.asp?articlekey=16736 (accessed July 20, 2006).
3. http://www.medterms.com/script/main/art.asp?articlekey=16736 (accessed July 20, 2006).
4. A more complete list of problems reported by rural physicians related to use of CTAS can be found in Canadian Emergency Department Triage and Acuity Scale (CTAS): Rural Implementation Statement (SRPC-ER Working Group 2002).
5. For an overview of CTAS and its application, see CTAS Implementation Guidelines (http://www.caep.ca/template.asp?id=B795164082374289BBD9 C1C2BF4B8D32#guidelines).
6. And often to ensure that appropriate billing and payment take place.
7. Vancouver Coastal Health. Briefing note: Patient Care Information System (PCIS) Enhancement Project. April 7, 2005. p. 1.
8. Notes taken at presentation about the system upgrade, March 18 2005.
9. It should be noted that at times the EMPI may be referred to as the ECIM—Enterprise Client Identity Management project. See for example Vancouver Coastal Health, Building the foundation for electronic health records: Recording accurate patient identity data is the key. In Medlink: The VCH physician newsletter. 2005, Sept. p.3.
10. Bingham, J. W. (2004, Oct. 29). Director, Health Reports and Analysis Canadian Institute for Health Information. *Personal Communication* (by e-mail).

REFERENCES

Balka, E. (2003) "Getting the Big Picture: The Macro-Politics of Information System Development (and Failure) in a Canadian Hospital," *Methods of Information in Medicine* 42(4): 324–330.

———. (2005) "The Production of Health Indicators as Computer Supported Cooperative Work: Reflections on the Multiple Roles of Electronic Health Records," in *Reconfiguring Healthcare: Issues in Computer Supported Cooperative Work in Healthcare Environments*, eds. E. Balka and I. Wagner, 67–75. Burnaby, BC, Canadian ACTION for Health.

———. (2005) "Socio-Technical Challenges in Constructing Continuity of Care: Reflections on Two System Development Projects," Proceedings of the Continuity of Care Workshop. Held in conjunction with Helseinformatikkuka (HelseIT). Trondheim, Norway: KITH. http://www.kith.no/templates/kith_WebPages_1469.aspx (No longer accessible).

Balka, E., and I. Wagner (2006) "Making Things Work: Dimensions of Configurability as Appropriation Work," Proceedings of the 2006 20th anniversary conference on Computer Supported Cooperative Work Conference 2006 (CSCW '06), Banff, Alberta, Canada: 229–38.

Balka, E., and S. Whitehouse (2007), "Whose Work Practice? Situating an Electronic Triage System within a Complex System," *Studies in Health Technology and Informatics* 130: 59–74.

Berg, M., and G. Bowker (1997) "The Multiple Bodies of the Medical Record: Toward a Sociology of an Artefact," *The Sociological Quarterly* 38: 511–35.

Beveridge, R., B. Clarke, L. Janes, N. Savage, J. Thompson, G Dodd, M. Murray, C. N. Jordan, D. Warren, and A. Vadeboncoeur (1998) Implementation

Guidelines for the Canadian Emergency Department Triage and Acuity Scale (CTAS)(Version: CTAS16.DOC December 16, 1998), http://www.caep.ca/template.asp?id=B795164082374289BBD9C1C2BF4B8D32#guidelines (accessed July 20, 2006).

Bjorn, P. and E. Balka (2007) "Health Care Categories Have Politics Too: Unpacking the Managerial Agendas of Electronic Triage Systems," in *ECSCW 2007: Proceedings of the 10th European Conference on Computer Supported Cooperative Work*, eds. L.J. Bannon, et al., 371–390. London: Springer-Verlag.

Bowker, G.C., and S.L. Star (1999) *Sorting Things Out: Classification and Its Consequences*. Cambridge, MA: MIT Press.

Calgary Health Region (2005) "Emergency Department Operational Overview," *Newslink*, http://www.calgaryhealthregion.ca/newslink/EmergencyDepartment_text.html (Accessed August 3, 2005, no longer accessible).

Canada NewsWire (2004) "Media advisory—Canada's emergency physicians want first ministers to address ED wait times," *Canada Newswire*, September 9: 1.

Capital Health (2004) "10-point plan for system improvement: Emergency overcrowding and long waits for hospital services," Capital Health Nova Scotia.

Chan, B., M. Schull, and S. Schultz (2001) *Emergency department services in Ontario 1993–2000*, Toronto: Institute for Clinical Evaluative Sciences.

Durojaye, L., and M. O'Meara (2002) "A Study of Triage Paediatric Patients in Australia," *Emergency Medicine* 14: 67–76.

Giuffre, M. (2005) "Progress is Being Made . . . But There Is a Long Way to Go," Vital Signs, January: 4–5.

Gouin, S., J. Gravel, D. K. Amre, and S. Bergeron (2005) "Evaluation of the Paediatric Canadian Triage and Acuity Scale in a Paediatric ED," *American Journal of Emergency Medicine* 23: 243–47.

Guardian (2005) "Saskatoon senior waits 52 hours in emergency," *Guardian*, June 4: A11.

Kirkey, S. (2004) "Heart patients dying in hospital emergency departments" [final edition], *The Vancouver Province*, November 26: A13.

McIntosh, T. (2005) "The taming of the queue II," *Colloquim Report*, March 31 and April 1, Ottawa: Westin Hotel.

Ottawa Hospital (2004) The Ottawa Hospital Performance Indicator Report. For period ending December 31, 2004. The Ottawa Hospital. http://www.ottawa-hospital.on.ca/about/reports/indicators/pir-dec04-e.pdf (Accessed August 5, 2005, no longer accessible).

Schuurman, N. and Ellen Balka (2009) "Alt.metadata.health: Ontological Context for Data Use and Integration," Journal of Computer Supported Cooperative Work 18(1): 83–108.

Sharman, Z. (2003) "Data Sharing Leads to Patient Caring?: Gender, Technology and Nurses' Caring Work," Master's thesis, Simon Fraser University.

Sin, L. (2005) "Kelowna hospital trial helps clear ED backlog: Diagnostic work: Nurses help cut wait times" [final edition], *The Vancouver Province*, May 22: A25.

Smith, D.E. (1990) *Texts, Facts and Feminity: Exploring the Relations of Ruling*. New York: Routledge.

SRPC-ER Working Group (2002) "Canadian Emergency Department Triage and Acuity Scale (CTAS): Rural implementation statement," *Canadian Journal of Rural Medicine* 7(4): 271–74.

Star, S.L. and J.R. Griesemer (1989) "Institutional Ecology, 'Translations' and Boundary Objects: Amateurs and Professionals in Berkeley's Museum of Vertebrate Zoology, 1907–39," *Social Studies of Science* 19(3): 387–420.

Star, S.L. and K. Ruhleder (1996) "Steps Toward an Ecology of Infrastructure: Design and Access for Large Information Spaces," *Information systems research* 7(1): 111–34.

Steyn, M. (2004) "Bad things come to those who wait: from C difficile to SARS, almost all scandals in Canadian hospitals boil down to the same thing: Sick people waiting longer in crowded room in dirtier hospitals," *Western Standard*, November 22: 69.

Suchman, L., and B. Jordan (1989) "Computerization and Women's knowledge," in *Women, work and computerization*, eds. K. Tijdens, et al., 153–160. Amsterdam: North Holland.

University Health Network (2005) University Health Network Wait Time Report Wait Time Summary. University Health Network. http://www.uhn.ca/patient/wait_times/docs/Wait%20time%20summary_0405.pdf (Accessed August 5, 2005, no longer accessible).

Wiener, C. (2000) *Elusive Quest: Accountability in Hospitals (Social Problems and Social Issues)*. Somerset, NJ: Aldine Transaction.

Young, M. (2005) "RIH adds staff to cope with emergency waits," [final edition] *Kamloops Daily News*, April 22: A5.

10 Making Sense of Numbers
The Presentation of Crime Statistics in the Oslo Police Annual Reports, 1950–2008

Heidi Mork Lomell

> It is not what you look at, but what you see
>
> (Henry Thoreau)

INTRODUCTION

Compared to many of the other statistics in this book, crime statistics is an old epistemological object. Crimes have been counted ever since society "became statistical" between 1820–1840, when an "avalanche of printed numbers" swept over many European countries (Hacking 1990: 1). During that period, various social phenomena were enumerated and publicized, especially social phenomena related to deviancy: suicide, crime, madness, prostitution, and various diseases. The underlying idea was to identify statistical laws and regularities in order to improve or control a deviant subpopulation. However, mere enumeration was not sufficient; the statistics had to be *interpreted* in order to make sense: "The laws had in the beginning to be read into the data. They were not simply read off them" (ibid.: 3–4).

One example is Adolphe Quetelet's (1796–1874) enumeration and subsequent interpretation of crime. Ian Hacking calls Quetelet "the greatest regularity salesman of the nineteenth century" (ibid.: 105), "fond of numbers, and happy to jump to conclusions" (ibid.: 106). With a background in astronomy, Quetelet hoped to discover the *social* mechanics, the underlying statistical laws governing crime (Beirne 1993: 77). When he interpreted the collected crime statistics, he found exactly what he had been looking for. Despite variations by up to 10 percent, he was stunned by an "alarming regularity" and a "surprising constancy with which the numbers of the statistics of crime are reproduced annually" (Quetelet 1996/1842: 15). He explained this "regularity" and "constancy" with a reference to the principle that "effects are proportionate to their causes, and that the effects remain the same, if the causes which have produced them do not vary" (ibid.: 16). He concluded that "the crimes which are annually committed seem to be a necessary result of our social organization, and [...] the number of them cannot diminish without the causes which induce them undergoing previous modification [. . .]. *Society prepares crime, and the guilty are only the instruments by which it is executed.*" (ibid.: 28, italics in original).

Like other statistics, crime statistics are always produced, by someone, for a purpose; crime statistics are not gathered from nonparticipant points of observation (Haraway 1991). The production of crime data is a social activity: "Rather than passively reflecting some objective world of crime, they actively construct a subjective world of crime" (Sacco 2005: 67). This view of statistics is not common in popular, public, and political discourse, where numbers are seen as superior to "mere" words, being taken to be scientific, independent and therefore accurate (Godfrey et al. 2008: 27). As Desrosières (1998: 325) puts it, "Statistical information does not fall from the sky." Statistical information is not a natural phenomenon, but a social product. Furthermore it is not "some pure reflection of a pre-existing 'reality'" (ibid.). In other words, in creating this social product, we set a social imprint on it—just as we have already set an imprint on, and socially constructed our understandings of, that which we use the statistics to describe.

In addition, as the example with Quetelet shows, the numbers themselves must be interpreted: "Numbers cannot tell a story on their own, they must be actively given a voice by others" (Haggerty 2001: 161). Statistics on crime and punishment are no exception. Who gives voice to the numbers, and with what agenda, are important questions to ask. In this chapter I will take as my point of departure that one always reads meaning *into* numbers, never simply *off* them. Not only enumeration, but also interpretation of crime statistics is not so much a descriptive as a productive, or maybe even a creative, activity.

I have analyzed how data on reported crimes have been presented and interpreted by the Oslo police in their annual reports from 1950 to 2008. All the annual reports contain crime statistics, but the presentations and interpretations vary. The study can therefore illustrate how the "same" statistics—data on reported crime—have been interpreted at different times and under different political and ideological circumstances. These different interpretations of the numbers help us to identify the rhetorics or discourses that have shaped the police's—and our—understandings of the numbers (Desrosières 1998: 322). Taking a cue from Henry Thoreau, it is not what crime statistics the police looked at, but what they saw—and *why* they saw what they saw—which is my focus in this chapter.

In a constructionist perspective, crime statistics can teach us just as much about the organization which produces them as about the activities they are supposed to describe (Maguire 2002: 333). This has led some to conclude that there is no point in analyzing crime statistics for the purpose of finding out anything about the extent of crime. Crime rates are "an aspect of social organisation," and by studying them one can better understand the agency producing them, not crime itself, making crime statistics "an end rather than a means of study" (Black 1970: 16). I will build on this approach, but instead of studying the processes involved in categorizing, processing, and recording crime in the police organization, I will focus on how the resulting statistics are interpreted and presented to the government and the public in the annual reports. My aspiration is to contribute to a better understanding of how both the crime problem and the police come to be interpreted or "framed" in particular ways under various governmental discourses.

METHODS

Annual reports are fascinating social documents, as they bear witness to the preoccupations of the time in which they were written. In addition, institutions draw their self-portrait when they report (Ericsson 1997, Munro 1996). By reading police annual reports over the last sixty years, we can see how this self-portrait has changed. My focus has been the placement and importance of crime statistics in this self-portrait, and how this has changed during the period.

For an institution, annual reports serve both performative and persuasive purposes. They do not simply present information to the government and the public; they also reveal an idealized image of the institution—their self-portrait. Unlike advertising, in which the purpose is openly persuasive, annual reports contain dual purposes—to deliver information about the institution's yearly activities and to convey an idealized image of the institution (David 2001). During the last two decades, police annual reports have increasingly been reshaped into PR instruments or marketing tools. More sophisticated design, striking covers, and graphics are used to promote a positive image of the police, to display their inclination to be accountable, and to increase their legitimacy in the eyes of both government and the public (Dixon et al. 1995). This development, which has had consequences for the description and interpretation of crime statistics, is part of a new management ideology, and I will return to this in the discussion. For now, it is important to bear in mind that the content, function, and purpose of the annual report itself have changed during the time period.

When reading the annual reports, I first looked at the placement, presentation, and description of the crime statistics. Where in the annual report (section, headlines, etc.) are the crime statistics presented? Are they prominent in the annual report? What words and phrases are used to describe the numbers—are they neutral, optimistic, or pessimistic? Most of the annual reports had a summary at the start of the report, often signed by the Chief of Police. I listed the content of every paragraph and transcribed the first sentence of the summary. The purpose of this more detailed analysis of the summary was to document the preoccupations of the police, with the assumption that the summary reflects the most important issues for the police—with the very most important listed first.

194 Heidi Mork Lomell

After this more descriptive mapping, I searched for explanations and interpretations of the crime statistics in the texts. Public discourse on social problems such as crime usually consists of both a diagnostic and prognostic component (Sasson 1995), and in my reading of the police annual reports, I searched for both. What explanations (causes, diagnosis)—if any—did the text suggest? Were the crime statistics related to policing activities in any way, and if so, how? Were the police part of the diagnosis—or the prognosis? How the police frame the crime problem in terms of causes (the diagnostic component) and remedies (the prognostic component) in their presentation of crime statistics, and how they situate themselves in the discourse—as part of the cause and/or remedy—influence (and are influenced by) not only crime policy, but also how we understand the crime problem—and the function of police in society.

By identifying the discourses surrounding crime statistics in police annual reports for nearly sixty years, we see how the numbers get filled with various contextual content, how the police read meaning into the numbers, and what consequences these various readings have. In the following section, I will describe how the placement, presentation, description, and explanation of crime statistics in the Oslo police annual reports have changed in the last sixty years.

CRIME STATISTICS IN OSLO POLICE ANNUAL REPORTS, 1950–2008

From a Secluded to a Paramount Position

When I began reading the annual reports from 1950 onwards, I had difficulties finding crime data at all. The annual reports were filled with an impressive amount of information about every conceivable aspect of the Oslo police—often illustrated with tables, graphs, and charts—but hardly any of them were crime data. The crime statistics lead a secluded life under anonymous headlines such as "Workload" or "Professional Activities" in the sections that described the Criminal Investigation Department's activities. Moreover, the department's number of officers, their field of work, organization, office, and technical facilities were thoroughly presented before the "workload"—the crime statistics. Even the Criminal Investigation Department led a quite anonymous life in the annual reports, in the shadow of the Administration, Management, and Personnel Departments.

Instead of crime data, the annual reports were filled with other data and details from a wide spectrum of the police organization and activities. To give a few examples, I found detailed column charts of the educational background of police recruits, line charts of the average monthly absence due to illness amongst clerical staff, charts of the monthly number of mentally disordered persons the police kept watch over, accounts of hunting rewards/bounty for carnivore, tables of the number of days police vehicles had been in for reparation, statistics on traffic accidents police vehicles had been involved in, and detailed accounts of festivity and performance licences. There were lots of charts and numbers documenting the police organization and all sorts of police activities, but only a few of them had anything to do with crime.

The annual reports from the 1950s and 1960s give very rich descriptions of the police and various police activities, but crime is not a major part of the police self-portrait in that period. In the 1950s and 1960s, crime statistics were peripheral, almost invisible in the annual reports. Some years they were not mentioned at all in the summary, but mostly they got a short paragraph at the end of the summary, where it was simply established that the crime statistics showed an increase or decrease last year. During the 1970s, crime and crime statistics gradually got more attention, mostly due to at times sharp increases. Crime was becoming more and more widespread in the annual report, at the expense of other topics. Gradually, crime got mentioned earlier and earlier in the summary. From 1985 onwards, crime was the first topic in the summary. During the 1990s, crime became the *only* topic in the summaries.

In the 1990s, the gradual vanishing of other police activities and their numbers from the annual report continued, and crime and crime fighting became dominant. From 1995, the Administration, Management, and Personnel Departments were no longer mentioned in the annual reports; also, policing activities other than crime control-related ones barely got mentioned.

From a peripheral position, crime and crime statistics have come to dominate the annual reports during the last sixty years. Other activities (and their statistics) have vanished from the reports. One important explanation is that reported crimes increased from 10,516 in 1950 to 64,800 in 2008. This has had a huge impact on the police organization, which is responsible for not only recording crime reports but also investigating crimes. Also, society, partly as a response to this huge reported increase, has become increasingly preoccupied with crime (Christie 2000, Garland 2001, Simon 2007).

However, this does not explain why all the other activities, which the police still perform, have been displaced from the annual reports, why they are no longer considered relevant features of the self-portrait. Why has the portrayal of the Oslo police become so *thin* in the annual reports? We will return to this question in the discussion. Let us first take a look at how crime has been described and explained in the annual reports during the period.

From Detachment to Attachment

Up until the 1990s, the crime statistics are mainly presented as *information*, not *knowledge* in the annual reports (Bell 1979). While information usually is described as a disembodied entity—independent of meaning and interpretation, knowledge implies a *knower*—somone who interprets and makes sense of the information (Brown and Duguid 2000). This "knower" is not that visible in the first decades. The presentation of crime data is often without any narrative, interpretations, and/or comments; both decreases and increases are established without any further explanation. It is left open to the reader to speculate and make sense of the numbers, on the underlying causes behind the changes in crime rates. The annual reports provide answers to "how many" and "what," but only at rare intervals do they suggest answers as to "why."

In the first two decades the language is dominantly neutral when crime statistics are presented in text. From the mid-1970s, pessimistic and/or concerned descriptions become more frequent, due to at times sharp increases. The increase is "alarming" in 1961, 1964, and 1974, and in 1978 the report states with dismay, "One really has to ask where this is going to end."

However, the police do not see themselves as having any power to influence the level of crime; they seem to be at the numbers' mercy. We find words such as "suddenly" and "luckily" describing decreases and "frightfully" describing increases—all indicating a detached attitude, not claiming any knowledge of the causes nor any responsibility for the development. When, in 1973, the number of reported aggravated larcenies decreases by more than 11 percent, this is called "remarkable," and further, "One should be very cautious of trying to explain why the statistics have taken such a turn. The Oslo police, at least, dare not draw any conclusions." They are very hesitant when it comes to explaining both increases and decreases; they often "dare not explain" and "don't know." In 1978, an increase in violent crimes "may be more of a random matter than a cause for fear." In 1985, we can read that "the crime level can remain constant for a long period of time, and then suddenly flare up in some parts of the city." Both in 1981 and 1982, the text suggests that the numbers of reported crimes seem to increase in correlation with inflation. All these examples show that the police see the crime statistics almost as living a life of their own, at least a life independent of the police and police activities. *Why* the number of crimes increases or decreases is not something the police are expected to know and/or report. This lack of explanations illustrates how crime statistics are merely presented as information that the police pass on to the government and the public; they do not make sense of the numbers or position themselves as knowledgeable when it comes to crime statistics.

When explanations *are* suggested in the reports, they point to conditions beyond the reach of the police. We find a few examples of socioeconomic explanations (e.g., inflation, growing unemployment, drugs) and a few examples of "blaming the youth" in the annual reports up to around 1980. In the 1963 report, a tripling of theft from 1953 is "caused by the large youth population," and, in 1960, the increase in crime "must partly be ascribed to the large youth population, whose activities mainly manifest themselves in minor larceny and car thefts." In 1977, a decline in car thefts is explained by the fact that "a large proportion of the youths now have their own cars."

Another interesting trend in the first three decades is "blaming the victim." In several of the reports the number of reported robberies is commented with this statement: "The complainants are often themselves to blame, since they, often under the influence of alcohol, make contact with strangers." In 1972, we can read that "Car owners are too careless and keep valuables, radios, and tape recorders in their parked cars." Also, burglaries are explained by inadequate securing of houses and apartments, and receiving stolen property is mentioned in several reports as an explanation of the increasing number of thefts. In 1971, we can read that "too many people are so eager to buy something cheap that they do not reflect on the origin of the objects." This explanatory trend disappears in the mid-1970s.

From 1990, we see a noticeable change in both describing and explaining the crime statistics. New policing initiatives emerge—community policing (Johnston 2003), problem-oriented policing (Goldstein 1979), and zero-tolerance policing (Harcourt 2001)—and declining crime rates are used as evidence of their success. The police begin to use themselves as an explanatory variable—both as part of the *diagnosis* and *prognosis*. They attach themselves to the crime rate by explaining decreases and increases as a direct result of policing activities. Furthermore, a new discourse appears in the annual reports which gains preeminence over other discourses surrounding the crime statistics. The annual reports get saturated with descriptions of a proactive police targeting crime—and succeeding—both when crime increases and decreases. "Targeted campaigns" (1990) and "targeted efforts" (1995) from the police now cause both increases and decreases in the crime rate: "After a period of problem-oriented policing directed at car thefts and thefts from cars, we have registered a substantial decline in the number of cases. Drug cases have increased substantially. The station's active efforts in order to fight drug crimes are one of the causes for this increase" (2001). Words such as "goals," "results," and "targets" now surround the crime statistics. As shown in the quote cited earlier, the police see the crime statistics as a direct result of their activities. Other socio-economic explanations, and the more hesitant approach that dominated up to the 1990s, vanish from the reports. The police *attach* themselves and their activities to the crime statistics to such an extent that we, in 2003, can read that, "the police station reduced the number of knife assaults with 24 per cent."

From 1990 the police not only present themselves as knowledgeable when it comes to crime, they also describes this as something *new*. The 1998 annual report states that "In the last years we have spent much time and energy on acquiring knowledge about the crime situation." The following descriptions are, however, based on the good old crime statistics that have been around in all previous reports. The statistics are not new, but the police make use—and sense—of them in new ways.

In order to sum up this brief examination of the presentation and interpretation of crime statistics in the Oslo police annual reports from 1950 to 2008, we can identify three different analytical translations of crime numbers: as *inputs*, *outputs*, and *outcomes*. In the following section I will explore these three translations in greater detail before analyzing the implications of these shifting translations for our understanding of crime and policing.

READING MEANING INTO CRIME STATISTICS: THREE TRANSLATIONS

Translation is an interactive process that involves both the translator and the translated (Hwang and Suarez 2005). It also involves transformation: "Each act of translation changes the translator and what is translated" (Czarniawska and Sevón 2005: 8). Latour defines translation as "the interpretation given by the fact-builders of their interests and that of the people they enrol" (Latour 1987: 108). I find the concept of translation a fruitful way to describe the various meanings that have been read into crime statistics in the police annual reports, as it highlights both the ideas and interests that are inscribed in various translations, but also the consequences for those actors that are enrolled. Crime statistics cannot tell a story on their own, they are given a voice—translated—by others. And, as we have seen, readings and interpretations have shifted during the time period. These multiple translations of crime statistics enroll different stakeholders and redefine and transform our understandings of crime, police work, and crime policy.

Crime Statistics as Inputs: Keeping Account of Crime

Up to 1990, crime statistics in the annual reports were mainly translated into *inputs*. The crime statistics headline and the text often used the word "workload," for example in the summary from 1960, where we can read that "While the personnel situation in 1960 was just as bad as it has been for years, the workload continues to increase: The number of reported crimes increased to 23,482 (21,871) [. . .]." When crime statistics are translated into inputs, the police need to relate to crime, administer it, and respond to it, but they do not position themselves as influencing the crime rate. The police keep account of crime in the same way as they keep account of personnel, offices, matériel, and so on.

When crime statistics are described as "workload," the connection made between the police and crime is that more workload (more crime) should lead to more police and/or more resources to the police. We see this connection in all the reports up to 1990. Crime is one of the major tasks for the police to deal with, and the crime statistics indicate the amount of this work (among other tasks). If crime increases, the police workload increases, and they need more resources allocated to this particular work: investigators, forensics, and so on. The amount of crime is beyond their control—socio-economic, cultural, and/or structural shifts influence the crime rate. Or, as we saw earlier, the police refrain altogether from explaining causes behind the numbers. When crime statistics are translated into inputs, the *responsibility* for crime remains general, communal, and not specific for the police.

Translating crime statistics into inputs coincides with the governance model of the time. The police, and indeed the public sector in general, was funded according to inputs, and the annual reports bore evidence of this. Increasingly, this mode of governance was criticized for being rule-driven and reactive, preoccupied with *inputs,* not *outputs* and *outcomes.* The justice system was seen as "uncoordinated, inefficient, and ineffective" (McLaughlin et al. 2001: 308). In a system with funding according to inputs, "police departments typically got more money when they failed," i.e. when the crime rate rose: "Because they don't measure results, bureaucratic governments rarely achieve them" (Osborne and Gaebler 1993: 139). What's interesting to note, however, is that the police have always *measured* crime; it is the translation that has varied. It wasn't lack of measurement, but translation and interpretation that were problematic, according to the proponents of the public management reforms. Until the 1990s, instead of results (outputs and outcomes) the police saw workload (inputs) when they looked at the crime statistics.

Crime Statistics as Outputs and Outcomes: Becoming Accountable for Crime

Under the label New Public Management (NPM), the public sector in many countries was reformed in the 1980s and 1990s (Christensen and Lægreid 2007, Garland 2001, Hood 2007). The emphasis shifted from *administering processes* to *achieving results*, setting explicit targets and performance indicators to enable the auditing of both efficiency and effectiveness (McLaughlin et al. 2001: 302; Long 2003). This reform took crime statistics into a new territory as a target and a performance indicator for the police. The police became *responsible* for reducing crime through reflexivity and self-monitoring (Garland 2001: 115), and this altered the translation of crime statistics from input ("workload") to output and/or outcome ("result").

In the annual report from 1992 we can read that "the personnel have been given direct responsibility for the development in crime." As crime statistics become directly linked to police activity, both decreases (20 percent decline in reported violence in city center "shows that targeted uniformed patrol at the right place and right time pays off" in 1995) and increases (increase in reported violence in city center because "increased efforts against street violence revealed a number of cases that otherwise wouldn't have come to police knowledge" in 1996) are explained with police activity. As the latter example illustrates, increases are within this new discourse translated into *outputs*. Here the police see themselves as influencing not the amount of crime in society, but the statistics. Their activities contribute to more crime getting reported and recorded. In the annual reports we find examples of this translation of increases for drug offenses, traffic offenses, and city center violence. If the police are visible and proactive in "finding" and/or "detecting" crime in society, or by encouraging citizens to report crime to the police (for instance in domestic violence cases), the crime rate will increase—as a result of policing activities. In the former example, declining crime rates are translated into *outcomes*. Declines are seen as the result—the preventive effect—of the policing activities. This translation sees the crime statistics as a *result* (outcome) of police work instead of a *starting point* (input). In the annual reports from the 1990s, declining crime rates are seen as an outcome, a result, of successful police work.

Since the 1990s, what the crime statistics in police annual reports purport to show is police activity. Other factors influencing the crime rate in society, such as alcohol consumption, education, unemployment, and so on, become invisible when the crime statistics are seen as the output or outcome of police activity. When crime statistics are translated into a target and are used as performance indicators for the police, the police become responsible and accountable for the crime rate—they become stakeholders. This is reflected in a new politicizing of the annual reports. Lobbying for increased resources now takes a different form. The police now use the crime statistics in a political struggle to look good, to be seen as always succeeding—both when crime increases and when it decreases. There seems to be less room for general descriptions and interpretations of the development in crime. Of course, this is not to suggest that the presentation and interpretation of the numbers was not politicized before the translation of crime statistics into a performance indicator. Rather, crime statistics were embedded in a different discourse up to the 1990s, a discourse where the police did not play an active role, neither as part of the diagnoses or the prognoses. When numbers are moved to a new mode of governance, they shift politics. I will explore further these shifts by using the notion of boundary object in relation to crime statistics.

CRIME STATISTICS AS A BOUNDARY OBJECT

We have seen that crime statistics can be used both as a measure of crime (input) and as a measure of the result of police work (both output and outcome). One way of understanding these multiple usages is to see crime statistics as a *boundary object* (Star and Griesemer 1989). Boundary objects are objects that are flexible enough to adapt to local needs and the constraints of the stakeholders employing them, yet specific enough to maintain a common identity across different interpretations. Crime statistics can be seen as an object that serves as an interface between different actors, organizations, and social worlds, allowing them to communicate, coordinate, and collaborate. The actors, organizations, and social worlds that relate to the crime statistics in the annual reports are the police, various governmental agencies (the Ministry of Justice, Police Directorate, Parliament, local government), the media, academics, and the public. Viewed in this way, one can describe the changes in the description and interpretation of crime statistics over the last sixty years as a case of shifting stakeholders. Crime statistics in the annual reports have during this period illuminated different actors, organizations, and social worlds. In the first decades, the numbers, if at all, illuminated the society in general and the public. Gradually, but most definitely from the 1990s, the crime statistics have come to illuminate the police.

By serving this interface, crime statistics can serve multiple purposes—as a measure (of society's well-being) and a target (good/bad policing), as both input, output, and outcome. However, what are the implications of this interpretative flexibility? I will first look more closely into implications for the numbers themselves, secondly into implications for the police.

From Workload to Result: Implications for the Crime Statistics

When a measure becomes a target, it ceases to be a good measure.
(Goodhart's Law)[1]

Michael E. McIntyre (2001) has described Goodhart's Law as a sociological analogue of Heisenberg's uncertainty principle in quantum mechanics: Measuring a system usually disturbs it. However, Goodhart goes one step further, in emphasizing the shift from "merely" measuring to turning that measure into a *target*. Applied to crime statistics, we can, following Goodhart, ask: By transforming crime statistics from a measure to a target, will it cease to be a good measure? This statement presupposes that crime statistics was *ever* a good measure, but "all forms of crime data are deficient in their various ways" (Coleman and Moynihan 1996: 68). Crime statistics have always been used to support specific policy aims. We can even state this more generally: Statistics have always been used to support specific policy aims (Levitas and Guy 1996, Starr 1987, Bayatrizi 2008). "There was *never* an idyllic moment when measures of human performance existed purely as measures, neutral and dispassionate" (Hoskin 1996: 266). Numbers are always produced, by someone, for a purpose. However, by turning crime statistics into a target, the police have become a stakeholder. The police are at the same time crucial in the production of the statistics. Until recently, more crime meant more money to the police, as crime statistics were used as a measure of demand for police services, not a measure of police performance. However, from having "a strong interest in keeping the number high and none in keeping it correct" (Reuter 1984: 136), the performance management reform changed the incentive to decrease recorded/reported crime. Turning measures into targets results in a novel conflation of the "is" and the "ought," using crime statistics simultaneously to describe and prescribe (Hoskin 1996).

A number of dysfunctional aspects of governing by performance indicators have been identified (Smith 1995). First of all, *misrepresentation*, the deliberate manipulation of data, may increase as a consequence of turning crime statistics into a target. The more pressure is placed on performing, the higher the temptation to actively manipulate the crime statistics. Not necessarily outright falsification, but techniques such as deceptive use of classifications and tolerance of methodological inadequacies. "Cuffing" crimes (not making a formal record of them) and "down-criming" (recording crimes as the least serious category possible) are examples of misrepresentation. There is a gradual transition from outright manipulation to strategic behavior, as the police still have considerable room for discretion. Numerous studies have demonstrated strategic behavior in crime recording and reporting long before crime statistics officially were turned into performance indicators (Bottomley and Coleman 1981, Young 1990, Black 1970). The police use their professional judgment to describe, categorize, and ultimately record an event. Performance indicators can influence this judgment by becoming a strong incentive to choose the category which maximizes the organization's performance result. If excessive reliance is placed on performance measurement, there is clearly an incentive for managers to manipulate the data under their control to show their organization's performance in the most advantageous light (Smith 1995: 292).

However, in a target driven culture, one can choose when setting targets both to increase the number of offenses (detection/output) but also to decrease it (prevention/outcome) (Gilling 1997). One example, from France, is that a *high* number of positive blood alcohol tests is a success criterion for the police (defined as "output"—a successful detection rate), while a *low* number of positive blood alcohol tests is a success criterion for the traffic police (defined as outcome—successful prevention [see Desrosières, this volume]). Either way, what we can observe is that when the resulting statistics are published, increases in crime often get interpreted as outputs ("what the police do": detection rates), while decreases get interpreted as *outcomes* ("what the police achieve": crime prevention and/or reduction). The same statistics, depending on whether they increase or decrease, can be interpreted both as the output and outcome of police work.

Performance measurement and performance indicators may also induce managerial *myopia*, which is the pursuit of short-term targets at the expense of legitimate long-term objectives (Smith 1995). Skewing practice to fit performance indicators, limiting the discretion of field staff, diminishing "real" effectiveness in order to maximize the practices that are most easily measurable are dysfunctional costs of performance management (Garland 2001: 189). There is a clear danger that what can be reduced in crime is largely synonymous with what can be counted, audited, and easily targeted (McLaughlin et al. 2001: 311). The police will be encouraged to focus on activities that give immediate, short-term results in the crime statistics. This can steer policing activities away from more long-term crime preventive activities.

From Input to Output and Outcome: Implications for the Police and for Policing

If managerial myopia increases, this will not only affect the crime statistics, but also the police itself. A related dysfunctional aspect is *tunnel vision*, which indicates an emphasis on quantified activities at the expense of unquantified or unquantifiable activities (Smith 1995). Performance targets may skew policing activities, priorities, and styles (Painter 2005: 311). The activities of the police have implications well beyond the immediate target of crime reduction. We saw in the annual reports, especially from the 1950s and 1960s, a very rich and detailed description of police activities, many of them quantified, however not as part of a performance measurement system. In a performance-driven culture, we can expect not only an increased emphasis on activities that can be quantified, but also a more narrow focus on the quantified activities that *matter*, i.e. a focus on crime statistics and other selected performance indicators. This might cause the police to avoid "high-risk activities," such as prioritizing crime investigations with low clear-up possibilities, crime preventive activities with low chance of immediate crime reduction effects, or noncrime activities. The tunnel vision effect will not only apply to an increased emphasis on crime reduction, but more narrowly an increased emphasis on *selected* crime performance indicators.

Misinterpretation (Smith 1995) is a dysfunctional aspect of translating crime statistics into a performance indicator which is also relevant for how we read meanings *into* numbers, never simply *off* them. Both crime problems and the functions of police in society are immensely complex, and interpreting trends in the crime rate is therefore equally complex. The number of recorded crimes from year to year provides no insight into the numerous variables that mold crime in society. By using crime statistics as a measure of police performance, one is at risk of misinterpretation. Two fallacies may result from using crime data to judge police performance. First, *the cosmetic fallacy*, which conceives of crime as a superficial problem of society, skin deep, which can be dealt with using the appropriate ointment, rather than as a chronic ailment of society as a whole (Young 1999: 30). Second, *the constabulary fallacy*, which is the tendency to think that the police are the key to crime prevention (Felson 1994: 8). Bayley calls this a myth (Bayley 1994: 3), and also Reiner concludes that "The police function more or less adequately as managers of crime and keepers of the peace, but they are not realistically a vehicle for reducing crime substantially. Crime is the product of deeper social forces, largely beyond the ambit of any policing tactics" (Reiner 2000: 215).

Rob Sindall has said that, "crime statistics are better viewed as a phenomenon in themselves than as a reflection of a phenomenon. It [is] on the criminal statistics, not the actual state of crime that both individuals and institutions base their beliefs about the actual state of crime" (Sindall 1986). I agree with Sindall, but not entirely. I would like to rephrase his statement. Crime statistics are a phenomenon in themselves, but they are also a reflection of a phenomenon. Looking historically at how the numbers have been interpreted, we can get a better understanding of how different discourses or ideologies affect how actors make sense of the numbers. It is nothing new that crime statistics are surrounded with interpretation, discourse, or "spin." What is new is how the numbers are spun today. Crime statistics are an old epistemological object, but as an object they are at the same time constantly reborn, as new modes of governance reconstitute and change how various actors read meaning into the numbers.

ACKNOWLEDGMENT

I am indebted to my coeditors, and in particular Ann Rudinow Sætnan for valuable comments on earlier drafts of this chapter.

NOTES

1. The original statement from Goodhart (advisor to Bank of England) from 1975 is: "Any observed statistical regularity will tend to collapse once pressure is placed upon it for control purposes" (Goodhart 1989). Keith Hoskin restated it into a more general principle (Hoskin 1996), which Marilyn Strathern restated into this more succinct and general principle which is the one I quote (Strathern 2000).

REFERENCES

Bayatrizi, Z. (2008) "From Fate to Risk: The Quantification of Mortality in Early Modern Statistics," *Theory, Culture & Society* 25 (1): 121–143.
Bayley, D.H. (1994) *Police for the Future.* New York: Oxford University Press.
Beirne, P. (1993) *Inventing Criminology: The rise of 'Homo Criminalis.'* Albany, NY: State University of New York Press.
Bell, D. (1979) "The Social Framework of the Information Society," in *The Computer Age: A 20 Year View*, eds. M. L. Dertoozos and J. Moses, 500–549. Cambridge, MA: MIT Press.
Black, D.J. (1970) "Production of crime rates," *American Sociological Review* 35 (4): 733–748.

Bottomley, K., and C. Coleman (1981) *Understanding Crime Rates*, Farnborough: Gower.
Brown, J.S., and P. Duguid (2000) *The Social Life of Information*. Boston, MA: Harvard Business School Press.
Christensen, T., and P. Lægreid (2007) "Introduction—Theoretical Approach and Research Questions," in *Transcending New Public Management. The Transformation of Public Sector Reforms*, eds. T. Christensen and P. Lægreid, 1–16. Aldershot: Ashgate.
Christie, N. (2000) *Crime Control as Industry: Towards GULAGS, Western Style*, London: Routledge.
Coleman, C., and J. Moynihan (1996) *Understanding Crime Data:. Haunted by the Dark Figure*. Buckingham, UK: Open University Press.
Czarniawska, B., and G. Sevón (2005) "Translation is a Vehicle, Imitation Its Motor, and Fashion Sits at the Wheel," in *Global ideas: How ideas, objects and practices travel in the global economy*, eds. B. Czarniawska and G. Sevón, 7–12. Malmö: Liber & Copenhagen Business School Press.
David, C. (2001) "Mythmaking in Annual Reports," *Journal of Business and Technical Communication* 15 (2): 195–222.
Desrosières, A. (1998) *The Politics of Large Numbers: A History of Statistical Reasoning*. Cambridge, MA: Harvard University Press.
Dixon, K., D. Coy, and G. Tower (1995) "Perceptions and experiences of annual report preparers," *Higher Education* 29: 287–306.
Ericsson, K. (1997) *Drift og dyd*. Oslo: Pax.
Felson, M. (1994) *Crime and Everyday Life: Insights and Implications for Society*. Thousand Oaks, CA: Pine Forge Press.
Garland, D. (2001) *The Culture of Control: Crime and Social Order in Contemporary Society*. Chicago, IL: University of Chicago Press.
Gilling, D. (1997) *Crime Prevention: Theory, Policy and Politics*. London: UCL Press.
Godfrey, B.S., P. Lawrence, and C.A. Williams (2008) *History & Crime*. London: SAGE Publications.
Goldstein, H. (1979) "Improving Policing: A Problem-Oriented Approach," *Crime and Delinquency*: 236–258.
Goodhart, C. (1989) *Money, Information and Uncertainty*. London: Macmillan.
Hacking, I. (1990) *The Taming of Chance*. Cambridge, UK: Cambridge University Press.
Haggerty, K.D. (2001). *Making Crime Count*. Toronto: University of Toronto Press.
Haraway, D. (1991) *Simians, Cyborgs, and Women. The Reinvention of Nature*. London: Free Association Books.
Harcourt, B.E. (2001) *Illusion of Order: The False Promise of Broken Windows Policing*, Cambridge, MA: Harvard University Press.
Hood, C. (2007) "Public Service Management by Numbers: Why Does it Vary? Where Has it Come From? What Are the Gaps and the Puzzles?" *Public Money & Management* 27 (2): 95–102.
Hoskin, K. (1996) "The 'Awful Idea of Accountability': Inscribing People into the Measurement of Objects," in *Accountability, Power, Ethos and the Technologies of Managing*, eds. R. Munro and J. Mouritsen, 265–282. London: International Thomson Business Press.
Hwang, H., and D. Suarez (2005) "Lost and Found in the Translation of Strategic Plans and Websites," in *Global Ideas: How Ideas, Objects and Practices Travel in the Global Economy*, eds. B. Czarniawska and G. Sevón, 71–93. Malmö: Liber & Copenhagen Business School Press.
Johnston, L. (2003) "From 'Pluralisation' to 'the Police Extended Family': Discourses on the Governance of Community Policing in Britain," *International Journal of the Sociology of Law* 31: 185–204.

Latour, B. (1987) *Science in Action: How to Follow Scientists and Engineers through Society*. Cambridge, MA: Harvard University Press.
Levitas, R., and W. Guy (1996) "Introduction," in *Interpreting Official Statistics*, eds. R. Levitas and W. Guy, 1–6. London: Routledge.
Long, M. (2003) "Leadership and Performance Management," in *Handbook of Policing*, Cullompton, ed. T. Newburn, 628–654. Devon: Willan Publishing.
Maguire, M. (2002) "Crime Statistics: The 'Data Explosion" and Its Implications," in *The Oxford Handbook of Criminology*, eds. M. Maguire, R. Morgan and R. Reiner, 322–275. Oxford: Oxford University Press.
McIntyre, M.E. (2001) "Goodhart's Law" http://www.atm.damtp.cam.ac.uk/people/mem/papers/LHCE/goodhart.html (accessed January 24, 2008).
McLaughlin, E., J. Muncie, and G. Hughes (2001) "The Permanent Revolution: New Labour, New Public Management and the Modernization of Criminal Justice," *Criminal Justice* 1(3): 301–318.
Munro, R. (1996) "Alignment and Identity Work: The Study of Accounts and Accountability," in Accountability, Power, Ethos and the Technologies of Managing, eds. R. Munro and J. Mouritsen, 1–19. London: International Thomson Business Press.
Osborne, D., and T. Gaebler (1993) *Reinventing Government: How the Entrepreneurial Spirit is Transforming the Public Sector*. New York: Plume.
Painter, C. (2005) "Managing Criminal Justice: Public Service Reform Writ Small?" Public Money & Management 25(5): 307–314.
Quetelet, A. (1996/1842) "On the Development of the Propensity to Crime," in *Criminological Perspectives: A Reader*, eds. J. Muncie, E. McLaughlin, and M. Langan, 14–28. London: SAGE Publications.
Reiner, R. (2000) *The Politics of the Police*. Third Edition. Oxford, UK: Oxford University Press.
Reuter, P. (1984) "The (Continued) Vitality of Mythical Numbers," *The Public Interest* (75):135–147.
Sacco, V.F. (2005) *When Crime Waves*. Thousand Oaks, CA: SAGE Publications.
Sasson, T. (1995) *Crime Talk: How Citizens Construct a Social Problem*. New York: Aldine de Gruyter.
Simon, J. (2007) *Governing Through Crime: How the War on Crime Transformed American Democracy and Created a Culture of Fear*. New York: Oxford University Press.
Sindall, R. (1986) "The Criminal Statistics of Nineteenth-Century Cities: A New Approach," in *Urban History Yearbook*, 28–36. Leicester: Leicester University Press.
Smith, P. (1995) "On the Unintended Consequences of Publishing Performance Data in the Public Sector," *International Journal of Public Administration* 18(2/3): 277–310.
Star, S.L., and J.R. Griesemer (1989) "Institutional Ecology, 'Translations" and Boundary Objects: Amateurs and Professionals in Berkeley's Museum of Vertebrate Zoology, 1907–39," *Social Studies of Science* 19 (3): 387–420.
Starr, P. (1987) "The Sociology of Official Statistics," in *The Politics of Numbers*, eds. W. Alonso and P. Starr, 7–57. New York: Russel Sage Foundation.
Strathern, M., ed. (2000) *Audit Cultures:. Anthropological Studies in Accountability, Ethics and the Academy*. London: Routledge.
Young, J. (1999) *The Exclusive Society: Social Exclusion, Crime and Difference in Late Modernity*. London: SAGE Publications.
Young, M. (1990) *An Inside Job: Policing and Police Culture in Britain*. Oxford, UK: Clarendon Press.

11 Statistics on a Website
Governing Schools by Numbers

Svein Hammer and Sigrunn Tvedten

In 2004, a new national quality assessment system for primary and secondary education was put into effect in Norway. As a vital element in this system, *Skoleporten* [the School Portal, or also the schoolyard gate] was launched, as a "website for quality assessment and quality development in the primary and secondary schools" (SP).[1] The main content of this website is statistical data on the performance of schools and municipalities across a range of quality indicators. The implementation reflects a breakthrough for a long evolving adjustment of the strategies for governing the Norwegian education system: from an input system of rules, regulation, and ambitious attempts at planning toward a more output-oriented system, combining decentralization with management by objectives and performance accountability.

In short, the website can be seen as a management tool composed of concrete instruments and techniques, related to various modes of thought and rationality, constructed as part of the national education system, and put to effect through acting subjects expected to interpret, assess, and make choices based on the numbers. Or, as it presents itself on the website:

> Skoleporten presents various data from the schools and school owners, and also resources for interpretation, assessment, and development in primary and secondary schools. This is a tool that in different ways can be used by school owners and leaders to evaluate and develop their own undertakings. The main target groups are decision makers in the school sector, but Skoleporten also provides useful information for parents, pupils, and other interested parties (SP).

These characteristics make it interesting to relate Skoleporten and its historical development to the analytical framework outlined by Hammer (this volume), where production and use of numbers were interpreted through an analytical model applying the perspective of "governmentality." In this chapter we explore how the public display of quantitative indicators on a website may be analyzed by applying this particular model. The strategy is to pay attention to the co-productive interplay between the following two dimensions:

1. Processes where concrete *technologies of governing*—such as the website itself, the data reported, and the procedures for using the numbers—interact with more abstract *governing rationalities*—such as overarching processes of discussions and conceptualizations of the system for governing education.
2. Processes where *vertical moves of governing* the educational sector relate to *horizontal practices of self-governing*—that is, the way different actors of the educational field develop their positions, make their choices, and assess their results.

These two dimensions are not discussed explicitly in the following analysis. Rather, they function as an underlying guide in our exploration of the website's design, content, intended usage, target groups, discussions, and reasons. The empirical analysis will be interpreted more explicitly through the framework of governmentality in the conclusion. Our main thesis is that statistical indicators and numbers not only tell a story about systematic knowledge, but can also be interpreted as a trend to operationalize and thereby standardize (even rationalize?) criteria for the (seemingly) free choices of humans. To analyze this we outline how the interplay between the understanding of quality, the statistics presented, and the guidelines for a constructive use of the information, point toward processes of subjectivation, where rational actors are encouraged to engage in active assessment and development of effective strategies for optimizing performance. To achieve this, the text analyzes both the website and some of the documents that constitute its genealogy—supported by secondary data on both the school system and the broader context of Norwegian government.

A challenge when studying a website is the constant revisions. Skoleporten was opened in 2004 as a free-standing website, operated by the Norwegian Directorate for Education and Training. In 2008, the first version of Skoleporten was closed down, and a new version was launched as an integral part of and service from the Directorate's website (http://skoleporten.utdanningsdirektoratet.no). All quotations are from texts published as part of the editorial content of the website, but with emphasis on the former version, as it was available until December 2007. An analytical argument for this is the heated public discussions related to its opening, with harsh criticism from the political Left and both teachers' and pupils' organizations. This may have influenced the former version, which shows more textual reflexivity, making it an illuminating representation of the discourse "in-the-making." Therefore we have decided to hold on to our 2007 material, but will end our analysis with a few words on the 2009 version.

THE GENEALOGY OF A QUALITY ASSESSMENT SYSTEM

In *The Politics of Large Numbers*, Desrosières (1998) describes how statistics has become more important in the governing of Western societies.

Pre- and post-WWII, "national spaces" went from being primarily political and judicial to also becoming statistical spheres of equivalence and comparability. In Norway new connections and interactions were established through intertwining discourses on national strength, social democratic government, and social scientific knowledge. Statistics were seen as providing systematic coding, monitoring, planning, and regulation of society at both macro- and micro-levels.

Since the 1980s, however, the discourses of the social interventionist regime have been challenged, especially by neoliberal discourses. In the public sector this has been visible through processes associated with New Public Management (NPM), and concepts such as *management by objectives/outcomes*, *decentralization*, more focus on *efficiency*, *quality*, *accountability*, and *performance*, and—not least—increased use of such techniques as *contracts*, *benchmarking*, *assessment*, and *competition* (see Naschold and von Otter 1996, Baldersheim and Rose 2005, Christensen, Lie and Lægreid 2007).

These shifts appear clearly when studying governing systems for the education sector, and the Norwegian case displays trends similar to several Nordic countries (see Telhaug et al. 2006), though local framing is apparent when global NPM reforms meet with local institutions and culture (Moos and Carney 2000: 12ff). The trend in question is a general convergence in the development of assessment systems in the Western countries, increasingly connected with various systems of accountability (Broadfoot 1996, Pont et al. 2008: 51). As Stephen Hopmann (2007) puts it, "We are entering 'the age accountability.'"

Looking at its history, the Norwegian education system went through an extensive development from the mid-1930s until the early 1980s (Telhaug and Mediås 2003). During these years, mainly under Labour Party governments, the political aims for the sector included (not without tension) social and cultural integration, democratic education, equality and individual self-development, and economic utility for the national economy (Telhaug and Mediås 2003). By the 1970s, the Norwegian school system was referred to as part of the Nordic model (Telhaug et al. 2006), with a comprehensive school, a state heavily intervening to ensure equal opportunities, and a faith in the potential of science in governing education. The education sector was governed through a centralistic rule-based system, detailed instructions, and earmarked remittances (Telhaug and Mediås 2003: 219). At the same time, however, a new Leftist radicalism, critical of positivism, instrumentalism, and strong central authority, teamed up with progressive, reform pedagogy, and challenged parts of the existing regime. Through these trends we can see several displacements, such as a pressure for a decentralization of the organization of the school activity (Telhaug et al. 2006, Telhaug and Mediås 2003: 219ff).

Regarding statistics, this reveals a break between, on the one hand, a focus on standardization and quantitative measurement, and on the other hand the reform pedagogy skeptical of statistical measures of educational outcomes (Sjøberg 2007: 112ff). This is reflected by Norway not participating in the first international comparative assessments of educational achievements in the 1960s and 1970s, even though neighboring Sweden participated, and even though aiming for a scientific development of the education system based on objective measurements would otherwise neatly follow the social democratic vision of rational societal development (Sjøberg 2007, see also Eide 1995: 75).

Not until the early 1980s did Norway participate in international comparative surveys, such as SISS (Second International Science Studies), and later TIMSS (Trends in International Mathematics and Science Studies) and PISA (Programme for International Student Assessment [Sjøberg 2007: 114]), and the results from these first surveys received only meager attention in the public sphere. However, since the late 1990s, the attitude toward achievement indicators has gradually changed, and after the results from PISA 2000 were made public, there has been massive public and political attention to the results from international surveys (Sjøberg 2007), paralleled with the introduction of national tests in basic skills, and a survey for measuring the pupils learning environment—all as part of a new national system of quality assessment and development.

These changes can be related to a new challenge of the social democratic regime from the early 1980s, now from *the new Right*, who also called for decentralization, but in combination with a focus on efficiency and the economic utility of education. As part of this came more use of block grants and the introduction of general plans instead of detailed instructions (Telhaug and Mediås 2003). An important event was when OECD in 1987 conducted an evaluation of the Norwegian education system, and revealed a concern with central authorities' lack of control over school activities, due to how the processes of decentralization were combined with an absence of a national assessment system (OECD 1989; Telhaug and Mediås 2003: 366–377; Eide 1995: 79). They questioned whether the central government was able to obtain a satisfactory picture of the functioning of the education sector at the local level, and particularly they requested a system clarifying the responsibility of different levels of authority. The report could be interpreted as partly legitimizing already existing ideas, but also initiating discussions, strategies, and projects the following decade, concerning national assessment of schools' performance[2] (Karlsen 2006: 201ff, Telhaug 2005: 50).

In 1991, the then dominant system of management by rules and inputs was partly replaced with management by goals as a national strategy for the educational sector (Stortingsmelding[3] 37/1990–1991). The central government was to formulate national goals for the educational sector, and then evaluate to what degree the schools achieved these goals. Actors at the local level were given considerable freedom to choose and organize strategies for realizing the goals. This implied that local initiatives and freedom could be combined with greater control by government through holding the schools accountable for the results of their activities (Telhaug and Mediås 2003: 366–377). This shift demanded an assessment system, but the process of bringing the new management rationality into effect was strained. In 1997, an official report on the issue of evaluation[4] concluded that routines for systematic evaluation of results in the school sector were still inadequate, but at the same time the report displays critical voices warning against external assessment of schools, which they feared would lead to over focusing on measurable results at the expense of broader educational goals.

Not until the new millennium was the process sped up, partly due to the initiatives of a new liberal-conservative government. The Søgnen Committee, founded in 2001, outlined a national quality assessment system and recommended that the long-standing practice of measuring school quality mainly through a focus on structure had to be supplemented by a stronger focus on systematic data on results, or performance. Among the suggestions was the introduction of compulsory national tests, systematic assessment of the learning environment, and the establishment of a website for publishing results (NOU 2002/10).

Concern about schools' performance swelled further when the results from the international examination PISA were made public in 2001 and again in 2004. Norwegian pupils had obtained mediocre results on learning outcomes and, though scoring high on well-being, Norway was among the lowest scoring countries when it came to learning environment, with noisy pupils and lack of motivation and attention (Telhaug 2006: 273). The results and ensuing debates heavily influenced the ongoing work on a new school reform, *Kunnskapsløftet* [Knowledge Promotion] (Kjærnsli and Lie 2006). In 2003 the path-breaking Stortingsmelding 30/2003–2004, *Kultur for Læring* [Culture for Learning], was published with frequent references to the results from international surveys. The new school reform was introduced, together with the new governance system and the new national assessment system. This brings us back to the website, Skoleporten, which was introduced as a central element of the new system.

SKOLEPORTEN AND ITS GOVERNMENTALITY

Analysis of Skoleporten reveals a website through which data on school quality is made public, and argued to function as a tool for local quality assessment and development. It mainly consists of statistical data gathered in an underlying database, presented in aggregated formats at school level or higher. The statistics are displayed in tables,[5] showing the arithmetic average of an organization (such as a municipality or a school) on a range of indicators. One can view comparisons over time, or compare organizations. In addition to statistics, the website offers resources for interpretation, usage and report making, as well as links to schools' and school owners' websites.

The main target groups of the website are local school agents, such as principals and local school administrations, who may log on to access more detailed information and make reports on their respective organizations. Additionally, both national authorities and the general public are assumed to make use of the website.

The content is structured around four assessment or quality areas: resources, learning outcomes, learning environment, and rate of secondary school completion.[6] These are areas pointed out by the government as the current main areas of focus in improvement of the quality of schools.

- *Resources* reports information on economic and material resources such as municipalities' spending on the school sector and availability of computers, as well as information on personnel, such as teacher:pupil ratios.
- *Learning dividends* show the results from various examinations and testing of basic skills: Norwegian reading, math, and English. These include the results from the (formerly) heavily debated national tests, first introduced in 2004.[7]
- *Learning environment* presents results from the questionnaire called the Pupils' Study. Through a compulsory, annual, web-based user survey, pupils are asked to evaluate their own school situation on topics such as motivation, teacher-pupil relations, bullying, and well-being.

Some of the statistics have been available to the organizations earlier, but Skoleporten opens up for new ways of making use of the numbers. As pointed out on the website:

> Most of the data on Skoleporten has been made public in other settings earlier, for example by the Statistics Norway. What is new with Skoleporten is that a great range of indicators are gathered in a comprehensive framework at the school level (SP).

The design and content of the website thereby facilitate new practices and techniques such as comparison over time and between organizations.

Governing Rationalities: Accountability and Quality

Following the analytical perspective presented introductorily, the materials and techniques of the website should be analyzed as co-productive with discussions and reflections concerning how to govern schools. In this way, the website can be seen as both outcome of and contributor to several intertwined processes. For example, related to the rationality of governing by objectives, the 1997 school reform was criticized for elaborating too many detailed goals in the curriculum, leaving the local level confused over what goals to focus on, at the same time as local freedom could be seen as threatened by national authorities (see Stortingsmelding 30/2003–2004). The new reform, *Kunnskapsløfet*, implemented from 2006, specified fewer and clearer goals in the national curriculum, reducing confusion at the local level as to national priorities while increasing freedom at the local level as to how to achieve those goals. Among the changes is a stronger focus on strategies for strengthening basic skills (Ministry of Education and Research 2007).

Through its self-presentation, Skoleporten claims to reflect the new national priorities:

> The content of the website (the data, the guiding resources, and the local examples) are structured around the four assessment areas—also called schools' areas of quality. These represent some of the main tasks of the school which the government has chosen to focus on now. The government's focus may change over time (SP).

And, although structural factors and process quality are addressed in the assessment areas of resources and learning environment, the website also conveys an increased focus on basic skills and learning outcome, as exemplified in the following:

> The supreme goal of establishing Skoleporten is to better the learning outcomes for the individual pupil by facilitating quality development on all levels in primary and secondary schools (SP).

Skoleporten as a website reflects the national governing rationality of management by goals, by facilitating the monitoring of goal achievement. By allowing members of the public to engage in their own monitoring activities, the website reflects the increased focus on public accountability. The shift toward management by objectives entails a shift of (some of) the pressure of evaluation from pupils to schools and teachers (Telhaug and Mediås 2003: 374). More generally, the new rationality involves a decentralization of responsibility and accountability, and the new national assessment system highlights the responsibility of the local level to systematically engage in school-based evaluation, as principals and local authorities are made responsible for school quality (see Stortingsmelding 30/2003–2004). The website was intended to serve as a tool for both control and development, and though the latter was to be given greatest attention, the website was also to enable national agents to maintain monitoring through the systematic presentation of data relevant for governing the sector (NOU 2002/10: 37, Stortingsmelding 30/2003–2004).

Accountable For What?: The Concept Of Quality

As discussed previously, the rationality of management by goals assumes that the national authorities set clear goals. A topic thus emerging when reading texts on Skoleporten is what characterizes the choice of educational goals and perspective on school quality presented through the website. An explicit definition or discussion of school quality is lacking, but an illuminating feature is that the assessment areas are all introduced by stating that the indicators are connected to quality:

- *Resources*: "The resources available to the school and how these are used affects the quality of the education" (SP).
- *Learning environment*: "The pupils' school and learning environment affect the quality of the education" (SP).
- *Learning dividends*: "There is a connection between the quality in the education and the learning outcome of the pupils" (SP).

In a national report, also discussing Skoleporten, three types of quality in education are defined: *structural quality, process quality*, and *result quality* (NOU 2002/10). These are readily related to the three assessment areas on the website. According to the report, there has in Norway for a long time been a focus on evaluating structural and process quality, and the recommendation was to put a stronger emphasis on results, justified by viewing process and structural quality as conditions for the ultimate purpose of school activity: learning.

This approach indicates a consensual and unambiguous picture of what school quality is: School quality is not specified, but is viewed as *the* desirable outcome dependent on all the three areas of quality assessment. This implicit perspective presented through Skoleporten resembles the outspoken claim put forward in Stortingsmelding 30/2003–2004, explicitly rejecting arguments on how broad educational goals such as tolerance or democratic education may be in conflict with emphasis on narrower academic goals as measured by test results. This nondefined, yet still unambiguous picture of what school quality is may be contrasted to more conflict-oriented perspectives, claiming that an inherent feature of school organization is broad, vague, and even incompatible educational goals. By dodging the challenge of specifying such vague and sometimes incompatible goals, it becomes (apparently) possible to establish one singular, neutral perspective on the ultimate characteristic of school quality (see Karlsen 2006: 138–139). The different aspects of quality are claimed as connected, and basic skills may be emphasized as conditional for broader goals. The design of Skoleporten effectively creates a common, public platform for the assessment of schools, undermining contradictions, and thereby neutralizing deliberations on priorities and what school quality may be (see Broadfoot 1996). The website does point out electronic links to the schools' homepages, which may present a more nuanced and comprehensive picture of their activities. This suggests a certain tension between the perspective promoting increased use of indicators and the perspective acknowledging the complexity of goal setting for the educational sector. This is further illuminated in the next section.

Discussions And Nondiscussions On Statistics

Skoleporten offers two reservations concerning whether the indicators adequately reflect national educational goals. Both are interesting here due to their explicit reference to the limitations of statistics. First, it is emphasized that not all national goals in education are reflected in the content of the website. While one reason may be the political priority of focusing on basic skills, it also appears due to there not being any methods for measuring a range of educational goals quantitatively. For example, the importance of both social and digital competence is stressed, so as to emphasize that there are not (yet) any tools for measuring these competencies.

> It is important to see the different areas of assessment in relation to each other, and at the same time know that they do not give a complete picture of school quality. There is a great deal the areas of assessment do not capture with the presented data. Work on social and digital competencies are two examples of this (SP).

This comment is particularly interesting if compared to Berge's (2007) discussion on the conceptualization of the term *basic skills* in the Norwegian curriculum. The concept was originally inspired by the work of OECD, where some broad, basic competencies for the postindustrial society were defined, later known under the term DeSeCo (Definition of Selected Competencies). The term was discussed in official reports, so that it could be introduced as a major priority in the new curriculum and quality assessment system (see Stortingsmelding 30/2003–2004). Berge (2007: 237ff) argues that the final definitions of the term in Norwegian official publications represents a simplified and instrumental interpretation of the OECD's DeSeCo, ignoring the comprehensiveness of competencies. The basic skills as presented on Skoleporten support this critical observation.

Second, a reservation is made on Skoleporten stating that the indicators do not reflect all there is to the qualities which are measured. The statistics will only show a score on a constructed indicator, leaving non-measured aspects in the dark. One obvious reason is to keep the number of indicators at a reasonable level. But it also reflects the challenge of certain aspects of quality being hard to measure quantitatively. The following example concerns how to interpret the results from the survey on learning environment:

> The questions in Skoleporten cannot possibly measure all there is to a concept, for example motivation. The questions measure what those who have developed the survey thought to be important and possible to ask about in the survey. It is therefore beneficial to discuss whether there is agreement between what one wants to measure and the user's own understanding of the indicators (SP).

This raises the question of how the challenge of developing quantitative indicators on broad educational goals regarded as important, has molded the design and functioning of this management tool. For instance, does the public display of indicators encourage a focus on goals which are quantifiable? This was of great concern to those critical of the national assessment system, fearing that schools might start "teaching to the test" and give lower priority to comprehensive aspects of quality, which may be harder to measure. The general point is how the strategy of management by objectives may lead to a displacement of objectives in public service organizations with complex goal structures, such as schools (see for example Sørhaug 2003).

As we have remarked on in this section, reservations are made in texts on Skoleporten:

> It is important to remember that the indicators only reflect a small part of the total result of the school's work. The indicators do not give an overall picture of the pupils' learning in the way the goals of education are formulated in the curriculum (SP).

The website is nevertheless intended to function as a tool for development, which ignores productive processes and indirect pressure mechanisms generated by the publishing of the indicators. This picture is further exaggerated when taking the context of decentralization of responsibility and accountability into consideration. According to Dahler-Larsen (2006: 123–47), evaluation of schools may not be considered simply as descriptive or intentional activity, but also as involving processes of construction. How actors make use of the data and evaluations can never be fully predetermined. This unpredictable use of data and management tools demonstrates a need to follow the developments closely through empirical research, and not to close the debates through nonbinding statements of compromises and reservations. The importance of continued debate is illuminated in a report by Engelsen (2008) on the implementation of the new governance system in the Norwegian education sector. It reveals how operationalization of educational goals, at the local level, tends to leave comprehensive, immeasurable goals as mere parrotings of the vague intentions formulated in national reports and curriculum.

Knowledge-Based Action: The Route to Better Schooling

We have now seen how the website reflects a national governing rationality of management by objectives and increased accountability, and also how it offers reservations that the statistics may not reflect all there is to school quality. This reveals a tension between two scenarios: on the one hand, the rationality of improved goal achievement, and on the other hand, the unpredictable outcomes as the technologies are interpreted and put to use by active agents. In this section we will continue the discussion as to how the website is intended to be used, and the purpose of providing information to decision makers in the educational sector. We thereby attend to the dimension of vertical and horizontal processes of governance.

We saw that critics fear that the educational goals represented through indicators will gain disproportional attention. In order to avoid this, there are no formal, direct sanctioning mechanisms associated with failure to follow up continued poor results. The outspoken intention of publishing results is for local development purposes mainly (NOU 2002/10: 37). But it is also emphasized that publishing the results will put pressure on the organizations to improve their performances (ibid.). The website can be interpreted as introducing an informal sanctioning mechanism grounded in the decentralization of accountability of results, combined with an increased focus on results, as the display of statistics indirectly creates pressures toward goal achievement. An important aspect of Skoleporten is thus how it is assumed to provide utility for local actors.

The intended usage of the statistics is noticeable in the texts on the website. With a sharp wording one is presented with a picture of data providing reliable information to rational actors involved in systematic school quality development at the local level.

> To be able to work systematically with development, one needs knowledge of what is being done well, and what can be improved (SP).

> The data on Skoleporten can give an indication of the situation within the areas of assessment. It can also over time give the users facts and opportunities for assessments indicating in what direction the development is leading in the various schools, in municipalities, counties, and nationally (SP).

We have already seen how the statistics on Skoleporten only capture a narrow aspect of school quality. Despite explicit reservations acknowledging this, the data are assumed to give actors an indication on which areas of school activity they are to focus on: which elements require extra attention and effort in their further quality development.

> One of the purposes of opening Skoleporten is to give decision makers on all levels in primary and secondary schools an efficient tool giving increased information about processes and results in their own undertakings. By enabling decision makers to follow developments in their area of jurisdiction, a more goal-oriented effort where it is really needed is made possible (SP).

Additionally, the website not only provides knowledge in the form of statistics, but also presents guiding resources for how to act on that knowledge:

Where the data on Skoleporten is to contribute to evaluating the present situation in the area of assessment, the guiding resources will be helpful in finding the way to accomplish both local and national goals of education. In this way interplay between quality assessment and quality development in the areas of evaluation is facilitated (SP).

When the consensus perspective on quality and the presentation of statistics to evaluate this quality is combined with the intention of providing guiding resources, the result is a distinct picture of the relation between subjects, knowledge, and quality development. Regarding the horizontal dimension of governing, we see how subjects are to engage in the regulation of school activities, according to the national educational goals. The subject now resembles the instrumental, rational actor of liberal, economic theories: The actors have a goal—in this case school quality, as measured through the indicators on the website. The actors are then given freedom to choose and develop the best suitable means to achieve this goal. For a rational evaluation of means, information is needed, and Skoleporten provides both statistics for evaluation of the present situation and guiding resources for the interpretation of the data. The actors are—so the website claims—provided with information required to make rational choices about where and how to improve the undertakings.

From 2007 to 2009

As mentioned previously, the analysis of the website is mainly based on its 2007 version. Since then the website has been revised and reopened, and we wish to make some comments on changes that have been made. Interestingly, in this new version the explicit reservations associated with quantitative indicators have been left out. There is a general call for the possibility of supplementing statistics with qualitative measures, but the explicit discussions directed at the validity of indicators and their limited scope has for now been reduced to general methodological discussions related to quantitative aspects—random variation, size of population, and challenges regarding comparisons. Additionally, in contrast to the first version of the website, the statistics have now obtained an even more salient position to visitors: Where in 2007 one had to actively look up the indicators of interest, presented in tabular form, graphic illustration of the latest results now dominates the opening page of the website. Based on the preliminary analysis of the new version, these changes in the content and appearance of the website may be interpreted as an indication of a further undermining of positions—previously quite prominent in the earlier very visible discourse and debate—which were critical of management by objectives in the public service sector, and in particular of quantitative measures in the field of education. The level of conflict concerning the website as a tool has subsided, and the governing discourse has seemingly become a more quiet force.

CONCLUSION: GOVERNMENTALITY AND THE SCHOOL

We started our journey by presenting Skoleporten as a technology of government, constituted by the website itself, the data presented through a database, and the textual resources for using the data to assess and develop schools. But when interpreted through the model of governmentality, we find that the website may be seen as more than this.

First, concepts like "school owners" and "decision makers" connote the vertical axis of government, as it is institutionalized in the school system of laws, objects, procedures, and formal roles. In other words: The website is part of a national system of quality assessment and development of Norwegian schools, where national authorities may follow and control the development, while local actors are given systematic information on their undertakings. As we have seen, the website is *assumed* to hold such a position, as the national authorities, county representatives, and school owners are invited to use the statistics to monitor the development, and thereby hold the schools accountable.

Second, the formulations about "resources for interpretation, assessment, and development" and "information for parents, pupils, and other interested parties" move our focus toward the horizontal axis, the interplay between active and constructive subjects expected to constitute school reality from below, who interpret, assess, and act in relation to each other and the website.

Lastly, through the discussions of this chapter, we have indicated how the website as a set of concrete instruments and techniques, constructed and used in relation to both vertical and horizontal processes, exists in a constitutive interplay with more abstract processes of questionings, discussions, conceptualizations, and solutions. We will argue that these can be categorized as part of at least three different forms of rationalization: *Instrumental* attempts to standardize practices through defining unambiguous goals and choosing or developing strategic means; *positivistic* ambitions in building objective knowledge through processes of precise operationalization, measurement, storing, and analysis of social reality; and *liberal* attempts to govern through freedom, competition, and self-assessment.

In other words, to understand the website, how it functions, what social reality it constitutes, or what kind of subjectivity it demands, we have to investigate the interplay between these processes of rationalization and the techniques they serve. Earlier we saw how the discourses on school reality were shaped by mediocre achievements of Norwegian pupils, according to the PISA test reports, which gave additional power to the demand for strategies and techniques directed at measuring learning outcomes. The reasoning was that "to be able to work systematically with development, one needs knowledge of what is being done well, and what can be improved." This indicates a shift in the practices of statistics, an adjustment in how we produce and use indicators and numbers, related to strategies and techniques for governing indirectly and from a distance.

In the empirical material from Skoleporten the combination of, on the hand, a concern with the central authorities' lack of control and, on the other hand, arguments for a new governing system highlighting local level responsibility, indicates how the vertical and horizontal dimensions are related. An example is the problem of too many detailed goals in the curriculum: The challenge was said to be that this leaves the local level both confused as to what goals to focus on, and wary that their autonomy is threatened by national authorities. The answer to this has been to specify fewer and clearer goals, and as we have seen to publish indicators on goal achievement on a public website. Through this, the national authorities can both reduce confusion and increase the possibility of strategic choices being made at the local level.

In other words, the goals, the data, and the guiding resources work to secure national standardization of what is seen as especially important, but this process of standardization is not primarily implemented through vertical, direct, top-down strategies. Rather, we see an increasing strategic use of horizontal, indirect governmental techniques, building on the active interpretation and interaction of self-governing actors. The function of Skoleporten in this governmental rationality is, through the combination of public presentation of data and decentralized accountability, to increase pressure towards improving quality.

These tendencies can be expected to have important consequences on many levels, including the day-to-day work of pupils, teachers, and other employees in the schools: As members of a seemingly more open and pluralistic field, every actor is invited to govern both herself and the school through complex practices of calculation—a decentralized governing through strategic choices in seemingly open situations, where the indicators and their measurements are part of technologies that work to optimize achievements, or outputs.

It is obvious that the processes of subjectivation here differ from what we had come to expect in social democratic society. The actor that now emerges is a rational exerciser of strategic choices, who through calculation is expected to maximize benefits to the best for himself, his organization, and the educational system—even society as a whole. As in liberal economic theory, a prerequisite for this is that the choices are enlightened; the numbers and the procedures for their production and use contribute to this in important ways. Perhaps we here are in touch with the central point of the expanded use of indicators and measures: They give a possibility to produce knowledge and manage practice *without* direct interventions.

ACKNOWLEDGMENT

This chapter was first presented at the workshop *Statistics as Boundary Objects between Science and the State*, Norwegian University of Science and Technology, May 14–16, 2007. We thank workshop participants for comments. We are also grateful for comments from Professor Lejf Moos, Norwegian University of Science and Technology and Danish School of Education, University of Aarhus.

NOTES

1. All quotations from the website are translated by the authors of this chapter and marked with (SP), as shown here.
2. For an example of an early study see the EMIL project (see Granheim and Lundgren 1990)
3. We will be referring to several forms of White Papers in this chapter. A "Stortingsmelding " is a Parliamentary report. An NOU is a public report on findings, often preliminary to a Stortingsmelding, which in turn may lead to a proposal for legal reform. Both are referred to by number and year, but often better known to the public by their titles or the name of the chair of the committee that prepared them.
4. Report from the Moe Committee: Nasjonalt Vurderingssystem for Grunnskolen [National Assessment System for Primary School], December 1997—see attachment in Stortingsmelding 47/1995–1996.
5. A feature of the revised Skoleporten is that data additionally may be displayed in a variety of ways, such as through bar graphs or line graphs, in addition to the tables, and which further accentuates the prominence of statistics in the appearance of the websites.
6. This fourth area will not be further commented in this chapter, as we focus on the governing systems for primary (elementary and middle) schools, which are owned and run by local authorities.
7. Due to harsh criticism of how the public display of results made possible an official ranking of schools, indicators on learning dividends are no longer made available to the general public at school level.

REFERENCES

Baldersheim, H. and L. Rose (2005) *Det kommunale laboratorium*, Oslo: Fagbokforlaget

Berge, K.L. (2007) "Grunnleggende om de grunnleggende ferdighetene" in Hølleland, Halvard (eds.) *På vei mot Kunnskapsløftet—Begrunnelser, løsninger og utfordringer*, Oslo: Cappelen Akademisk Forlag.

Broadfoot, P.M. (1996) *Education, Assessment and Society*. Buckingham, Bristol: Open University Press

Christensen, T., A. Lie, and P. Lægreid, eds. (2007) *Transcending New Public Management*. Burlignton, VT: Ashgate.

Dahler-Larsen, P. (2006) *Evalueringskultur—et begreb bliver til*, Odense: Syddansk Universitetsforlag.

Desrosières, A. (1998) *The Politics of Large Numbers*. London: Harvard University Press.

Eide, K. (1995) *OECD og norsk utdanningspolitikk—en studie av internasjonalt samspill*, Oslo: Utredningsinstituttet for forskning og høyere utdanning

Engelsen, B.U. (2008) *Kunnskapsløftet. Sentrale styringssignaler og lokale Strategidokumenter* Rapport no. 1. Oslo: Universitetet i Oslo.

Granheim, M. and U.P. Lundgren (1990) *Målstyring og evaluering i norsk skole*, Sluttrapport fra EMIL-prosjektet, NORAS/LOS-i-utdanning, LOS-notat nr. 7-90

Hopmann, S. (2007) "Epilogue: No child, no school, No state left behind: Comparative research in the age of accountability" in Hopmann S. (Ed) *PISA according to PISA, Does PISA keep what it promises?* Wien: LIT Verlag.

Karlsen, G. (2006) *Utdanning, styring og marked: Norsk utdanningspolitikk i et internasjonalt perspektiv* Oslo: Universitetsforlaget (2. utg.)

Kjærnsli, M. og Lie, S. (2006) "TIMSS og PISA: Prinsipper og hovedfunn fra studiene i 2003" in Brock-Utne, Birgit, og Bøyesen, Liv (Eds.) *Å greie seg i Utdanningssystemet i nord og sør*, Oslo: Fagbokforlaget.

Ministry of Education and Research (2007) "What is the 'Knowledge Promotion'?" http://www.regjeringen.no/en/dep/kd/Selected-topics/andre/Knowledge-Promotion/What-is-the-Knowledge-Promotion.html?id=86769 (accessed March 26, 2010).

Moos, L., and S. Carney (2000) "Ledelse i den nordiske kontekst: Realiteter, muligheder og begrensninger" in Lejf Moos, Stephen Carney, Olof Johansson, Jill Mehlbye (eds.) *Skoleledelse i Norden*. Report from the Nordic Council of Ministers.

Naschold, F. and C. von Otter (1996) *Public Sector Transformation: Rethinking Markets and Hierarchies in Government*. Amsterdam: John Benjamins.

NOU 2002/10 *Førsteklasses fra første klasse*, Oslo: Ministry of Education and Research.

OECD (1989) *OECD-vurdering av norsk utdanningspolitikk. Norsk rapport til OECD. Ekspertvurdering fra OECD.* OSLO: Aschehoug.

Pont, B., D. Nusche, and H. Moorman (2008) *Improving School Leadership Volume 1: Policy and Practice*. Paris: OECD.

Stortingsmelding 37 (1990-91) *Om organisering og styring i utdannelsessektoren*, Ministry of Education and Research.

———. 47 (1995-96) *Om evaluering, skolevurdering og nasjonalt vurderingssystem*, Ministry of Education and Research,

———. 30 (2003-2004) *Kultur for læring*. Ministry of Education and Research.

Sjøberg, S. (2007) Internasjonal undersøkelser: Grunnlaget for Kunnskapsløftet? in Halvard Hølleland (ed.) *På vei mot Kunnskapsløftet—Begrunnelser, løsninger og utfordringer*. Oslo: Cappelens Akademisk Forlag.

Sørhaug, T. (2003) "Fra plan til reformer: Det store regjeringsskiftet" in Neumann, Iver B. and Ole Jacob Sending (eds.) *Regjering i Norge*, Oslo: Pax forlag.

Telhaug, A.O. (2005) *Kunnskapsløftet—Ny eller gammel skole* [The Knowledege Promotion—A new or an old school], Oslo: Cappelen Akademiske Forlag.

———. (2006) *Skolen mellom stat og marked—Norsk skoletenkning fra år til år, 1990-2005* (2.utg) [The school between state and market—Norwegian educational thinking from year to year, 1990-2005 (2nd edition)], Oslo: Didaktika Norsk Forlag.

Telhaug, A. O., and Mediås, O.A. (2003) *Grunnskolen som nasjonsbygger—Fra statspietisme til nyliberalisme*, Oslo: Abstrakt forlag.

Telhaug, A.O., O.A. Mediaas, and P. Aasen (2006) "The Nordic Model in Education: Education as Part of the Political System in the Last 50 Years," *Scandinavian Journal of Educational Research* 50(3): 245–283.

12 Locating the Worths of Performance Indicators
Performing Transparencies and Accountabilities in Health Care

Sonja Jerak-Zuiderent and Roland Bal

> Transparency on performance is a strong incentive to improve healthcare. We hope that all institutions score well. That is in everybody's interest, but especially in the interest of the thousands of patients that can enjoy a longer life
> (Dutch Council on Public Health and Healthcare 2005).[1]

PERFORMANCE INDICATORS—TRANSLATING MARKETIZATION AND SAFEGUARDING QUALITY?

The fanfares of marketization and privatization in health care echo as a common theme through recent welfare state discourse. Since the 1980s, public services such as health care and education have been increasingly invaded by neoliberal conceptualizations of the State, market, and society. This neoliberal sociomaterial context has shaped the rationale of how to frame and govern public policies, with performance indicators being a direct consequence of these conceptualizations. However, what is it that is "performed"? What is "accomplished" or "achieved"? The definitions of performance vary and the emphasis has been placed—at different moments—on economy, efficiency, effectiveness, outputs, quality, outcomes, and social impacts (Freeman 2002: 126). "Consequently, the indicators themselves have also varied widely, in terms of what they purport to measure, their presentation and their intended audiences" (Freeman 2002: 126), as have their justifications.

The justifications for introducing and working with indicators center on the idea of providing data to facilitate transparency, but they are rooted in different logics. One justification is that performance indicators[2] make health care more transparent and, thereby, accountable, while another claims that we need transparency to make market mechanisms (e.g., free choice, price mechanisms as competition) work. Yet another logic highlights the preventive effect indicators purportedly have if things are not going according to an expected standard. Further strands of justifications, also implied in the introductory quote, cluster around facilitating learning and improving by measuring or by making and breaking reputations. In all these justifications, indicators are purported to deliver the "right" information on healthcare products and services. They seem to be *the* tool to facilitate the growth of knowledge for diverse purposes; *the* tool through which patients, citizens, policymakers, stakeholders, and professionals can understand what is achieved and accomplished in health care.

In this chapter we want to unpack this unproblematic take on information. Following the indicators in a hospital we analyze what these indicators "actually" do. Do they fulfil these ambitious, multiple, and sometimes contradicting promises? If so, in what ways? If not, why not? We focus our analysis on a specific set of indicators in health care, the "Basic Set of Performance Indicators" for hospitals, developed under the aegis of the Dutch Health Inspectorate (*Inspectie voor de Gezondheidszorg*, or IGZ; IGZ, NVZ, et al. 2007).

The "Basic Set of Performance Indicators" is the first example and set of indicators in health care in the Netherlands that has been rolled out by a governmental body on a national level. The Inspectorate is an independent agency of the government and their *raison d'être* is acting in and protecting the public interest.[3]

The introduction and development of the "Basic Set of Performance Indicators" for hospitals in the Netherlands in 2003 was strongly related to the history of the Inspectorate. In 1999, the National Audit Office (*Algemene Rekenkamer*) published a devastating report on the Inspectorate, arguing that the Minister of Health "could not, on the basis of information provided to her by the Healthcare Inspectorate, form an image of the quality of Dutch healthcare" (Rekenkamer 1999: 4). This was actually a direct critique of how the Inspectorate dealt with the consequences of a 1992 airplane crash southeast of Amsterdam. The Dutch Inspectorate failed, in the eyes of the Ministry, in one of its main responsibilities: advising and counseling on issues concerning public health. A direct measure in response to this critique was the reorganization of the Inspectorate and the development of the "Basic Set of Performance Indicators" as part of a new mode of supervision based on the idea of foreseeing and managing risks. The indicators, as so-called preventive measurements, are one part of the "layered and phased" supervision, grounded in the logics of risk management (IGZ, NVZ, et al. 2003: 3).[4] These indicators are intended to be used by the hospitals for purposes of external accountability *and* internal quality improvement. The combination of external and internal aims is a rather unique endeavor. Usually, the two principal uses of indicators systems relate to *either* (internal and external) control and accountability *or* formative quality improvement (Freeman 2002: 128).

For us, the interesting question here is not so much the pros and cons of these two traditions in theory, but rather how expectations for indicators are shaped in the materialities and work related to the indicators in practice. In our analysis we map the successes and failures of indicators "in action" (Latour 1987). By empirically analyzing the "visible and invisible work" (Suchman 1995, Star and Strauss 1999) around the indicators, and by exploring the "worths" (Boltanski and Thévenot 2006) and investments at stake, we specify how the indicators of the Inspectorate relate to the practices they are supposed to measure and in which sense the indicators result in improvement and controlling. Or not.

In the following section we elaborate on our methodology. In the third and fourth sections we zoom in on the work undertaken to "make indicators work." Based on our empirical analyses we conclude with some critical reflections on the notions of transparency and accountability resounding throughout the discourses on public issues in the recent welfare state discourses.

PERFORMANCE INDICATORS—PROMISES OF SUCCESS, FAILURES, AND THE ACTUAL WORK

By grouping the arguments raised previously on the value and use of indicators around two different, but complementary, analytical lines in theory—"promises of success" and "focus on failures and fundamental disagreement"—we open up another analytical perspective: "indicators in action."

Promises of Success

The principal justifications for the value and use of indicators are either control and accountability or learning and quality improvement. Following these two justifications, the literature refers to summative and punitive "performance regimes" when used for external or internal control or to positive, supportive and formative ones when used for learning. However, the vast literature of believers in indicators—either in their internal or external use—comes together in being mainly practice-oriented, prescriptive, and optimistic of the values and use of performance indicators in health care. The literature of such believers focuses mainly on selecting the "right" indicators, on notions of valid and reliable data collection, and on sensitivity in the analysis and use of data for improving healthcare delivery. Interestingly, although being largely discursive and lacking an empirical base for their claims (Freeman 2002: 127–128), such initiatives mushroom the health care world with claims of smart solutions and expertise on how to measure and thereby manage and improve performances.

The indicators developed under the aegis of the Inspectorate can be seen as an interesting experiment in combining the logic of external control with the logic of learning within an organization reproducing the promises of success—ironically, without proof:

The set would enhance the transparency of the hospital sector, and stimulate individual hospitals to improve their scores. Bridging some of the classic distinctions between "internal" and "external" indicators, the Inspectorate's vision was to rapidly produce a feasible set of indicators that would fulfil these aims, while maximally preventing "side effects" such as misinterpretations, defensive or perverse reactions. Explicitly avoiding the trap of searching for exhaustive validity of the indicators, the Inspectorate's motto was "feasibility first" (Berg et al. 2005: 59).

"Feasibility first" can be seen as a compellingly enthusiastic manifesto for the use of performance indicators. However, the authors fail to problematize the investments that must be undertaken to "make these indicators work," thereby missing the opportunity to point out how to trace successes or failures in practices around and beyond the very figures of the performance indicators. In this contribution, we address the loci of investments that are at stake when producing indicators and that are not captured by these very numbers.

Failures and Fundamental Disagreement

The literature we refer to as focusing on failures deals primarily with incidents and controversies related to the value and use of indicators. Concepts like "decoupling," "colonization," and "gaming" are based on the observation that indicators do not work as they should (Power 1997). But the reasons for failure are usually interpreted from a social-institutional perspective (e.g., "cultures of resistance," power games, or path dependencies) or from a social-psychological perspectives (e.g., that "people are creatures of habit and do not want to change," or referring to obstructions due to conflicting stakeholder strategies) without zooming in on the actual practices. Explaining failure from these perspectives precludes analysis of the investments at stake as well as analysis of if and how these "failures" could also be turned into making these indicators a success.

Besides this literature on "perverse effects," we found literature fundamentally disagreeing with the value of performance indicators in health care, claiming that the "faith in measurement may be misplaced" (Freeman 2002: 129). Power argues that indicators do not replace the need for trust, but rather displace existing informal and unrecorded modes of quality assurance, thereby generating suspicion and fear and ultimately undermining the conditions of trust required for quality improvement (Power 1997). Strathern refers to the ubiquitous pervasion of "audit cultures" as a way for the State to evade other (ethical and political) responsibilities (Strathern 2000), while Wiener fundamentally disagrees with the format and idea of indicators by pointing out the complexities of healthcare practices and critiquing the superiority of market forces in the mentality of performance indicators (Wiener 2000). We argue, however, that, by disagreeing so fundamentally authors blind themselves to moments of successes in relation to the values of indicators, whereby they also miss the opportunity to analyze the failures in a productive way. Again, this calls for more detailed analysis of what happens when performance indicators are introduced into sectors such as healthcare.

Indicators in Action

What kind of investments and work must be undertaken to "make indicators work" and who or what does such work? Who is held accountable, and for what? What consequences do these investments have for the practices in a hospital? We traced the consequences of the IGZ indicators by following them "in action" (Latour 1987) as participant observers in the work around the indicators in a hospital from April until June 2006. We primarily followed the quality management team and their endeavors by attending meetings, filling in data on overviews, and assisting in preparing a presentation to the medical staff, but also observing the quality management team in action in relation to their other responsibilities and activities. We also followed the indicators on their way through the hospital. This allowed us to observe the indicators as actors in the daily exigencies of the hospital. We were allowed to join any meeting of interest and to interview any hospital employee directly or indirectly involved in the investments in and work on indicators. In addition to the quality management team we also interviewed and observed the "coders" in the hospital, the IT team, the head of the Department of Internal Medicine, a "division manager," diabetes nurses, a cardiologist, and a representative of the board of directors. We were particularly interested in if and how the "external indicators" of the Inspectorate were "internalized" into the social and material fabric of the hospital.

MAKING WORK ON INDICATORS VISIBLE—
DELEGATING WHAT TO WHOM ACTUALLY?

In the hospital the stories of the IGZ indicators are stories of perambulations, delegations, and translations that result in work. The Inspectorate places the order of delivering knowledge of the current status of health care in the format of the "Basic Set of Indicators" on the desks and agendas of the board of directors of the Dutch hospitals, and delegates thereby the task of proving the quality of health care practices to the hospitals themselves. The board of directors of the hospital where we did our observations (a general hospital with around 400 beds, henceforth referred to as "the hospital"), in turn, delegated the task in the first year to a long-time employee, later on to a management team, the management team to a project group set up specifically for the task of collecting data on the IGZ indicators, and the project group, once set up, to the IT Department and the so-called "problem owners"—a doctor, nurse, or someone from the quality management team who has been assigned the responsibility for gathering data on a specific (set of) indicator(s). In the following we follow the "blue booklet"[5] and trace the work and investments undertaken in the hospital to collect the requested data.

At first, the blue booklet did not do much. The order for the IGZ to gather data on the "Basic Set of Indicators" needed much specification and locating work. Not only did the "network of informants" need to be detected, but the indicators also needed to be translated into "something meaningful for the people doing the real work," as indicated by the quality manager. She meant the people on the work floor—doctors and nurses working in various wards and departments. Every indicator has its own story of "translating" (Callon 1986, Latour 1986, Law 1986, Wiener 2000) and "articulating" (Corbin and Strauss 1993) various values when balancing priorities of practicing health care "in the drive of daily exigencies."

In spite of many difficulties, there are a few successes to be booked in the search for measurable performances. But successes for whom? And what are the costs? And who is bearing them? The Dutch Inspectorate succeeded in making the hospital gather data that matched their new mode of supervision: supervising by numbers. The hospital succeeded in submitting the requested data, but failed to a certain extent in the Top 100 ranking that is published since the introduction of the indicators in 2003 based upon the publicly available Inspectorate data. Also, the project group in the hospital succeeded in designing a network of informants, but failed to make the "informants as problem owners" accountable for their "problems," the indicators. Ownership was attributed, but not accepted. In the following sections, we review and specify the work and investments that have taken place to make these successes happen since the first indicator year.

Investing in Management, Administration, IT

The acquisition of a quality management team, a senior and a junior position, was the largest visible investment after the first two years of collecting data on performances in the hospital. In the third year of data gathering, the year we observed, the quality management team took the overall lead in gathering the data on the indicators via the established net of informants. The quality manager experienced the task of gathering the data as an "experiment and research" and one aspect of her tasks as "conveying a message people do not want to hear"[6]: "Quality is per definition paper work and something no one is really keen on, something that is forced by strangers who have no idea about the real work."[7]

As she explains, work on collecting the requested data is mainly characterized by overcoming the distance that is enacted by the apparently unambiguous and standardized format of the indicators. This standard does not mean much "to the people doing the real work" and needs to be translated into something "meaningful."[8] She translates her most important task and her raison d'être into

> "making sure that they [quality initiatives and projects on quality (improvement)] don't conflict with their daily work; I creep as much as possible into their daily work, so that I understand what the connection is. I then take care of the annoying part and they only have to tell how they are doing it. That is, I bridge."[9]

She herself engages with the indicators only partly. In the search for accountabilities she engages with the assignment given by the Inspectorate and the board of directors: She invests time, meetings, and infrastructures in assigning and enrolling problem owners for the various indicators, thereby diffracting and spreading accountabilities across the net of informants. However, she does not engage with the dimension of "being held accountable for the outcomes"; this she delegates, together with the Inspectorate and board of directors, to "someone else." They all make a clear distinction between being responsible for gathering the requested data (to the public, the Inspectorate, the hospital directors) and "others" being accountable for the outcomes, thereby contributing to a "diffraction" (Haraway 1991) of accountabilities. They all hereby evade reflecting upon and questioning the unambiguousness of such data in informing or contributing to knowledge about the performances of health care practices. This task is reserved for the practitioners at the end points of the network of informants who are supposed to juggle and balance the multiple ambiguities of the requested data in a meaningful way.

Catching and Betraying Data and Accountabilities

The quality management team began its mobilization work on the collection of the requested data within the hospital by working out an action plan. The action space and the accountability of the project group in this plan was clearly demarcated and limited to the mere collection and delivery of data. The assumption that there is an essence of performances somewhere out there in health care practices that can be retrieved and collected in an objective and unambiguous way seems to be an instantiation and reproduction of the "myth" referred to by Haraway (1988). We show, however, that the steps and the investments undertaken to collect the requested data are not neutral at all and instantiate specific values that are prioritized (or left out). What values were at stake in our hospital when investing in IGZ indicators?

The action plan consisted of various steps (meetings, e-mails, follow-up, etc.) and the project group agreed upon a deadline and a responsible person or group for each step, which, from the perspective of the quality management team and the project group, was a clear and reasonable plan. The retrieval of the data on many indicators located "somewhere" (or actually, "nowhere") in the grey area between the planned steps, however, turned out to be much more difficult and ambiguous than envisioned by the project team. The "betrayals" (Callon 1986) that the quality management team experienced caused delay and, eventually, extra work.

As the deadline for one of the last steps of the action plan—to present the gathered data to the medical staff for feedback and approval—loomed ever closer, much of the requested data was still missing. The medical staff is a crucial partner in the running of Dutch hospitals, alongside the board of directors.[10] The presentation to the medical staff was therefore considered an opportunity to get the blessings of the medical experts for the collected data and a way to engage the medical staff with the indicators. The quality management team sent reminders via e-mail, asked for feedback and data via phone, trying to "catch"[11] the betraying problem owners. Furthermore, it invested time in organizing meetings with problem owners and the IT Department to clarify what kind of data could be retrieved, as the standard format of the IGZ indicators turned out to be in many instances not unambiguous at all. Some betrayals by problem owners were framed as "merely technical," requiring the investment of time; these were left under the responsibility of the ICT Department, which was to develop a smart query or combinations of queries. Other betrayals were directly related to the allocation of accountabilities and discussions on who or what can be held accountable for a specific outcome.

Figure 12.1 reflects the messiness and ambiguities of the data collected following the action plan approximately one month before it was submitted to the Inspectorate. The figure displays an overview of performance measurements within the hospital for 2004 and 2005 and with the national average for 2004. The last column contains the handwritten data collected for 2006, which was accompanied by various notes and questions related to the comparison of the handwritten data with the measurements of previous years. For example, in relation to the percentage of unplanned reoperations of risky interventions such as *aneurisma aorta abdominalis* and *cholecystectomie* we find: "Quite high in comparison to previous years. The year before measured in relation to the total amount of risky (?) interventions?" In relation with the average blood sugar value we find: "high? check with lab if values normal?" and "LMR [the national medical registration system] ➔ only clinical values; no registration of type I and type II, in DGR not registered." Or references like "check with so and so."

The presentation of the outcomes that the quality management team prepared for the medical staff and as displayed in the working document (Fig. 12.1) went differently than expected. In summary, the doctors raised concerns about the value of the data—the operationalization of the measurements, the interpretation of the outcomes, and the definition of the indicators themselves. These were questioned, discussed, negotiated,

232 *Sonja Jerak-Zuiderent and Roland Bal*

rejected by various doctors during the presentation, instantiating yet another "betrayal."

In the following section, we first unpack how the aforementioned betrayals and extra work instantiate in action, that is, in the work of collecting and interpreting data. We specifically focus on work that has been done to

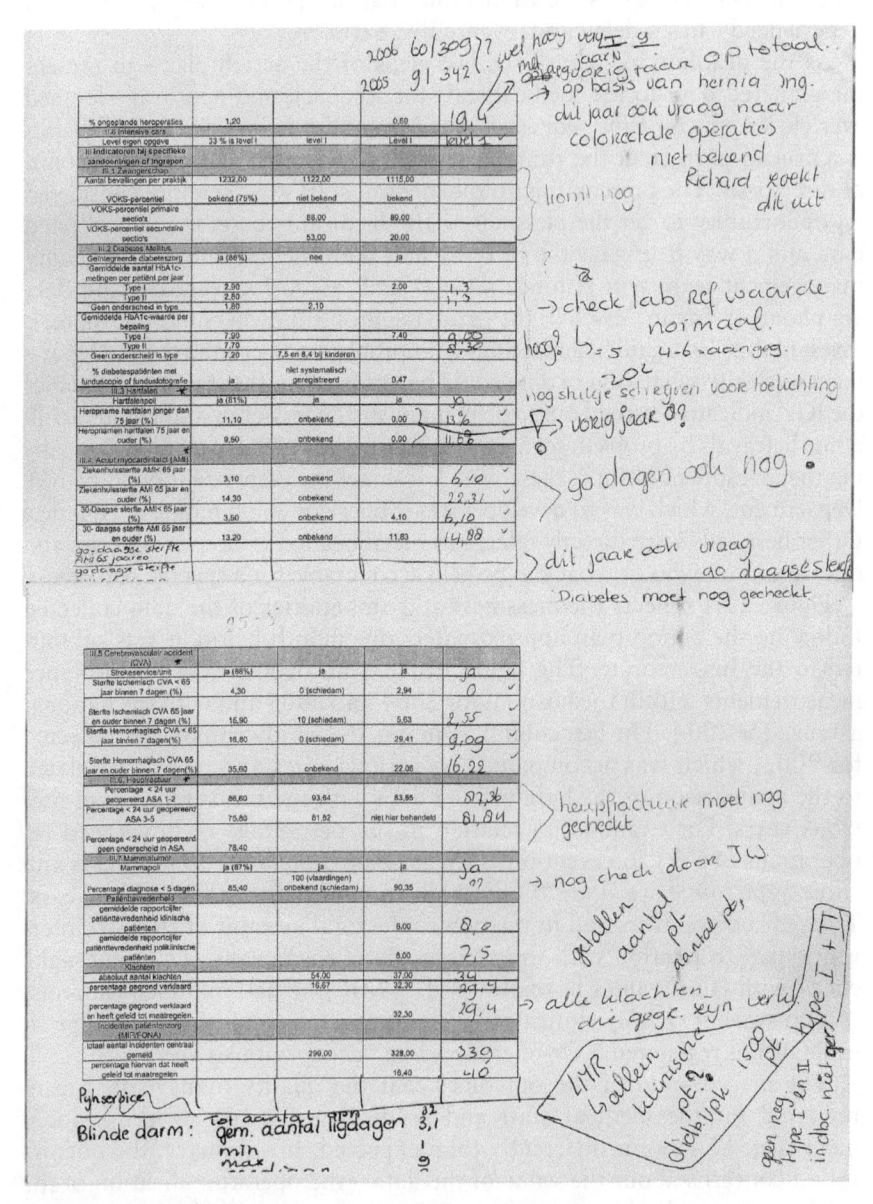

Figure 12.1 Working document of quality management team for collecting the data on performance indicators requested by the Health Inspectorate.

get the "pure number" and "objectify" (Desrosières 1998) for the figures collected on the blood sugar value of diabetes patients treated in the hospital. The collection of the data triggered questions from the project team with respect to figures from previous years and the national average. Did these questions result in learning or improving within the hospital as suggested by the endeavor of the Inspectorate?

LOCATING THE WORTH OF THE BLOOD SUGAR VALUE

Various interviewees referred to indicators as a driver to act and to do "something." Members of the quality management team, doctors, members of the IT Department, as well as representatives of the Inspectorate, the Ministry of Health, the Federation of Patients and Consumer Organisations in the Netherlands (*Nederlands Patiënten Consumenten Federatie*—NPCF) all agreed that the indicators do "something." But what? Does this "something" equate to accountability and improvement in health care practices? These questions, we claim, can only be answered in a situated way and by conceptualising indicators as a process. In the following sections we elaborate on the investments and efforts in composing a network of informants and accountabilities for one specific indicator: the indicator on the average HbA1c value of diabetes mellitus patients.

Diabetes mellitus is increasingly becoming an issue in the so-called Western world as one of the prominent chronic diseases. Diabetes is the most significant cause of adult blindness in the non–elderly and the leading cause of non–traumatic amputation in adults. People who suffer from diabetes mellitus have a too high blood glucose level due to a diminished production of insulin (in type I) or resistance to its effects (in type II). The long-term consequences of this imbalance and its treatments can cause many complications. Juggling and balancing the blood sugar value by measuring it regularly becomes one of the core activities of diabetes patients and their doctors or nurses. Being such a prominent health care issue, the inclusion of a set of indicators on diabetes care in the blue booklet of the Inspectorate seems justified.

The blue booklet emphasizes that "good" diabetes care involves multiple organizations, persons, moments, and locations. A hospital has therefore "good" care in place if it takes the variety and plurality of care into account by facilitating integrated care trajectories. Another indicator used by the Inspectorate to trace "good" diabetes care is the blood glucose value, measured as the HbA1c value. Diabetes patients with a "good" adjustment of the HbA1c value have a lower risk of complications and according to the booklet, the value is easily measured and calculated including the following components:

- sum of all HbA1c values of diabetes patients,
- amount of HbA1c measurements done,

- average of HbA1c value per measurement,
- average of measurements per patient and year,
- norm value used by the hospital lab, and
- database (national, regional, or own) used by the hospital to retrieve data (IGZ, NVZ, et al. 2005).

The responsibility to collect the information on the diabetes indicators of interest to our analyses was eventually assigned to the head of the Department of Internal Medicine, who describes her responsibilities in administrative and managerial terms, i.e., not being involved in health care practices herself. All she knows about diabetes care is based on theoretical studies or based on the information she receives as head of the diabetes outpatient clinic.

Her first difficulty in fulfilling her new role as the problem owner of the diabetes indicators was that the requested data were not traceable in the existing registration systems without translation and "articulation work" (Strauss et al. 1997). The numbers initially retrieved by the IT Department were too high according to the current professional standard[12] for a healthy average (around 7 mmol/l)[13] and compared to the numbers from previous years: 9,00 mmol/l for diabetes type I and 8,3 mmol/l for diabetes type II.[14] Supported by the quality management team, the problem owner tried to figure out how to provide the requested information on blood sugar values. The IT Department did considerable detecting work to understand how the existing registration systems could be combined in running a query that would provide the requested information—without success.

> You can take things from different packages. In order to see which patients have diabetes, you can have a look in the National Medical Registration (LMR). But there you can only trace those who have been admitted to the hospital [not those who come to the outpatient clinic] . . . and this is coded by, well, the people in the basement. How reliable this is, I don't know. (. . .) [Interviewer: Who is sitting in the basement?] (. . .) We have special coders. They get the status and they code, they indicate and define the codes. [Interviewer: Are they educated and qualified to do so? Do they have a background in health care?] No, maybe some, but in principle not. They did a course in medical terminology, some sort of basic course in disease, therefore, the reliability, I don't know.[15]

To start with, already the differentiation of diabetes type I and type II was difficult to trace in the digital information systems in the hospital. Only patients who have been admitted to the hospital are registered and differentiated as diabetes type I or II patients in the hospital-wide systems. The regular diabetes patients in the outpatient clinics are registered only locally via a locally conceptualized registration system. The conceptualization of the electronic diabetes outpatient registration system was based on the needs of the nurses practicing in the Diabetes Outpatient

Clinic, with its own logics and values. The differentiation in type I and II diabetes patients was done differently and was not easily compatible with the data traced via the hospital-wide information system. From the perspective of the problem owner there was no need to register the type of diabetes and

> (...) in none of the registration systems is a difference made between Type I or II. In the DRG registration they are all thrown on one heap. Only the National Medical Registration (*Landeijke Medesche Registratie* - LMR) does it; which in its turn is done by the coders who are literally sitting in the basement with us. I wonder how is it done in the [rest of the] country? How do they know what is Type I or Type II? Do the doctors keep up with it? Or have they appointed an administrative employee to do so? You know, I can organise it and ask the nurses in the outpatient clinic to do so. But at the same time, next year it will become clear that all the hospitals have difficulties in coming up with a valid differentiation. And this after we have just organised everything to make it happen.[16]

Several meetings with the quality management team, IT Department personnel, and problem owner were organized to clarify the issue of the rather high HbA1c value of 9,00 mmol/l for type I diabetes patients.[17] It was not the first meeting that had taken place to "sort things out" (Bowker and Star 1999) in the hospital around the IGZ indicators. The contribution of the IT Department in "sorting things out" was major. The IT team often found itself in a situation of creatively connecting queries from different information systems, based on the logics of these systems, but in absence of the medical background. They did what made sense to them, not knowing that diabetes type I and II were inconsistently registered in the various systems. After going through the data with the problem owner and after her specifying which data was and was not consistent, they decided to leave out the data on type I completely, although data on both indicators are requested by the IGZ. The figure the problem owner felt was representative and that was submitted in the end for type II was retrieved from the LMR and based on the work of the coders. It basically only reflected patients who were admitted to the hospital and not the outpatient clinic visitors. What does this story say about accountabilities and possible improvement in the quality of care?

> The idea is nice, you know, the idea, we want to compare hospitals and the care they deliver. But it is rather recalcitrant and difficult and the consequence is that the numbers you get, do not say everything (...). [T]here are so many things that distort this measurement (...) The patients, but (...) assuming that we have organised it very well with the general practitioners; that is, that all people that are well adjusted just stay with the family doctor, the general practitioner, where they can be taken care of; there is no need for them to come to the hospital.

> The consequence is that (. . .) we will only get the problematic cases; and of course the value will increase! Because those that are well adjusted will go out of our system. (. . .) [T]hose hospitals that are not so well organised, and are doing it all themselves, will have a lower value than those that are well coordinated and organised. (. . .) Our work is so well integrated that the standard range of the HbA1C value does not work anymore as an indicator of "good diabetes care." The "good" patients do not come to the hospital anymore, but they are examined and checked by their GPs.

This example shows that the issue is more about how to set up a smart query with the existing registration systems than actually mirroring the diabetes care taking place in the hospital. The evidence-based, anonymous, and unlocated knowledge of the fact that a blood sugar value around 7 is a healthy value, is of little meaning without the situation around the value 7 or the story of the number 7.[18] It sketches, however, one possible translation of the initial endeavor of "making hospitals accountable for their health practices": accountability remains a matter of working out intelligent and smart queries in existing data systems and, at best, results in designing a different registration system.

When the focus is only on the figures, then the stories and work behind them are omitted. Interestingly, the diabetes nurse in the outpatient clinic was not even aware that there was a project group on indicators that was collecting the aforementioned data. In our observations of the consultations at the outpatient clinics, one crucial asset of the indicator on the average blood sugar value is only enacted when the value is related to the stories around the figure of a patient. The indicator gets meaning and "comes into life" through the work of the diabetes nurse who relates the measurements to the specific situation and stories of a patient. For a lady whose blood sugar values generally vary strongly, an average of around 10 can be a "good sign." The diabetes nurse was not so much interested in the absolute number of the average of that lady, but much more in the range of the variations. Focusing only on the blood sugar value as an average outcome implies that the blood sugar value changes from having a relative value that allows the patient and the nurse to construct a story around the patient and the disease into an absolute value in itself that needs to be reached under all circumstances.

The initial focus on the structure and the process of diabetes care entailed improvements translated as an integrated care trajectory and as a sign of having "good diabetes care" in place. The introduction of the clinical outcome indicator on the HbA1C value jeopardized the previous efforts and attainments. A possible consequence might be the depreciation of the worth of coordinating care by discussing HbA1C values with patients and exchanging it for the worth of compliance and having a good outcome.

What our analyses show is that the successes and failures of performance indicators are not reflected in the numbers alone. Only in relation to the loci and situations where the data are ascribed meaning and only by relating the figures to the stories around them do moments of success and failure become traceable. Moreover, there are many moments of successes and failures that are not captured by the figures at all: the collaboration between the management team, the problem owner, and the IT team; the lack of a consistent hospital-wide information system. It all depends on the question: success or failure for whom or what?

Locating the Worth of the Indicator on Acute Myocardial Infarct

Our second example refers to data on the mortality rate for patients diagnosed with heart attacks. Acute myocardial infarction (AMI), commonly known as heart attack, is the leading cause of death for both men and women worldwide. AMI occurs when the blood supply to part of the heart is interrupted causing some heart cells to die. The data requested for two indicators on AMI is related to patient's mortality rate within a specific period of time after having been diagnosed with AMI:

(a) patients dying within thirty days after admission or patients dying while being at the hospital based on the diagnosis of AMI;
(b) patients dying within three months after admission based on the diagnosis of AMI regardless of the actual causation of death.

Furthermore, the requested data is split into two categories of patients: The first category comprises patients younger than sixty-five years of age, the second category patients sixty-five years of age or older (IGZ, NVZ, et al. 2005).

AMI outcomes as presented by the project group on indicators to the medical staff before submitting the data to the Health Inspectorate were strongly questioned. One month after the presentation of the results and six days before the June 1 deadline, the medical staff sent a letter to the hospital's board of directors indicating that they did not support the results presented to them and that they would need to check the data properly before it could be submitted. The medical staff stated explicitly that they would prefer missing the deadline rather than risk submitting incorrect data. The letter was a formal manifestation of the medical staff's "dissidence" (Callon 1986) prioritizing the value of valid data to the value of obeying the rule enacted by the Health Inspectorate. The quality management team felt bypassed as they had not been informed of the letter beforehand. They interpreted the letter as a devaluation of their work on facilitating learning and improving within the hospital and as an assault against all their efforts to make indicators work. However, delaying the submission of the data was no option for the board of directors who insisted on combining the value of validity, respecting the authority of the Inspectorate, and investing in the worth of reputation at stake. Not submitting the data on time would have been immediately translated into zeros in the ranking prepared by the Dutch magazines resulting in the lowest possible position for the hospital. But what was wrong with the figures according to the cardiologist and head of the medical staff?

After disapproving the figures presented to the medical staff one of the cardiologists went through the data himself. He requested the respective patient records and proved that the outcomes detected via the National Medical Registration database [*Landelijke Medische Registratie* (LMR)] were not correct. The figures of the AMI indicator as presented by the quality management team to the medical staff were based on the interpretation work of the coders, who process the information contained in a patient record into the LMR once the patient is discharged.[19] However, the cardiologist showed that in some of the cases the diagnosis was coded as an AMI although the cause of death was not clear. Similarly, he showed that in some cases patient records had been coded as AMI although patients had been admitted and treated for another disease. Through this correction work the mortality rates improved dramatically. The mortality rate thirty days after cardiac treatment decreased by about 6 percent for patients under sixty-five years of age and about 10 percent for patients sixty-five years of age or older.

How to explain the differences in coding? The cardiologist responsible for the figures on AMI considered this a relatively complex diagnosis and elaborated on the complexity of diagnosing and treating AMI. If AMI is diagnosed immediately after someone experiences symptoms, the treatment can be immediate and the accountabilities of these treatments of the patient would be clear. But if AMI is diagnosed after a week or more, the chances for successful treatment diminish. Similarly the accountabilities of the treatment are not as easily "located" anymore. Is it the responsibility of the specialist in the hospital if the treatment is not successful or perhaps a general practitioner who missed the initial diagnosis? Who or what will be able to tell? According to the cardiologist the figures on AMI have therefore little to no explanatory power on the performances of care practices.

Circumstances external to the hospital also influence indicator outcomes but are not discernable in the data. One example the cardiologist mentioned was the extremely low mortality rate in Sweden after operating aneurysms in the stomach. In Sweden it seemed that the percentage of patients with aneurysms in their stomachs surviving the operation was 80 percent higher than in other countries. The reason, however, was not the higher quality in diagnosing and treating patients with an aneurysm. The explanation for this low mortality rate was that the distances to the hospitals were comparably big in Sweden, so that the patients with a higher mortality risk had already died on their way to hospital.[20]

In the case of the AMI indicator in the hospital, the Inspectorate succeeded eventually in getting a figure after quite some extra work and betrayals within the hospital. But what does this figure actually show? And what do this extra work and the clashes mean? How do they relate to the figures? Is the indicator on the mortality rate of patients diagnosed with AMI an example of making care more accountable? Who is actually held accountable for what? And what about improvement? Who is learning and improving here? And what is learnt and improved?

PERFORMANCE INDICATORS—PERFORMING "EVERYBODY'S INTEREST"?

Do performance indicators make health care more transparent as suggested by the Dutch Council on Public Health and Healthcare in the quote at the beginning of this chapter? And is their use in "everybody's interest"?

As we stated, the literature on performance indicators in health care can be divided into two traditions: the believers and the promises of success on the one hand, and skeptics and the claims of failures on the other. We started our endeavors differently. Following the performance indicators developed by the Dutch Health Inspectorate in 2003 "in action" (Latour 1987), we problematized the notion of unambiguous information that is reproduced in both traditions and their evaluation of indicators. We did so by tracing the "invisible" (Star and Strauss 1999) work and investments that are undertaken to tackle the multiple ambiguities indicators actually bring along. By conceptualising indicators as processes and by studying what was "actually going on" when the indicators entered the world of a hospital, we traced how successes or failures materialize in practices that are not (only) captured by numbers.

We discerned two phases: Firstly, the primary work of setting up a project group and network of informants—a "primary" infrastructure—who took up the task of translating the order given by the Inspectorate to the hospitals. The project group in the hospital succeeded in designing a network of informants, but failed to make the informants, the so-called problem owners, accountable for their problem, the indicators. On the search for "locating accountabilities" (Suchman 2002) ownership was attributed but not accepted. Interestingly, there was a clear distinction made by the problem owners between being responsible for "merely" gathering the requested data and others being accountable for the outcomes thereby diffracting accountabilities.

Secondly, we closely examined two specific indicators: that on the blood sugar value of diabetes patients and that on the mortality rates of myocardial infraction patients. We explored how possible successes or failures of delivering data instantiate when zooming in on the "secondary work" that has been invested in submitting the final data.

As we show, there is neither *one* transparency for all nor is their *one* use or value of performance indicators at stake. Assuming one transparency for all and assuming something that needs to be discovered, investigated, and evaluated "out there"—that is, beyond the very methods, technologies, ideas, and values at stake when evaluating health care practices—obstructs our seeing the (sometimes surprising) successes that might exist when performing performance measurement.

Conceptualising indicators as processes and "draw[ing] on the theoretical vocabulary" (Pols 2004) of Boltanski and Thévenot (2006) by thinking in terms of investments is a first step, we suggest, toward opening up the dominant discourses on performance indicators as alluded to in the beginning of this chapter. Numbers have their own value, but they have to be seen in relation to the ambiguities of the work and investments they entail. Mapping the investments and the values at stake in the work on indicators allowed us to extend the loci of successes or failures beyond the number as such. From our analyses, we argue for the importance of developing discourses—in the Foucauldian sense, languages and materialities together—not only to reflect and talk about which performance and knowledge we want to invest in, but also discourses that allow us to design evaluations that take into account that goals shift in the course of tinkering toward them (Zuiderent-Jerak et al. 2009).

Our analyses showed that the promises of the use and value of indicators for making health care transparent and accountable and for facilitating improvement are not taking place as and where they are promised or expected to be. At the same time our analyses show that we cannot talk about failure or decoupling or gaming in a general sense either. Indicators do have consequences: an emerging network of informants, meetings between doctors and nurses who otherwise have limited communication about patient care, the betrayals of the doctors, the discovery of the people in the cellar, new queries in the information system, reflections on the values of the blood sugar value, and so on. We promised in the introduction to specify the successes and failures of the indicators in action, but we refused—hopefully successfully—to come up with an absolute judgment of the use and value of performance indicators in health care. In this sense we have attempted to productively engage in theoretical reflections in a practically oriented environment, where terms are not so much discussed in terms of underlying or implicit values but used (Mol 2008: 103 N 8) for "everyone's interest."

ACKNOWLEDGMENT

This chapter would have not been possible without the generous support of the quality management team and the staff in our hospital of analyses. Next to that our chapter was part of a collaboration with Steve Harrison, Christopher Pollitt, and George Dowswell on an ESRC-funded project "Performance Indicators in Health Care: A Comparative Anglo-Dutch Project" (RES-166-25-0051). The final version of this chapter benefited from feedback by Jessica Mesman, Teun Zuiderent-Jerak, and Samantha Adams.

NOTES

1. All Dutch quotes have been translated by the authors. See http://www.rvz.net/cgi-bin/nieuws.pl?niew_srcID=159 (accessed July 28, 2008).
2. In the following we use "performance indicators" and "indicators" as synonyms.
3. See www.igz.nl/uk (accessed February 15, 2009).
4. This layered supervision is divided into three phases: Phase 1 involves the collection of data in the form of indicators; Phase 2 is contingent on Phase 1 and entails gathering supplemental data; Phase 3 entails intervention in the form of on-the-spot inspecting and investigating if Phase 2 provides reason to do so.
5. We refer here to the A5-sized hard-copy edition by the Health Inspectorate that is sent to the hospitals every year since 2003, containing information on the "Basic Set of Indicators".
6. Interview with quality management on January 27, 2007.
7. Ibid.
8. Ibid.
9. Ibid.
10. Historically, Dutch medical specialists have been self-employed within hospitals and have a strong economic and organizational autonomy and efforts at integrating medical staff differently within the internal governance arrangements of the hospital have largely failed so far (Scholten and Grinten 2002).
11. Interview with quality management on January 27, 2007.
12. The standard changes according to the development in societies, technologies, and sciences (Mol 2002).
13. As specified in the professional online handbook for doctors, "Compass for Diagnosing" (diagnotisch kompas 2003), and as used by the quality management team as point of reference.
14. Following the information of a standard in clinical diagnostics in the Netherlands, the "Compass for Diagnosing," the values of orientations for the blood sugar average in the format of the HbA1c-value for non-diabetics is < 4–6 mmol/l. For diabetics anything < 7 mmol/l is considered to be a well-adjusted value, a value between 7 and 8 mmol/l is considered to be fairly well-adjusted and anything that is higher than 8 mmol/l is considered a bad adjustment.
15. Interview with problem owner on diabetes on July 17, 2007.
16. Ibid.
17. The average of diabetes type I in the hospital of our analyses in 2005 has been 7.4 mmol/l and the national average in comparison in 2004: 7.9 and 8.3 mmol/l. In 2005 no information had been available on the average for diabetes type II in the hospital of our analyses and the national average of diabetes type II in 2004 indicated 7.7 mmol/l.
18. Interview with problem owner on diabetes on July 17, 2007.
19. The data of the National Medical Registration database is regularly forwarded to Prismant, an organization providing national statistics on hospital care.
20. See http://www.ncbi.nlm.nih.gov/pubmed/1555072?dopt=Abstract (accessed April 18, 2010).

REFERENCES

Berg, M., et al. (2005) "Feasibility First: Developing Public Performance Indicators on Patient Safety and Clinical Effectiveness for Dutch Hospitals," *Health Policy* 75: 59–73.
Boltanski, L., and L. Thévenot (2006) *On Justification—Economies of Worth.* Princeton, NJ: Princeton University Press.

Bowker, G.C., and S. L. Star (1999) *Sorting Things Out: Classifications and Its Consequences.* Cambridge, MA: MIT Press.

Callon, M. (1986) "Some Elements of a Sociology of Translation: Domestication of the Scallops and the Fishermen of St. Brieuc Bay," in *Power, Action and Belief: A New Sociology of Knowledge?,* ed. Law, 196–233. Boston: Routledge and Kegan Paul.

Corbin, J.M. and A. Strauss (1993) "The Articulation of Work Through Interaction," *The Sociological Quarterly* 34(1): 71–83.

Desrosières, A. (1998) *The Politics of Large Numbers—A History of Statistical Reasoning,* Camridge, MA: Harvard University Press.

Freeman, T. (2002) "Using Performance Indicators to Improve Health Care Quality in the Public Sector: A review of the Literature," *Health Services Management Research* (15): 126–137.

Haraway, D.J. (1988) "Situated Knowledges: The Science Question in Feminism and the Privilege of Partial Perspective," *Feminist Studies* 14: 575–599.

———. (1991) *Simians, Cyborgs and Women; The Reinvention of Nature.* New York: Routledge.

IGZ, NVZ, et al. (2005). *Prestatie-indicatoren ziekenhuizen—Basisset 2006.* Utrecht.

———. (2007). *Prestatie-indicatoren ziekenhuizen—Basisset 2008.* Utrecht.

———. (2003). *Basisset prestatie-indicatoren ziekenhuizen 2003.* Utrecht.

Latour, B. (1986) "The Powers of Association," in *Power, Action, and Belief—A New Sociology of Knowledge?,* ed. Law, 264–280. London: Routledge.

———. (1987). *Science in Action: How to Follow Scientists and Engineers through Society.* Cambridge, MA: Harvard University Press.

Law, J. (1986) *Power, Action and Belief—A New Sociology of Knowledge?* Boston, MA: Routledge.

Mol, A. (2002) *The Body Multiple: Ontology in Medical Practice.* Durham, NC: Duke University Press.

———. (2008). *The Logic of Care: Health and the Problem with Patient Choice,* New York: Routledge.

Pols, J. (2004) "Good Care—Enacting a Complex Ideal in Long-term Psychiatry," PhD dissertation, Twente: University of Twente.

Power, M. (1997) *The Audit Society. Rituals of Verification.* Oxford, UK: Oxford University Press.

Rekenkamer, Algemene (1999) Inspectie voor de Gezondheidszorg. The Hague: Algemene Rekenkamer.

Scholten, G. R. M., and T. E. D. Grinten (2002) "Integrating Medical Specialists and Hospitals: The Growing Relevance of Collective Organisation of Medical Specialists for Dutch Hospital Governance," *Health Policy* 62(2): 131–139.

Star, S.L. and A. Strauss (1999) "Layers of Silence, Arenas of Voice: The Ecology of Visible and Invisible work," *Computer Supported Cooperative Work* 8: 9–30.

Strathern, M., ed. (2000) *Audit Cultures—Anthropological Studies in Accountability, Ethics and the Academy.* New York: Routledge.

Strauss, A., et al. (1997) *Social Organization of Medical Work.* London: Transaction Publishers.

Suchman, L. (1995) "Making Work Visible," *Communications of the ACM* 38(9): 56–64.

———. (2002) "Located Accountabilities in Technology Production," *Scandinavian Journal of Information Systems* 14(2): 91–105.

Wiener, C.L. (2000) *The Elusive Quest: Accountability in Hospitals.* New York: Aldine de Gruyter.

Zuiderent-Jerak, T., M. Strating, et al. (2009), "Sociological Refigurations of Patient Safety: Ontologies of Improvement and 'Acting With' Quality Collaboratives in Healthcare," *Social Science & Medicine* 69(12): 1713–1721.

Part IV

Reporting and Acts of Resistance

13 Constructing Mirrors, Constructing Patients—with High Stakes Statistics

Gunhild Tøndel

The situated, contingent character of medical decision making has been thoroughly documented (see, for instance, Posner et al. 1995, Geest and Finkler 2004), so much so that it might begin to seem a banal truth. And yet, if we observe everyday medical work from a distance, it can also appear an unproblematic process where determinative patient conditions are identified by procedures based in scientific knowledge, resulting in uncontroversial diagnoses and treatment options (Hughes 1977: 128). These truths live in parallel, simultaneous "universes." This means that—dependent on perspectives, audiences, and uses—a diagnosis can embody different characteristics. In one such nexus we might find diagnoses as tentative hypotheses. In another we find a system of governance where diagnoses are viewed as conclusive truths translatable into budgetary units. These two (and more) parallel universes coexist—both of them legitimate, at times as if mutually exclusive, at times explicitly or implicitly acknowledging one another, thereby intermingling, interfering in one another's self-creation. This chapter will explore the creation of diagnosis data where such universes meet.

When we observe the construction of data—irrespective what kind of data—the construction is choreographed by technical and technological solutions, "tools" for efficient and valid data collection—at least defined as efficient and valid by system designers and end product users. One such tool is *Diagnosis Related Groups* (DRG), a classification system which groups in-hospitalized patients on the basis of primary diagnoses, secondary diagnoses, and complications, surgical procedures, age, and discharge routines (Aas et al. 1989). It is a management-oriented case mix system which aggregates upwards (away from the details of medical practice; Bloomfield 1991). In Norway, DRG has since 1997 figured as a payment system in activity-based financing (ABF) of four regional national health enterprises.[1] ABF effectuates the financing of specialist health services with a 60 percent basis allocation and a 40 percent activity-based allocation (Norwegian Directorate of Health 2009). The individual DRG groups, about 500 altogether, are meant to mirror the average cost of treatment in that group. The primary diagnosis is key to the grouping outcome.

This interactive ABF-DRG arrangement gives diagnosis decision making a high stake identity, since the registered diagnosis codes affect the financing of the service which the reporter (in Norway, primarily the treating physician) works within. In this chapter I draw upon participant observations of and qualitative interviews with physicians about medical diagnoses and DRG statistics in-the-making to explore how high stakes numbers interact with what and whom they quantify. The data collection (or, more precisely and reflexively, the data *construction*) took place in a Norwegian internal medicine ward during 2005. I focus on the (re-)shaping of medical diagnosis practice which took place during physicians' trivial, fleetingly brief yet high stakes work of translating patients' conditions into quantifiable units: diagnoses and diagnosis codes. How is this translation undertaken? How does this affect the patient? What responses appear "at below," among producers of DRG statistics, in this case physicians? And how do those responses affect the diagnoses produced? In order to address these questions, I begin with a brief contextualization of the DRG system. In the second section of this chapter I enter the medical ward and recount a medical diagnosis decision and formalization process as directed toward and simultaneously against the practical production of DRG.

DIAGNOSIS RELATED GROUPS: CLASSIFYING, COUNTING HOSPITAL ACTIVITY, REIMBURSING PATIENT STAYS AND TREATMENT COSTS, GENERATING A CONTROVERSY

DRG plays a major role in the transformation of the hospital sector in countries that have implemented it (Torjesen and Byrkjeflot 2007). Still, the system seems to function mostly behind the public scenes, as a boring body of numbers, codes, technical terms, and texts. As Bowker and Star (1999) point out, classification systems are serious matters indeed, full of consequence and conflict however dry and formal they may appear at the surface. They are also pervasive. DRG is only one of many classification systems physicians confront during everyday work. In the Norwegian health sector we find at least twenty-one medical and administrative code systems (KITH 2002)—including the well-known International Classification of Diseases, but not counting local code systems such as those activated through research projects.

DRG originated in the United States during the 1970s as a technological concept at Yale University, where a group of researchers developed a classification system aimed at measuring hospitals' productivity and efficiency as well as managing resource usage (Torjesen and Byrkjeflot 2007: 7; Jørgenvåg and Hope 2005; see also Slåttebrekk 1990). The designers' point of departure was an industrial system for among other things monitoring quality and cost (Torjesen and Byrkjeflot 2007, Jørgenvåg and Hope 2005).

In the 1970s and 1980s, DRG was portrayed as a technology for the future. Aas (1985) writes in a literature review of existing DRG literature that DRG renders it possible to slow down budget growth in the hospital sector. DRG was quickly assigned many further roles—producing better working conditions for professionals and administrators, producing data for studies of patient compositions, planning hospital personnel compositions, and forming a common language for administrative and medical personnel (Aas 1985: 55, 70)—in other words, a multitasking system.

Despite challenges, DRG is still defined as an outstanding innovation in health planning and politics (Torjesen and Byrkjeflot 2007). And yes, if we measure its success by the number of countries using it, DRG does appear as an outstanding technology in its ability to diffuse. Today, it exists in several different public editions adjusted in accordance with the formal organization of the given country's health system. A quick count through the Advanced Medical Technology Association's (2009) listing of International Diagnosis Related Groups ends in twenty-four industrialized countries. Diagnose that.

One vector of DRG which contributes to turning hospital patients' diagnoses into high stakes units for their reporters—beyond them being personal high stake representations for the patients—is the approach to hospital productivity measurement inscribed into the system's technical design. In his review, Aas (1985) concluded that DRG opened up the possibility of measuring productivity and efficiency and simultaneously considering the hospitals' "patient composition": From measuring hospitals' output defined as the goods and services which a hospital gives to the patients—production, product, and performance—DRG defined hospitals' output as "the discharged patient." This patient status, characterized by patient-related and resource use factors, was seen as indicating hospitals' work load and progress (Aas 1985). However, this high stakes relation is not unproblematic. In Norway, the relation between formal patient representations and budget composition has turned DRG into a controversial object. Critics have questioned physicians' capacity to diagnose patients without bringing economic considerations into the diagnosis-making and registration process. Headlines in Norwegian media have fronted cases such as:

"The hunt for profitable diagnoses" (Hafstad 2003)

"The patients believe that the advice is professionally based. They may receive an offer of operation because it is more profitable for the hospital . . . " (Schjerve 2003)

"The medical summary—invoice or medical document?" (Glomsaker 2005)

(my translations from Norwegian).

Some of the tensions surrounding DRG can be seen to reflect the fact that there are simply different ways of classifying diseases and medical procedures, and a variety of ways of inscribing and calculating costs (Bloomfield 1991). In addition, some physicians may have acted too enthusiastically in a situation where the high stakes relation between diagnosis and procedure reports and reimbursement rates became established and transparent. However, in accounting for the competence of physicians, we tend to focus on their intellectual qualities, and perceive "medical decision making" as a mental process, located in the individual physician's brain (Berg 1996: 504). This also accounts for much of the critique of physicians' medical integrity. These potential situations are not interesting in this case. The hue and cry over individual physicians' purported medical-administrative report work failures (I refrain from passing judgment here) communicates rather with the questions of good or bad, correct or wrong statistics. Let us instead take a closer look at DRG's inscriptions of and into medical diagnosis practice.

A Formal Modeling: What's in a Statistical Code Production Process?

The Norwegian department and directorate responsible for health issues have issued several articulations regarding how physicians are to manage medical diagnosis categorization and registration as a quite straightforward procedure. Physicians are instructed to categorize patients' conditions and the medical interventions undertaken "into diagnosis and procedure codes on the basis of the documented information about the patient" (Jørgenvåg and Hope 2005: 17). To make a "correct choice of principal diagnosis is important with regard to statistics, correct steering data, research, DRG grouping and ABF reimbursement" (Norwegian Directorate of Health 2009: 37). A correct, conclusive diagnosis is assumed to be extractable from the available patient documentation, and the physician's choice has been given a direct effect upon important nonmedical macro-conditions such as budgets.

The DRG instructions model the diagnosis' course of production and transportation as a process consisting of three consecutive steps (Tøndel 2007):

1. The physician decides "what this patient suffers from."
2. The physician finds the relevant diagnosis term in the World Health Organization's (WHO) *International Classification of Diseases* (ICD) and registers it.[2]
3. This term is translated into DRG groups which enter reporting and budgetary processes.

Notice the temporal and social ordering of the steps. In this command, step 1 shall be accomplished before step 2, and be precisely reflected by step 2, which is seen as merely mirroring the more important and professionally qualified step 1. Step 2 shall be independent of and completely untouched by step 3, which can be translated into DRG grouping and the DRG system in its entirety, including the system's role in budgetary frames of references. Step 2 can be relegated to lower level medical staff. Step 3 can even be accomplished by secretarial staff. Already at step 2, the process is intended to be separated from the patient. Once the diagnosis is made (step 1) the patient should not be affected by any further coding (steps 2 and 3). DRG are intended to describe hospital activity without changing either the situational construction of the diagnosis or the repetitive production aspects of the underlying decision-making process. The clear-cut boundary between diagnosis making and diagnosis coding becomes a necessity for the system's functionality, since the statistics are presented as a product of reliable medical documentation unaffected by economic interests.

INTERACTING INSTITUTIONS

Following DRG's description of the diagnosis decision as a three step process, I will portray this process through three sections that respectively illustrate (a) that DRG affects the production and formalization of itself and hence what it counts in the first place, (b) how DRG may affect the counted one, and (c) different ways physicians face DRG and its' report instructions in order to do their work and simultaneously control the interaction between medicine and DRG. Introductorily, the next vignette illustrates contextual conditions of the diagnosis decision and registration which both diagnoses and DRG statistics depend upon as products and "outputs."

Production Context

> The physician is apparently totally alone when she formalizes the medical diagnoses. The patient is apparently not in the same room. "Well, to make a diagnosis is to make a diagnosis. You do that by summing up all information, as clinical and laboratory findings." The physician uses her index finger to draw an imaginary plan on the office table as she explains the process. "I work with a list that I construct, containing what is important to include, about the patient, and what is done, what others think is important to include" (field notes including physician comments, during an observation in field).

The physician is "apparently alone" according to my notes, and yet in her comments, she is clearly aware of multiple audiences, "what others think is important." She knows what product she shall produce: a formal medical diagnosis translatable into a statistical code. However, she does not yet know for sure what diagnosis term she will end up with. As she conducts her almost invisible work of producing a term and a code, she is aware of my presence, and that of my possible future readers, even though we are only marginally relevant professionally. As we shall see in the following pages, she directs her activity toward formal standards set in advance for how to perform the documentation work. The standards are located within the medical ward due to interests of, among others, state and administration representatives such as the Directorate of Health and the local hospital management, to get periodical overviews of the "activity"[3] undertaken at the ward. The patients have an acknowledged interest in good quality reports. The physician knows that her decisions during the act of registration may be taken to generate or defend forthcoming acts upon the patient. Her translation from something not yet formally mapped and controlled into a formal representation of the patient is not only a declaration of a current working hypothesis; it may affect future hypotheses and actions. The diagnoses, whether formal or informal, may at some point in time result in direct medical and therapeutic interventions into the patient's flesh, bones, nerves, and muscles; social interventions into the patient's relationship to insurance companies and social security services; and personal interventions into the patient's self-biography.

Due to the administrative organization of diagnosis outputs, collected and used as information about resource use, the diagnosis registration situation generates a paradox: The national DRG system groups the information she gives away about her patient and how she and her colleagues have treated the same patient. The resulting code takes on a life of its own, operating independently of and external to the physician's medical awareness, at the same time as it is made dependent upon her awareness. The system ascribes a monetary sum to the categories she reports which acts back on her workplace through feedback budgeting processes. The physician knows that if she does not report as her employer has prescribed, she may be suspected of sloppy work, even disloyalty. Her employer has underscored the importance of including secondary diagnoses, which increase the DRG values. But she is also aware of the media where accusations of cheating have been prominent. She tries therefore to project, make, and register a medical diagnosis decision, without involving other institutions.

At this point in time, the act of deciding a medical diagnosis seems characterized by control (the information is already produced, she only needs to sum it up), few interactors (she is the only person in the room, although aware of others), emotionally balanced (none are confronted here and now, face to face, with the diagnosis she registers), and of very brief duration (although the production of the patient's medical identity may have spanned over a long time period due to testing and treatment). By only observing this moment, one may be led to believe that the making of a diagnosis is decided—yes, nearly determined—by the reporting physician and that there are no audience or discerners for her act. It may seem as if she dictates the diagnostic truth about the patient through her categorization of the "real" cause of the patient's in-hospital stay. In sum, the diagnosis seems to be determined by conditions *outside of* and *before* her existence, which she slavishly follows and encodes. This kind of interpretation follows a means-end scheme. It is partly a lengthening of the ivory tower idea of medicine as instrumental and truth generating, with "correspondence" as the operative criteria, not "usefulness" (see Strauss 1993).

An alternative interpretation is rather that the process toward a crystallization of one term that describes the supposedly necessary medical experiences about, with, and on the patient, is embedded in the physician's coordinating movements during the formalization of the diagnosis. This interpretation is the one I follow. As described in the vignette at the beginning of this section, the physician connects many communicative spots through her action: her office, the clinic, the labs, myriad collegial venues, and any others who influence her understanding of the practical act of thinking, whoever they may be. These movements, supplemented with her theories of what their interaction produce and shape, shape the registration process. This makes a diagnosis into a product which never gets its umbilical cord cut, and a structured distributing and collecting device (Berg 1996: 510). The diagnosis' connection to still ongoing and upcoming processes which interact with its production context may more or less legitimately define its characteristics in broad terms, whether their scope is ideologically internal or external, medical, or administrative, political, or economic.

Included in this decision-making activity, we find a guardian of closure due to physicians' commitment to the natural scientific worldview: They must manage Occam's razor to produce explanations of the patients' condition(s) that include the most, but not too many, of the variable and sometimes diverging accessible reported observations about the patients. Occam's razor is a special case of the principle of verifiability—that no theory is meaningful unless it leads to some observable consequences (Feuer 1957: 112). The best theory is in principle the one that holds the simplest explanation of most of the observations, at the same time as it points out further actions for exploration, treatment, and so on. The physician will be left with such a theory when she shaves away any observations—transformed into and temporarily termed as information—not necessary to consider in the very moment her actions have brought out the product: the best theory of what she assumes is medically wrong (or not!) with the patient.

Many conditions can speak to Occam and through this affect or even stop the physician's management of the diagnosis decision. In the previous vignette, several internal conditions bear rather immediately on the initiation and projection of the action (Strauss 1993: 78), among others represented by the physician's aim of making a medical diagnosis decision: her reasoning that to make a diagnosis is equivalent to summarizing all existing information which she has available—and that others think is important to include in the decision—is important for her. She is forced into a juggling situation, where she must manage several potentially incompatible requests of representation at the same time. She is stuck in a process where she is required to translate work defined as already undertaken into a term and a code directed at multiple audiences, simultaneously as this work actually continues through her management. It is not ready made. It is not ended. She ends it, for a while. And during this process, local worries, work routines, contingencies, certainty, uncertainty, and so on, get melted into, refracted, modified, and codified into a socially accepted truth in the shape of an output. In this situation, questions about what went (or not) into the products take unavoidable steps forward from their background position to the explicit foreground, like undiscovered yet visible threats on the chess board which the player has overlooked, ignored, or underestimated during the play. How can we trust that the outcomes describe something that has happened? How do the outcomes interact with the still ongoing production of the diagnosis? An entire mapping of conditions that affect this work is not the aim here. Such a mapping would include such conditions as the medical technology, the course of the illness itself, and the patient (Strauss et al. 1985). My point is that at this moment we are apparently not yet introduced to the DRG system. Or are we? What conditions bear materially on her performance of this plan or list construction? What does she include when considering what others might think is important to include? Who are these others, what are their purposes, and how do they intervene with her performance of the medical diagnosis decision? To illustrate these questions we must enter another phase in the processual production of the diagnosis.

Simultaneously Producing Statistics and Medical Patient Representations

The Norwegian Directorate of Health writes (2009: 12, my translation and emphases): "The *actual* disbursements through the ABF-system shall be based upon *real* activity, and shall be *determined* in line with this in connection with the final settlement." In other words, central authorities take the relationship between patient conditions, medical diagnoses, and statistical codes as naturally occurring and based upon a direct derivation of form from flesh. I described earlier how DRG instructions rely upon a clear-cut boundary between working diagnoses and formal diagnoses. Physicians I followed during fieldwork did not express such a distinction between terrain and map,[4] practice and stylization, medicine and state. Let us go back to the physician from the earlier section and her administration of what *she* denotes as a medical diagnosis decision:

> The physician says she starts with the diagnosis. She enters a Norwegian version of ICD-10 at the hospital's internal webpage. In this case she says she does not find something she recognizes as an exact

diagnosis code. "Some symptoms are not registered in under the diagnosis here." She looks at a list on the wall that contains many codes and terms. The list says it consists of a sample of complicated diagnoses from ICD-10 and also the most usual diagnoses (with related codes) for the internal medicine ward. I count the units on the list; about 70 usual diagnoses, 25 complicating diagnoses, 15 procedures, and around 20 supplemental codes. The physician says she does not change the patient's diagnosis to adjust it to ICD-10. "So then we start from an unspecified diagnosis. Let's say that [mm] is an illness, and that its specification is [mm] plus another term, [mm-mm]. I overlook the latter term, and write only [mm]". I ask whether this is a standardized procedure in that every physician does as she does. She says no, she does not think so. "I think everyone has their own way of doing this, their own solutions and routines. I do how I have found best for me, to first get an overview over what to include, and find a diagnosis that fits the best . . . I go into the DRGs then (. . .) go in and look for a diagnosis that fits, eh usually I go into the chapter that describes such and such organ system, and find the diagnosis that fits best. Eh yes . . . And then you do that with every diagnosis . . . You have knowledge about the patient and his illness biography . . . " (field notes including physician comments, during an observation in field).

Something happens when the physician translates between the different terminological systems—medical practice, ICD-10, and DRG—that figure in this vignette. First, she illustrates that the translation act changes what is translated. Second, we see that DRG, ICD-10, and the medical diagnosis decision making—codes, numbers, and medical diagnoses—are not separate entities. The physician has a categorical understanding of the patient. This understanding is most likely medically acceptable and compatible with local practice and tradition. But she is not able to transfer this understanding to a category in the formal code system that bears upon the same understanding as she wishes to express in the first place. During her effort to find a term that fits the projected phenomenon she thinks describes the patient's illness, both her self-reflection and physical hand movements are interrupted because the anticipated needed diagnosis category is lacking. The physician attempts to articulate the illness in medical-administrative terms without success due to, respectively, the code systems' administrative requirements and her knowledge about the patient and his illness biography. She does not find a formal diagnosis code which she thinks includes every symptom of the patient. She is somehow forced to conclude with another diagnosis specification because of the structural requirements of the formal classification system.

The theory she ends up with is less complex than the theory that is "more right" in accordance to the conglomerate of medical symptoms earlier portrayed in and on the patient. However, this is not the simplest theory even though we may presume it to be so due to it being expressed through one term instead of two. The simplification she makes masks a more complex understanding behind a simpler term, but this does not simplify later physician's work with the patient. The medical knowledge she practices supports a more complex theory, [mm-mm]. The simpler theory, [mm], becomes a more difficult one, as it, in its hereby documented shape, cannot explain itself for the next physician who encounters it in the same patient's medical record in the same way as [mm-mm]. The physician *says* she does not adjust the patient diagnosis in accordance to ICD-10. However, when she ends up registering an unspecified diagnosis instead of the specified one she would have preferred to report in the first place, she actually does report another diagnosis constellation because of the formalities that tie up this report situation.

But, is it because of DRG that the physician ends up registering this other constellation rather than the diagnosis she means fits best medically? Physicians have coded ICD due to international obligations for many years. How can we know that the presence of DRG alters its own formalization and the diagnosis registration in ways different from the alternative—a medical diagnosis decision making and registration process not connected to DRG, only ICD? And, does it matter if DRG affects the diagnosis output? Does it matter how a patient is represented at this formal stage, which seems so detached from the patient him or herself? Let us repeat the observed situation described earlier:

> The physician says she does not change the patient's diagnosis to adjust it to *ICD-10*. "So then we go out from an unspecified diagnosis. Let's say that [mm] is an illness, and that its' specification is [mm] plus another term, [mm-mm]. I overlook the latter term, and write only [mm]." I ask whether this is a standardized procedure in that every physician does as she does. She says no, she does not think so. "I think everyone has their own way of doing this, their own solutions and routines. I do how I have found best for me, to first get an overview over what to include, and find a diagnosis that fits the best [. . .] I go into the *DRGs* then (. . .) go in and look for a diagnosis that fits, eh usually I go into the chapter that describes such and such organ system, and find the diagnosis that fits best."

Obviously, the physician mixes ICD-10 and DRG at the terminological level. She co-mentions the two systems as the same object. Does this imply not only a co-mentioning, but also a co-construction? The co-mentioning seemed to be typical for physicians at the ward. Other physicians described the same situation in similar ways:

> I see the point with the DRG system, but medically I think it is mostly disturbing. You use so much time on it, and it is difficult to place the diagnoses in categories that are too theoretical . . .

I only think about DRG during discharge. I only think about medical diagnoses in "the DRG way" when I'm in my office. I only think about DRG every time I write a case summary. Otherwise: never.

Physicians' identification of DRG as present in the course of medical diagnosis registration actualizes the question: Why? The situation seems not only to be defined by mixed terms and systems. Rather, the systems seem to have become practically intertwined. The physicians seem to be configured to think about ICD as DRG and practice DRG and ICD as one package, simultaneously as their debates seem to focus on the boundaries of the systems. At first glance, we may presume that the physician's last utterance is a proof for the opposite; that DRG acts only in the restricted area of his office cut off from patient contact. Yet, due to the social organization of medical diagnosis practice, this is also a description of how DRG gets direct access to the clinical work which is undertaken before any formal summary of a patient is made. Theoretically, it is "the ICD way" physicians are meant to follow during the writing of case summaries. According to central authorities' way of knowing DRG, the system has neither the potential nor the ability to "disturb" medical practice; it only groups ICD codes which physicians have already produced. Most likely there *is* a boundary between these systems in the medical diagnosis decision practice. The important point to notice is that since physicians drag DRG into the formalization of the medical diagnoses, the systems' meeting point(s) are not policed neatly into the formal documentation act, that is, step 2 in the DRG instructions. But, is it the physician who invites the system beyond those bounds? Does it invite itself? Or, is it perhaps parked there by another institution? Let us go from the medical-administrative work and pay the medical reading of the patient a flying visit.

Being Counted, Being Framed, Becoming Another?

During a walk down a medical ward a physician emphasized that "before [DRG was implemented] secondary diagnoses would not have been reported." As a budgeting system, DRG rewards secondary diagnoses. In response, physicians record multiple diagnoses more frequently. This situation goes straight into the counted ones' affairs. Through DRG's involvement with the formalization of medical diagnoses at the case summary stage described earlier, DRG may also affect a physician's reading and interpretation of a patient's past illness trajectory, which in turn may affect the patient's future medical management.

> On many occasions we must manage patients with headache for many years, and then we must go back in the record to study their illness history, and then you use every detail in the record for many years back. Then you study every diagnosis that the patient has had that may

overlap and that may be interesting for the total picture of the patient. I read every diagnosis and every illness description reported from the patient's birth (Physician, interview).

Diagnosis code registration is not an isolated event at the medical ward, but episodically repeated actions which, due to physicians' documentary duties, play an obligatory part of the patients' medical biographical trajectories (see Strauss 1993: 53–54). Diagnosis recordings enter physicians' evaluations of patients' medical situation, and thus their coming interactions. Small details can be taken into account to explain patients' symptoms. Classification codes act forward and backwards in time, just as Berg (1996: 519, 517) notes regarding the role of the medical patient record as part and parcel of the ongoing (re)construction of the present, actively shaping the very events it "represents." Earlier documented formal diagnoses are important tools for physicians in their medical mapping of for them "new" patient situations. This means that when DRG affects which diagnoses physicians register, DRG has the power to affect the one that the system counts aspects of, the patient. The counting generated by central authorities through DRG inevitably (re)frames the patient as the patient is ascribed diagnosis characteristics which physicians perhaps would not have bothered to register earlier. The question is how physicians manage this potential situation, and how they control DRG as practice.

Physicians' Strategies at the Ward: Protecting the Patient

Physicians' different ways of performing DRG have a varying impact on the situation of the counted ones. Common for their acts is that they try to protect patients from being turned into DRGs that imply them being documented in ways unfavorable for their future patient trajectories. I have chosen to structure this section according to Table 13.1.

The tensions follow two axes. On the one side we can describe physicians' strategies at a practice level (compliance vs. resistance) and on the other side we can use their direction (active vs. passive) to try to explain and further categorize practice at a more detailed level. These strategies are not separate ways of acting. They all communicate with the same background framework. Because physicians know that diagnoses can affect the coded ones, they try to manage the patient characteristics registration and

Table 13.1 Physicians' Strategies and Their Characteristics

	Active	Passive
Compliance	Co-opting	Adapting
Resistance	Sabotaging	Ignoring

statistics production in ways they more or less actively identify as protecting the patient from the potential harm formal medical categorizations can invoke.[5] Physicians I followed during fieldwork seemed to practice the distinctions between work diagnosis and formal diagnosis as tentative, but simultaneously robust. Roughly speaking, they shifted between a skeptical and a robust position depending on the position's practical value in their everyday work. They seemed to use these positions actively to sort out and control the interaction between diagnosis practice and DRG, between diagnoses-in-the-making and diagnoses-declared (see also Latour 1987: Ch. 1). The previously outlined strategies are all expressions of these positions and different ways of managing this position play.

First of all, the co-constructive interaction between DRG and medical diagnosis practice is not necessarily harmful for the patient.

> I see it [the increased focus on diagnosis coding] as an advantage for the patient. Formerly, secondary diagnoses would not have been put up for instance. For the patient this should make existence safer, as it gives more information about the illness picture. For instance a patient with a kind of cancer as primary diagnosis. Secondary diagnosis anaemia. The patient bleeds a lot. The anaemia comes naturally from the primary diagnosis. When the anaemia is specified, this may increase the impression that it is important to consider the anaemia where the patient comes the next time. The coding structures the trajectory and the patient's condition in a more lucid way (Physician, interview).

This physician experiences DRG as a trust generator as it improves physicians' descriptions of patient conditions. He thinks this is confidence inspiring and risk repressive. For him, medical documentation is interpretatively inflexible. When diagnoses are registered they represent facts. He depends on other physicians' work, and he trusts other physicians, who read the documentation he has produced, to depend on the diagnoses he has registered. This implies that he uses earlier reported diagnoses to construct theories of patient conditions regardless of the potentially shaping presence of DRG—because he thinks they promote valuable information. He actively complies with the formal rules as they are given. By emphasizing the importance of increased coding attention he co-opts DRG. He colonizes it and portrays it as a valid tool. In situations where the diagnoses he considers and uses are wrong, due to their being reported in response to purely economic considerations, he may make medical decisions that are out of order. Nevertheless, this potential disorder will probably be negotiated and counteracted by actions of his colleague collective, since his kind of trust to registered diagnoses and diagnosis codes is only one of several strategies negotiated at below.

Another physician working at the same ward described his own relationship to diagnoses and the potential pitfalls following the compliance strategy as such:

> I never use the diagnoses when I read through the journal. Never. I read the text. Earlier sicknesses most of all. And sometimes the case summary text. I never consider the actual diagnoses. (. . .) I know it has been negligent through all years and has been different systems and been written about that . . . (. . .) There are many who should have had [mm] and who seemingly should have had it during this in-hospital stay too, but who have not. That it has been coded as chronic but it has only been atypical or an isolated incidence. I have seen many times that diagnoses have chased themselves, even though they haven't been relevant. Then the patient has been branded as more sick than he is. There are many who have got the diagnosis hypertension or . . . there are lots, lots, many incidences. Absolutely. Well, eventually sometimes one diagnoses wrong and then the wrong diagnosis is maintained. That's why I don't read such case summary headlines for diagnoses because there are lots of errors. And words I am not familiar with. Never trust diagnoses that have been registered before . . . (Physician, interview).

This physician tells about earlier experiences of managing the intervention of different "systems" in diagnosis documentation practice which have affected physicians' reports. He says that his knowledge of this condition affects how he projects and reports the patient. He is both skeptical and pragmatic to the interaction. He implies that he is superior to those who read the diagnoses. He says that some do, but that he dissociates himself from that. In one way this physician also complies with the rules that are given about diagnosis coding, and he also uses DRG as a tool in his medical diagnosis work—as a tool to demarcate and legitimate that he does not trust nor presumably use earlier reported diagnoses. He draws these demarcation lines actively, simultaneously as he takes a passive position to the information that medical diagnoses can communicate. He passively resists DRG, and ignores its potentially positive contribution to his decision-making work. This position interacts neatly with the compliance strategy, simultaneously as it may function as a point of departure for more or less active sabotage of the formal DRG instructions.

> You adapt to what is the local law in the starting point. This is not necessarily correct in accordance to DRG guidelines. I try to adapt easy and fast, so I don't get negative feedback on my work. Whatever the law is, I will adapt [. . .] they [some physicians and especially the older ones] haven't got the energy or interest to sit, not a doubt. That one, when one sits there and ok the patient has [mm]. And the hospital can get so and so much if we code [mm]. It is not given that we bother to code heart failure as a secondary diagnosis if [the patient] comes in with [mm]. Because it implies that we must find the codes and that is stress (physician, interview).

Constructing Mirrors, Constructing Patients 259

The physician uses his interpretation of other colleagues' ways of knowing and managing DRG as an excuse to legitimate his own resistance and refusal to follow up the managerial and political authorities' instructions. He views his colleagues, rather than the authorities, as authoritative, but is selective in which colleagues' views he values. He allies with colleagues through positioning DRG as a burden on medical work and a potential threat for the patients. His emphasis on secondary diagnoses reflects the pressure he feels from hospital management to code patient conditions as detailed as possible so as to generate a larger reimbursement from central authorities. Note that the physician informs that he actually does not know whether the diagnosis codes he and his colleagues report are correct according to DRG. He goes on to tell how he and his colleagues construct representations of what DRG consists of to which they remain ambivalent:

> I don't know what they do down on the clinic. We try to coordinate the coding rules, we try to follow the advice they (hospital ward administration, colleagues etc.) give. We never talk about coding on a regular basis. I don't know how the other physicians code. Of course, some are more concerned about including secondary diagnoses and small stuff like that, because that implies more money (physician, interview).

Perhaps is it normative not to talk about DRG and not to show detailed knowledge about diagnosis coding. This entails portraying oneself as either resistant or passively compliant to DRG. Physicians do not want to signal that they negotiate actively, supportively with the DRG system. At the same time as the physician says he does not know how his colleagues' code diagnoses, he knows what they think it is important to include. At the same time as the physician says he does not know how his colleagues code diagnosis, he knows what they think is important to include. At the same time as he says he does not talk about diagnisis coding on a regular basis, paradoxically, normative rules for how to connect numbers, words, and medical worlds are established at the mediacl ward.

No matter what strategy physicians choose to control DRG, they construct theories of diagnosis coding, for instance which codes they assume (or have noted) that hospital administration wants them to code. But what do they need such theories for? Perhaps they structure them into routines and norms for how to manage the presence of uncertainty that they experience during situations where they cannot simply derive a diagnosis from a condition and a code from a working diagnosis. In this way, they can always actualize structural remedies during the practicing of medical diagnosis decision making to position DRG at a comfortable (in their eyes) distance from the patient. This applies both in situations where physicians must interpret and manage the potential presence of DRG in earlier reported documentation and when they themselves shall produce and shape a person into a counted and classified one during diagnosis and diagnosis code systematization and registration.

SUMMARY

The fuzzy boundaries between medicine and DRG and the negotiations concerning the ICD and DRG boundaries suggest that the examined debate concerns not only the technical systems involved, but the multiple parallel, collaborating and competing "universes" where diagnoses play a role. Of these, we have focused most on medical diagnosis practice and state regulation practice. The case seems to be one of social world membership and representation, with some physicians being especially significant transmitting agents (Strauss 1993). As the active parties negotiate this blurry, conflicted boundary, they seem to be improvising a complex choreography (Cussins 1996) in which it is unclear who has the lead, what shape and size the dance floor has, or even what the dance is about (precision? care? skill? knowledge?).

What is at stake with high stakes diagnoses? What would happen if the diagnosis had not embodied a high stakes identity? How can the medical diagnosis-making process be maintained in the face of the potential influence of DRG? Do the differences that DRG brings to this process make a difference? DRG adds up as another layered convention in the local history of diagnosis documentation management. The social world does not stand still. The interaction between DRG and medical diagnosis practice is not a problem per se as far as the patients are not represented in unfavorable terms for future intervention and the financing of the health services is coordinated in a way accepted as just (as is one of the main goals with ABF). In this context "an unfavorable term" implies a representation of the patient's identity over time which does injustice to the patient when faced with medical assessments and treatments, social security benefit applications, or insurance policies. So DRG's effects on how the patient is presented at diagnosis code level are partly counteracted in practice through other segmented routines at the medical ward, such as physicians' multiple collective personal-professional strategies for managing diagnosis documentation. Through their active focus on protecting the patient during their reading, interpretation, and further presentation of the patient in the formal documentation systems, physicians stabilize the blurrily bounded arena where DRG and medicine meet. However, the boundaries stay unavoidably blurred. Turning to the financing aspect, this means that there will always be a sliding relationship between diagnosis and number. If there is some "best answer" for how a just financing of the health services should be, this is a moral question which cannot be simply delegated to a technology.

ACKNOWLEDGMENT

This chapter is based on an earlier published op ed piece in Tidsskrift for Den norske legeforening, see Tøndel (2007), as well as my Master Thesis in Sociology from the Norwegian University of Science and Technology from 2006.

Constructing Mirrors, Constructing Patients 261

It was also presented in 2007 at the Annual Meeting of Society for Social Studies of Science under the title "The Mutual Shaping of Medical Diagnosis Practice and Statistical Knowledge-building: The Case of Diagnosis Related Groups." Thanks to Ann Sætnan and Aksel Tjora at the Department of Sociology and Political Science, NTNU, for proofreading and comments.

NOTES

1. Services included in ABF are day and night treatment, day surgery, and day medical treatments, as well as day surgery undertaken by specialists in private practice which the enterprises buy (Norwegian Directorate of Health 2009: 13).
2. The World Health Organization (2008) defines ICD as an "international standard diagnostic classification for all general epidemiological, many health management purposes and clinical use. These include the analysis of the general health situation of population groups and monitoring of the incidence and prevalence of diseases and other health problems in relation to other variables such as the characteristics and circumstances of the individuals affected, reimbursement, resource allocation, quality and guidelines." In 2009 it is the tenth version of ICD which is in use in Norway.
3. "[A]ctivity" is quote-marked, because my use of the term here refers to the DRG instructions definition of "activity" as measured by patient characteristics, and not "activity" in interactional terms.
4. "The formal is the representation, the *map*; the empirical is the represented, the *terrain*" (Berg 1997: 406). According to Berg (1997), this dichotomy is, caricatured, dominant in our understanding of formal tools and what roles they play in organizing work within workplaces.
5. I have constructed Table 13.1 for practical reasons, not because its categories correspond systematically to my data material. However, in the material I find tendencies that support this schema adequately for me to trust it for use within this case.

REFERENCES

Aas, I.H.M. (1985) *DRG: Diagnose Relaterte Grupper. En litteraturoversikt.* NIS-rapport: 3. Trondheim.
Aas, I.H.M., et al. (1989) *The Making of Norwegian DRGs.* NIS-rapport: 3. Trondheim.
Advanced Medical Technology Association (2009) *International Diagnosis Related Groups,* http://www.advamed.org/MemberPortal/Issues/International/internationaldiagnosis.htm (accessed September 2, 2009).
Berg, M. (1996) Practices of Reading and Writing: The Constitutive Role of the Patient Record in Medical Work," *Sociology of Health & Illness* 18(4): 499–524.
———. (1997) "Of Forms, Containers, and the Electronic Medical Record: Some Tools for a Sociology of the Formal," *Science, Technology & Human Values* 22: 403–433.
Bloomfield, B.P. (1991) "The Role of Information Systems in the UK National Health Service: Action at a Distance and the Fetish of Calculation," *Social Studies of Science* 21(4): 701–734.

Bowker, G., and S.L. Star (1999) *Sorting Things Out: Classification and Its Consequences.* Cambridge, MA: MIT Press.
Cussins, C. (1996) "Ontological Choreography: Agency through Objectification in Infertility Clinics," *Social Studies of Science,* 26(3), 575–610.
Feuer, L.S. (1957) "The Principle of Simplicity," *Philosophy of Science,* 24(2), 109–122.
Geest, van der, S., and K. Finkler (2004) "Hospital Ethnography: Introduction," *Social Science and Medicine,* 10, 1995–2001.
Glomsaker, T. (2005) "Epikrisen—faktura eller medisinsk dokument?" *Tidsskrift for Den norske legeforening,* 125(7), 866.
Hafstad, A. (2003) "Jakten på lønnsomme diagnoser." *Aftenposten,* http://www.aftenposten.no/meninger/kommentarer/article570550.ece?service=print (accessed August 2, 2005).
Hughes, D. (1977) "Everyday and Medical Knowledge in Categorising Patients," in *Health Care and Health Knowledge,* eds. R. Dingwall, C. Heath, M. Reid, and M. Stacey, 128–140. London: Croom Helm.
Jørgenvåg, R., and Hope, Ø.B. (2005) "Kvalitet på medisinsk koding og ISF-refusjoner. I hvilken grad er journalgjennomgang et nyttig verktøy?" SINTEF Helse, Norsk pasientregister, http://www.kith.no/upload/2178/Kvalitet_medisk_koding_ISF_ref.pdf (accessed September 2, 2009).
KITH (2002) *Kodeverkstabell. Oversikt over norske og internasjonale helsefaglige kodeverk og definisjonsgrunnlag. Administrative kodeverk, veiledere og terminologiske hjelpemidler,* http://kith.no/templates/kith_WebPage____1135.aspx (accessed January 6, 2010).
Latour, B. (1987) *Science in Action: How to Follow Scientists and Engineers through Society.* Cambridge, MA: Harvard University Press.
Norwegian Directorate of Health (2009) *Innsatsstyrt finansiering 2009.* Oslo, http://www.helsedirektoratet.no/vp/multimedia/archive/00092/Innsatsstyrt_finansi_92249a.pdf (accessed October 2, 2009).
Posner, K.L., W.M. Gild, and E.V. Winans (1995) "Changes in Clinical Practice in Response to Reductions in Reimbursement: Physician Autonomy and Resistance to Bureaucratization," *Medical Anthropology Quarterly,* 9(4), 476–492.
Schjerve, H. (2003) "Pasientene tror rådene er faglig basert. De får kanskje tilbud om operasjon, fordi det er mer lønnsomt for sykehuset . . . " Dagbladet.no, 14. juni, http://www.dagbladet.no/nyheter/2003/06/14/371168.html (accessed September 1, 2004).
Slåttebrekk, O.V. (1990) "Hva koster pasienten? En beregning av norske kostnadsvekter til DRG". NIS-rapport: 1. Norsk institutt for sykehusforskning: Kommuneforlaget.
Strauss, A., S. Fagerhaugh, , B. Suczek, and C. Wiener (1985) *Social Organization of Medical Work.* Chicago, IL: University of Chigaco Press.
Strauss, A.L. (1993) *Continual Permutations of Action.* New York: Aldine de Gruyter.
Tøndel, G. (2007) "Hvordan DRG-systemet påvirker medisinsk praksis". *Tidsskrift for Den norske legeforening.* 127(11), 1532–1534.
Torjesen, D.O. and H. Byrkjeflot (2007) "When Disease is Being Priced: The Translation of the American DRG System to the hospital sectors in Norway and Denmark". 2nd Nordic Workshop on Health Management and Organization, December 6–7, 2007.
World Health Organization (2008) *International Classification of Diseases,* http://www.who.int/classifications/icd/en/ (accessed October 5, 2008).

14 GIS in Practice
Domestication of Statistics in Policing

Helene I. Gundhus

In this chapter I reflect on what end-user responses to statistics in policing, particularly aggregated data derived from geographical information systems (GIS), can tell about co-construction of statistics in policing. It will point to how the users of technology are important contributors to the further development of that technology, by adjusting it to their own needs and "lifestyles" (Lie 2003: 20). Fieldwork undertaken at a police station in Oslo is the starting point for the discussion. Prefacing that, I present shifts in knowledge discourses in policing. I end the chapter by reflecting on the implications of findings for debates about shifts in police knowledge practices.

GIS AND SHIFTS IN KNOWLEDGE DISCOURSES

GIS are computer systems for capturing, storing, integrating, analyzing, and displaying geographical data. GIS can exist in several forms, for example as a desktop software package, as a web-based system, or on mobile handheld computer devices (Home Office 2005). GIS is used in different sectors: in business, industry, government, and education. Since the mid-1990s, its use has been emerging in the field of criminal justice. The technology can be used in both simple and complex ways. Various referenced data about crime, policing, and socio-economic context are captured, and can be stored, linked, analyzed, and displayed for different policy purposes. GIS can be used for crime mapping, the direct application of considering the inherent geography in crime. Similar mapping techniques can also be applied to other police data, such as incidents, offenders, victims, intelligence, efforts, and so on. For example, mapping can be used to identify why incidents may have occurred in one location and not another, and to analyze how offenders traveled to the incident location.

The use of GIS in policing is a strategy for gaining more of a focus on the geographical dimensions of crime, risk, and disorder. As Maguire (2000: 316) has pointed out, this interest is motivated by a change in knowledge discourses in policing, which means a shift from "reactive investigation of individual crimes" to a "strategic, future-oriented and targeted approach to crime control." The shift represents a vision about a "new" knowledge regime, based on evaluating aggregated knowledge and statistics about risk as more important than knowledge about individuals involved in crime. Knowledge-led policing is an umbrella term for these knowledge discourses in Norwegian policing (Politidirektoratet 2007). It includes philosophies such as problem-oriented policing (POP), directed toward "problems" (a broader concept than crime), and intelligence-led policing, directed toward serious crime using law enforcement instruments. The aim of these philosophies is to implement methods based on the systematic use of information sources to analyze various patterns of crime or risk. These strategies[1] intend to improve the analysis functions in police organizations, so that better crime reduction efforts are implemented in police work. The National Police Directorate's intention is that all police districts are familiar with and work in accordance with the principles of knowledge-led policing. Since these changes are dependent on changes in organizational structures, this is an ongoing process that is implemented quite differently in the twenty-seven police districts in Norway.

The analytical approach to police work is intended to change traditional police patrolling by intervening on the basis of knowledge about what causes crime and what the possible future risks are. The aim is to widen the traditional knowledge base in policing, which is built on individual police officers' experiences and so-called street knowledge produced by observation during patrols. Knowledge-led policing is a strategy for managing potential risks more effectively, by governing the police officers' tasks more in detail and by managing the police organization directly through the POP process.[2] POP is a knowledge-led strategy usually applied to reducing everyday crime and disorder. It employs a process called SARA, which is an acronym for the four stages in POP[3]: scanning, analysis, response, and assessment. SARA is a practical guide for better allocation of resources, by applying an analytical approach for dealing with risks (Politidirektoratet 2002). By introducing analytical responses to a problem, the objective is to ensure that the problem is effectively identified and tackled in such a way that even future risk can be reduced. In addition, since knowledge-led policing indicates a police role that is subordinate to the knowledge processes, the aim of, for example, a police patrol is changed by implementing the SARA model. The patrol becomes standardized on the basis of scanning and analyses of possible risks and trends. Between incidents, patrolling becomes an instrument for these processes, and certain patrols are dedicated to POP. Police officers have to integrate a more analytical view into their knowledge at street level, and to a higher degree become information retrievers for police databases, by collecting and registering more than just crime incident information into computers for analysis purposes. The police officers have to follow a plan prescribed in the POP manual, and can to a lesser degree decide what to do autonomously while patrolling.

To a remarkable extent, police strategies like POP and intelligence-led policing focus on how statistics and aggregate data can make policing proactive and future oriented. In these knowledge discourses, traditional police practices are described as passive and managing incidents in a reactive way. The police officers are considered to be biased by particular experiences and human weaknesses. The aim of becoming less incident-led and more knowledge-led is to become more rational in decision-making processes through knowledge management, thus making police practice more predictable. Since the late 1980s, performance indicators, management measures, financial auditing, and accountability have been tightly integrated within the mentioned police reform projects (Garland 2001, Johnston 2000).[4] But as Reiner (2000: 77) has shown, measuring police work is not a unilateral process, because "police effectiveness is a notoriously slippery concept to define or measure." Interestingly, the implementation of POP has two objectives. First, the aim to become more effective in managing budget control, and second, to make police work more effective by managing the police officers. There is an ongoing tension between increased demand and decreased resources. The intention is to target future risk, more serious and more covert crime. It offers an opportunity for the police to extend the background for better prioritizing, and become more just by moving "beyond traditional narrow focus upon highly 'visible,' but in many cases relatively minor, delinquent and anti-social behaviour of the poorest and most marginalised members of society" (Maguire 2000: 329). The problem, however, is that there is no guarantee that the system will operate in this way.

GIS—A FLEXIBLE TECHNOLOGY

GIS has been highlighted as a tool supporting the future-oriented approach to crime control, less focused on offenders than considering possible future victims and risks. Its mapping feature can be an important tool in strategic planning, where problem-solving maps are used as management tools for the decision-making process. However, it is important to stress that GIS also can be used within the "old" reactive knowledge practices. The development of GIS in criminal justice in the United States seems to be more advanced than in Europe, but it is increasingly attracting decision makers in the latter. For example, the U.K. Home Office (2005) has made "a good practice guide for front line officers," describing how crime mapping can improve front line officers' performance. In public policing, GIS is used to inform deployment changes, for example regarding changes in hot spots over time and the types of crime occuring.

Theories about how GIS can improve policing are connected to the earlier mentioned shift in knowledge discourses. Home Office (2005) points to studies which show that police officers' perceptions of where crime and disorder happens don't necessarily match up with where such problems actually occur. Crime mapping is used to better inform and overcome inaccurate and limited perceptions of where problems are located. The police officers' restricted perceptions have several causes—for example, they can be insensitive or biased about newly emerging patterns, or have a bias in the case of historical problems. The "facts" and "truths" about the area that the perceptions are based on will always be limited, and can be weighted by anecdotal information. The last argument can be seen in the light of the sociological observation that stories about police work in the streets are central components of police occupational culture. As Shearing and Ericson (1991: 491) have pointed out: "In their street talk police officers use stories to represent to each other the way things are, not as statements of fact but as cognitive devices used to gain practical insight into how to do the job of policing." Police officers have long traditions of "knowledge of particulars" (Shearing and Ericson 1991: 492). The criteria for evaluating the truth in the stories is not scientifically grounded, "but rather whether the knowledge they capture 'works'" (Shearing and Ericson 1991: 491). The declared intention with crime mapping is to challenge this value of "knowledge of particulars," and make the police officers more aware of its biased character.

As Terence Dunworth (2006) has pointed out, crime mapping has three main functions: operational maps identify and disseminate specific crime information, strategic maps identify and analyze long-term crime data, and administrative maps provide the administration with general information about the community.

Mapping is also used to inform citizens about risks. For example, maps are published on the Internet, to inform citizens so that they can take precautions. The purpose is to involve citizenry as coproducers of law and order. As early as 1995, the California police published crime maps on the Internet, and this has become a more and more common practice. This is part of a preventive strategy, where the purpose is to influence citizens to avoid hot spots or give important tips to the police (Stoe 2003). But as Stoe and others point out, this practice raises ethical issues, especially when the identities of former offenders get published on maps on the Internet. To reveal the home addresses of "potential criminals" can be stigmatizing and lead to social exclusion, which in turn may exacerbate a tendency toward crime.

GIS is a flexible technology, wherein different types of users are invited to interpret and shape data, or interpret the presentations of the included stories about crime, risk, and the community. Also, within the police organization, maps are presented as a visually accessible format for diverse audiences, with the aim of providing a basis for discussion, for example in deployment decisions. However, since mapping is so tightly knitted to the new knowledge discourse and the goal of improving police performance, how it is measured and for what it is used are important dimensions shaping the discussion within the police.

Co-Construction of GIS at a Police Station

Grand theories concerning risk management in policing (Feeley and Simon 1992, Harcourt 2007) have been criticized for being too abstract, decontextualized and for predicting dystopian futures without room for maneuver and resistance (Zedner 2009). The problem with such grand theories is that they tend to take the advent of some technologically advanced future for granted, i.e., they are typically technological determinist[5] stories. Such stories can be technophile or technophobe: Advocates want the systems that critics fear. Both agree on what the systems can really do, and how extensively they can do it. The problem with determinism is that it closes the space for a criticism that might constructively try to see other realities.

In order to examine the use of information and communication technology (ICT) in policing on a small scale, I have empirically explored how these new knowledge-led policing strategies are domesticated[6] in two Norwegian police organizations. Fieldwork has been undertaken in two different workplaces within the Norwegian police service, a crime intelligence division and a local police station. I have also interviewed police officers and nonpolice employees at the two workplaces, and analyzed different types of relevant documents. In this chapter I will focus on findings from the second case—the implementation of POP/GIS at the police station. This police station was observed from January to July 2004. I conducted sixty-eight hours of observation in patrol cars, meetings, and daily briefings, and thirteen hours of observation at the police station's front desk. During the fieldwork, the total number of employees at the police station was 240. I interviewed sixteen police officers working at the station, and carried out twelve focus group interviews with twenty-nine police officers working in the station's patrol unit.

THE POP/GIS STORY

The GIS story started in 1999 and the objective was to form an analytical basis for the senior police officers (i.e., the managers) in their effort to make better decisions about long- and short-term priorities (Tolloczko and Rimstad 1999). Since GIS can be used in multiple knowledge discourses—both in terms of "knowledge of particulars" and in a more abstract, analytical knowledge discourse, and in reactive and proactive knowledge-using practices—GIS can be used in two ways at a police station. Firstly, it can be used to improve efficiency in responding to crime, i.e., reduce the response time. Secondly, GIS can be used among other efforts to change traditional policing—by implementing GIS as an analytical tool in a POP context. At the police station GIS was clearly part of a POP reform project, and this was the first station in Norway to implement POP and GIS.

The managers, in cooperation with enthusiastic and engaged young police officers, who had learned about POP at the Norwegian Police University College, were the driving forces behind the project. The vision of POP at the station was to prevent crime by implementing a more analytical approach. POP and GIS were integrated at the police station in the following way: In contrast to traditional policing, POP in this project indicates a more sensitive police role, delivering community safety via "partnerships." One intention is to reduce the use of punishment as a crime reduction tool. Another is to make crime the responsibility of other sectors of society than the police. To prevent crime in this context means to intervene early before crimes occur. The purpose of using GIS is to support the implementation of POP, by using the earlier mentioned SARA model, referring to the four stages in POP: scanning, analysis, response, and assessment. GIS is first used to enhance the police's ability to identify and scan hot spots; the maps show where problems are located. Secondly, it is used to analyze spatial patterns of crime and risk behavior; the maps help to test hypotheses about problems. Thirdly, it is used to find new ways of intervening early by making practical responses so that these problems are less likely to occur in the future; the maps show how problems are handled. Finally, it is used to assess the impact of the interventions; the maps show changes in problems (Clarke and Eck 2003).

By integrating POP/GIS, project proposers hoped to change crime-fighting practices from the traditional focus on the individual committing crimes to a focus on risk reduction and problem prevention (Oslo Politidistrikt 2000: 4). The individual offender himself/herself is usually unknown and difficult to identify. By seeing crime not just in relation to the offender, but in relation to other variables such as location and victim, it becomes possible to reduce risk. The police connected police computer databases to GIS, and made it feasible to analyze different variables. This made it possible to get an overview of hot spots, where crime is most often committed. In addition, they connected overviews of Oslo municipality databases of bars and restaurants, in order to explore the possibility that hot spots are close to nightlife activities.

So, GIS technology is a tool to enhance the police officers' observation and ability to identify the hot spots, analyze spatial patterns of crime and criminal behavior, and to share the information both across the organization and with the community outside. To value just the "knowledge of particulars" should not be enough for a police officer.

These systems were governed by the strategic planning department at the station. The intention with the use of these communication technologies was to maintain close contact with problems citizens report to the station, and stay in touch with what they experience as difficult in their everyday lives, to reduce so-called "repeat victimization" and "prolific offenders."

The management and strategic planning department also used statistics to communicate with potential victims about possible risks through different types of media. For example, one Saturday afternoon, the planning department sent out a text message to 15,000 youths aged between twenty to thirty-five years of age in Oslo, instructing them to take precautions against pickpocketing and violence. Another example was warning posters at shopping malls, with printed texts such as "take care of your wallet—pickpocketing in the area." These were among several strategies used to distribute responsibility for crime and disorder by bringing the problems back to the so-called "problem owners," such as companies, shop owners, regulatory agencies, financial institutions, health and welfare agencies, and ordinary citizens. POP/GIS was implemented at the police station in a community safety context, combined with knowledge-led policing strategy—the intention was to move away from the punishment paradigm toward crime prevention by using risk communication and partnership strategies.

After I completed the fieldwork, I found an advertisement from Geodata (2009) on a website, about delivery of GIS software to police stations. It is reported from the participators that GIS has contributed to powerful crime reduction in the streets of Oslo. By collecting historical data of crime, visualizing them on maps, looking into patterns and hot spots, and allocating resources on the basis of the maps, it has become possible to be a step ahead of future crimes. This version of the POP/GIS story at the station is in line with the larger story about technological determinism I referred to previously. By managing the police patrols in hot spots, it has directly given results in crime statistics. Such success stories were not presented to me during the fieldwork. It is interesting how presenting images of the technology itself was seen as more important in the "glossy" version of GIS in the publicity material at the station.

STREET KNOWLEDGE VS. STATISTITCS

It is important to examine how GIS was co-constructed and domesticated by the end users at the police station. According to Mazerolle, Bellucci, and Gajewskij (1998: 133), it is possible to identify two primary crime mapping system "end users": crime analysts and street-level problem solvers. During my time of observation at the station, the end users were mainly the street-level users. The maps showed the exact locations for "criminogenic" zones, problem owners, and so on, but they were far from thematic, and therefore not very useful for crime analysts. It was all about scanning and identifying hot spots, based on citizens' reports to the police.

The enthusiastic police officers at the police station talked about POP as a process that made them better at professional policing and creating community safety. They were proponents of the "new" knowledge practice, with attitudes toward a more cooperative, just, and analytically derived policing practice. Front line officers are important in community safety strategies, where the aim is to reassure the citizens in the coproduction of law and order by both police and community (Innes 2004). But when I observed daily life at the station, GIS was primarily used to identify crime and allocate resources for where and when to patrol. It was used more like a management system.

The end users at street level turned out to be the most skeptical about GIS at the station. They saw the charts as tools that would yield greater internal surveillance of their individual police discretion. The front line officers were concerned with dividing between "need to know" and "nice to know" information, and statistics were merely viewed as "nice to know." Street statistics from GIS were also seen by some as a good example of "not nice to know." They saw it as rather useless in their daily work because the maps visualized facts they already knew, or that were seen as wrong compared with "the truth" from their own experiences in the street. The very act of counting as a specific way of understanding was in conflict with the weight given to experience as a source of truth. The scanning was for example interpreted literally, and criticized as if the indicators were mistakenly presented decontextualized. Rather than perceived as context driven, the end users had a literal interpretation of the GIS data. So in spite of the intention, the crime maps did not broaden the police officers' perception of street crime.

Useful information was considered by the front line police officers to mean "informative," and connected to individuals. There was a huge interest in searching police databases for information about individuals and their relationships in criminal milieus, particularly among the street-oriented intelligence officers. Personal knowledge about criminals was seen as important in daily work. That type of individual information was talked about by officers with the assumption that it would enhance their prestige within the police organization in the future.[7] Knowledge that was highly valued was inductive, something that grew out of on-the-job experiences. Valuable information was described as short-term based, useful, effective, freshly produced, and practical. It was knowledge that makes the police officers vigorous, active, and efficient (i.e., information that can be used for traditional reactive crime fighting). Statistical data were less transferable to this type of work. This pointed to uncertainty when it came to domestication of statistical data (i.e., data produced for statistics used as data for information about individuals).[8]

The "informative" use of knowledge tended to work at the expense of abstract, statistical, and analytical understandings of future risks. The police officers I interviewed saw themselves as using craft knowledge and found it difficult to police "by books and statistics." They interpreted the maps as the extension of their own experiences, and in that way they co-constructed the maps in a traditional framework. In spite of the new knowledge discourse, the finding revealed a primacy for street knowledge, personal data, intuition, experience, "grounded" knowledge, instead of analytical, abstract, statistic analysis in the organization. Knowledge types at street level were also evaluated differently from management's knowledge. It was still based on face-to-face interaction, "street sense," and it was contextual, particularly to places and time, and intuitive as opposed to management's more abstract type of knowledge. This can be connected to the different types of knowledge circulating in the organization, which are shaped by the employee's position in the organizational structure. This illustrates that, as knowledge derived from GIS is office based and computer driven, GIS represents the antithesis of action-oriented police work. It also indicates that features of police culture—e.g., the value attached to action-oriented police work vs. office work, along with experience, grounded knowledge, and the use of discretion by the police—are shaping the integration of GIS in policing.

Among front line officers, the new practices were potentially rather than fully integrated into their daily work. The situation was also far less rational, and much more "messy," both in terms of risk management and of interagency cooperation. Statistical "codes" were not transferrable between different occupational cultures at the police station. Efforts toward standardizing the images of the city collided with local values and the street-level police officers' ways of interpreting crime by their own street experience. This indicates that GIS may contribute to making police work more standardized and oriented toward management-driven targets, but this would be far from making policing analytical and knowledge based. In an occupational culture with resistance to new knowledge practices, aggregated data are used within the framework of the "old" knowledge discourse. Data produced for statistics are used as information about individuals. This makes it difficult to understand how statistics can disprove the police officers' subjective perceptions about truth, contexts, and risk trends.

Street Level vs. Management—"Firewalls of Resistance"

The findings also indicate tensions between POP as an instrumental management tool with focus on resource allocation, and POP as an analytical approach to crime prevention. The POP/GIS project was started at the initiative of a few enthusiasts at the street level in the organization. It met with positive response from senior police officers, who comprise the management, but with resistance from the rank and file. Front line officers preferred not to be placed in a POP car. They experienced the implementation of POP as being a hindrance to their freedom and discretion as police officers, since management created performance indicators for what they should do at particular places, at specific times, and with particular selected targets. The front line officers did not report back on POP work in the police computers. This had a negative effect on the crime mapping and its evaluation.

The end users at street level were rather critical about the maps, and saw them as only relevant for management decisions. This could actually be seen as in line with the intention of the POP/GIS project at the station:

to have better management because of better objectives. But during the fieldwork the management expressed worries about "out-of-control" patrol projects. The technology's flexibility made it easy for the end users to co-construct the data and the management and strategic planning department were worried about what they called "hobby analysts," with their own Top 10 lists of the worst criminals, which were not in accordance with the targets prioritized by management, but reflected their own individual obsessions. The social landscape that ICT was used in is far from seamless and free from conflicts. The combination of interpretatively flexible data and the need for flexibility, discretion, and freedom at street level for doing good POP also opens up for vain heroics.

In the analysis of the POP/GIS project, I found it important to separate between GIS users at different levels in the police department. Understandings of risks differed between front line officers and management. For example, the uniformed police at the station experienced pickpocketing as a much smaller risk than the resources allocated to prevent it. Some of the police officers I interviewed were rather irritated at the resource allocation, and tried to resist it during patrolling. The management saw it as important to solve the pickpocketing problem in order to reduce fear of crime and increase safety in the community. This differing perception of what risks should be selected illustrates Ericson and Haggerty's (1997: 100) argument that it is necessary to focus on the interaction between, on the one side, institutional constraints and routines, and on the other side, conflicting and dominating political and corporative interests to understand the selection of risks.

Since GIS was presented as mainly a management tool, the maps were met with an open resistance from the main end users of the maps, i.e., the patrol units. The end users' main concerns were about GIS restricting their professional judgments and autonomy. They had concerns about working in such a knowledge-led way when resource allocation was considered inferior to "real police work," and criticized as "management stuff." It was also associated with femininity, because it was connected to routine-based office work.[9] "Thick professionalism" was mobilized against the management's press toward new knowledge practices. The reform project was conceived as a negative "scientification" of police work's traditional connection to "knowledge of particulars."

Since the GIS/POP project was a reform project, the implementation was hindered because the most important end users resisted, and interpreted the statistics within their traditional framework. The co-construction of the maps made apparent the different occupational cultures at the police station. So the findings indicate the existence of two cultures inside the organization, revealing conflicts between management and front line officers. This quote from a police officer at the station illustrates this:

> You might call it an internal struggle—for that's what it is—and that is making it difficult to implement management's objectives at the police

station. It is a macho system making people play their cards close to their chests.

This also indicates that end users at street level are marked by a distance from management. They use technology bottom-up instead of top-down. This can be seen in what Lipsky calls street-level bureaucrats, who are often characterized with a style of occupation marked by autonomy, independence, and distance from management (Carlström 1999: 165, Lipsky 1980). As Wright (2002: 142) puts it, "measures to ensure accountability have often been resisted with a heavy preference for self-regulation." Knowledge-led policing at street level is in conflict with strategies, plans, and management, because the police officers struggle for recognition, position, material benefits, and legitimate authority. In this struggle it is possible to see a tension between individualistic and collectivistic cultural ideals, norms, and values within the organization. The paradox can be described as how to create latitude for knowledge-led policing as a reflexive, street-level practice, at the same time as discretion is standardized and management-by-objectives is introduced in the police organization.

Aggregated data are introduced to change knowledge and learning processes. But this has been perceived as control, and "firewalls of resistance" are preventing the implementation. Both operative management and front line officers function as "firewalls of resistance" against transformation. This resistance underpins Kemshall's (2003: 98), Lynch's (2000) and O'Malley's (2000, 2001) argument that occupational culture and the role of both workers and staff are important sites for both mediation and resistance to new policy implementations. Since the reform project was perceived as a management project, the resistance can also be understood as an answer from those who are being monitored by the new way of "counting" police work.

To conclude, the finding revealed a tension between "old" and "new" ways of policing. To a great degree, this tension reflects conflicts between "old" and "new" knowledge discourses. The field of policing has changed, but not police habits (Chan 2003). The management worked toward standardization of challenges and the introduction of more scientific ways to define "truth." But the front line officers insisted on local values and traditions for constructing the truth. In a way, it can be understood as counteraction against unwanted surveillance from management.

Domesticating GIS into Everyday Life

GIS data in policing can serve as audit tools that can be subject to quality assurance and regulation, introduced in order to replace the vagaries of individualized decision making. But the findings in this chapter point to the fact that such tools are domesticated, shaped, and adapted in occupational culture and practice. One illustration of this is that the police officers at the station responded to management-driven targets and the use of analytical knowledge as a threat to professional judgment, experiencing management's message as a signal of distrust. At the same time, the analytical knowledge challenges their view of "real" policing. In the new management strategies there are, according to Kemshall (2003: 58), "also skepticism in the professionalism and expertise of criminal justice personnel, and a lack of trust that front-line workers could manage risk appropriately if left to their own devices." The newly imported working strategies and analyses of risk could either replace the biased and inconsistent clinical judgments of workers, or merely structure them toward managerially desired ends. This points to obstacles for implementing "new" knowledge discourses in police practice.

This also indicates different conceptions of professional policing. GIS is part of the new standardization of professionalism in policing, which is associated with knowledge-led policing. The "firewalls of resistance" make it difficult to apply GIS to the occupational culture of the front line officers. This indicates that the newly imported working strategies *cannot* replace professional judgment and thereby structure police work toward managerially desired ends. Since implementing tools like GIS requires that they become embedded into particular agency cultures and objectives, the translation of knowledge-led policing into police practice reflects processes that are not univocal but multivocal. These findings support O'Malley's (2001: 100) argument that disciplines are politically multifaceted and pragmatic. The findings also indicate that there is dissonance between management objectives and practitioners' views on valuing risk knowledge. As Kemshall (2003: 23) has pointed out: "managers and policy-makers valu[e] actuarially based knowledge for its consistency and accountability, and practioners valu[e] professional, individualized judgment for its flexibility and responsiveness to individual factors."

The practice of new knowledge discourses is to a large degree shaped by bottom-up perspectives, not only by top-down management. The technology is *domesticated*: a perspective on technology that emphasizes that it is not enough to make the technology available for the employees; it is the interaction between availability and use that can tell something about practice (Lie and Sørensen 1996). By treating the GIS statistics as *not* "need to know" information, police officers can limit the discrediting of their street knowledge. In theory, POP could be about "flattening" hierarchies in the police organization and making police officers more reflective. However, many of the employees working with analysis and community contact call for a much stronger governing and control of the POP process, if the new strategies are to become more than mere rhetoric and symbolic in the everyday work of the police.

CONCLUSION

By using examples from the fieldwork connected to the use of GIS technology, I have shown that there is resistance to the new tools, and this is resulting in certain "firewalls of resistance" that work against a full integration of new knowledge practices in policing. Although it was not the intention, GIS was perceived as a tool for fixing numbers and acted upon as an obstacle to discretion and personal freedom. Findings in the case underpin Kemshall's (2003: 123) argument that "at the micro-level, change in police practices can be variable, with national policy and objectives subject to local context, front-line interpretations and workers resistance." In criminological theory, a technology like GIS is presented only as a tool for analyzing risk because the use of it has the intention to predict future crimes (Kemshall 2003). However, the material from the police stations indicates that evidence for a new risk-based penology is by no means clear-cut, and risk is perhaps more readily characterized by discontinuity, resistance, and mediation than by continuity and any "inexorable logic" (O'Malley 2001, 2004). Both front line and management decisions on risk need to be placed within this broader context of workplace demand and discretion. It was within this context police practitioners mediated and diluted the managerial imperatives. Tool use and risk classification in particular have proved to be important sites of worker and management mediation, and traditional practices have been adapted to new demands rather than replaced. In this area of practice, risk-based criminal justice has had an adaptive and evolutionary impact rather than a transformative one (Kemshall 2003: 98).

Significant police sociology with roots in Giddens' sociology (1990, 1991), describes how trust in policing is eroding (Loader and Mulcahy 2003: 21, 112–19; Reiner 2000: 47–80).[10] The ebbing away of confidence in social institutions, such as the police, has also in part triggered the introducing of policing reforms that are aimed at building or restoring the public's trust, for example community safety and "reassuring" policing, and recently the new buzz phrase in England and Wales, "police community co-production of law and order" (Innes 2004). The implementation process at the police station illustrates that trust is not only an issue when it comes to the relationship between police and the public. Trust between management and employees is also an important issue when it comes to improving police performance and implementing these police reforms in practice.

What type of data is collected, how the visualization is interpreted, which theories are incorporated in the data maps, and how the complexity of the crime problem is managed and operated are all crucial practical and ethical aspects of GIS. The focus on the technical aspects of GIS technology and management's aim of reducing crime numbers is driven at the expense of the more cognitive, analytical, and interpretative aspects of POP. This indicates that the implementation of GIS also depends on how criminological theoretical perspectives are integrated in the implementation process. As Hope and Sparks (2000: 180) argue, situations in situational crime prevention are too narrowly defined. In that type of theory, situations happen naturally or accidently, a theory which differs from sociological understandings of situations.

If policing is to become more knowledge led than incident led, it depends on the mobilization toward change of both organizational structure and police culture. Problem-solving policing is not only about technical issues, because the technology is co-constructed in the structures and cultures of the police organization. The findings also strengthen the argument that statistical tools like GIS both can broaden and limit the policing outlook, depending on how it is domesticated and co-constructed. Although GIS can reduce complexity, because it is at a remove from the understanding of an individual's motives for committing crime, GIS must also be discussed in light of the organizational framework it is used within. Problem-solving maps and GIS technology may help bring new perspectives on crime and the city into an organization otherwise extremely focused on "knowledge of particulars," but how perspectives change will be an outcome of struggles, negotiations, and cultural processes.

NOTES

1. In the literature, knowledge-led policing is called a strategy when describing a plan of action and knowledge-led policing when describing a policy.
2. Crime science (Smith and Tilley 2005) and evidence-based policing (Sherman et al. 2002) are two distinct schools built on this new knowledge practices.
3. The first step is to identify and scan hot spots; the second is to analyze spatial patterns of crime and criminal behavior and to make hypotheses about problems. The third step is to find new ways of intervening earlier and in a way that best fits the problem, to make it less likely to occur in the future. The fourth step is to assess the impact of the interventions (Clarke and Eck 2003).
4. The introduction of national police performance indicators that can measure crime prevention and community safety is relatively new in the Norwegian context, especially compared to Britain. For example, national priorities in police patrol work are missing, and crime prevention has never been measured. Performance contracts have first and foremost been connected to the Director General of Public Prosecutions, clear-up rate, detection rate and crime-reduction statistics.
5. See Smith and Marx (1994) for a thorough discussion of the concept of technological determinism.
6. See Lie and Sørensen (1996) and Silverstone, Hirsch and Morley (1992) for an introduction to the domestication concept.
7. See Carlström (1999: 64, 97–99) for similar findings.
8. In many ways the emerging tension between statistics-based knowledge and "knowledge of particulars" in policing is similar to the tension between epidemiological and clinical knowledge in medical practice. One important difference, however, is that in medicine it is clinical knowledge that has higher status, not only in day-to-day medical work but also at the (medical professional) managerial level and even in many research contexts. See for instance Sætnan (2000).
9. For another study where computerization was found to be at odds with working-class masculinity, see Lie (1995).
10. Others, however, have found that trust in the police tends to be higher than trust in many other public institutions, and is consistently particularly high

in Norway and other Scandinavian countries (Listhaug and Wiberg 1995, Listhaug personal communication 2010).

BIBLIOGRAPHY

Carlström, A.K. (1999) *På spaning i Stockholm. En etnologisk studie av polisarbete.* Stockholm: Institutet för folklivsforskning, Stockholms universitet.

Chan, J. (2003) "Police and New Technologies," in *Handbook for Policing*, ed. T. Newburn, 655–679. Collumpton: Willan.

Clarke, R.V., and J. Eck (2003) *Become a Problem Solving Crime Analyst in 55 Small Steps.* London: Jill Dando Institute of Crime Science.

Dunworth, T. (2006) "An Overview of Policing Use of Geographic Information Systems (GIS) in the United States," presented at the Crime, Justice and Surveillance Conference, Sheffield April 5–6, 2006.

Ericson, R., and Haggerty, K. (1997) *Policing the Risk Society.* Toronto: University of Toronto Press.

Feeley, M., and J. Simon (1992) "The New Penology: Notes on the Emerging Strategy for Corrections," *Criminology* 30(4): 449–475.

Garland, D. (2001) *Culture of Control: Crime and Social Order in Contemporary Society.* Oxford, UK: Oxford University Press.

Geodata (2009) "Lovens arm over hele kartet". http://www.geodata.no/GIS-i-praksis/Politiet/ (accessed February 5, 2009).

Giddens, A. (1990) *The Consequences of Modernity.* Cambridge, UK: Polity Press.

———. (1991) *Modernity and Self-Identity: Self and Society in the Late Modern Age.* Cambridge, UK: Polity Press.

Harcourt, B. (2007) *Against Prediction: Profiling, Policing, and Punishing in an Actuarial Age.* Chicago, IL: University of Chicago Press.

Home Office (2005) *Crime Mapping: Improving Performance: A Good Practice Guide for Front Line Officers.* London: Home Office.

Hope, T., and R. Sparks (2000) "For a Sociological Theory of Situations (or How Useful is Pragmatic Criminology?)," in *Ethical and Social Perspectives on Situational Crime Prevention*, eds. A. von Hirsch, D. Garland and A. Wakefield, 175–189. Oxford, UK: Hart.

Innes, M. (2004) "Reinventing Tradition? Reassurance, Neighbourhood Security and Policing," *Criminal Justice* 4(2): 151–171.

Johnston, L. (2000) *Policing Britain: Risk, Security and Governance.* Harlow, England: Longman.

Kemshall, H. (2003) *Understanding Risk in Criminal Justice.* Maidenhead, UK: Open University Press.

Lie, M. (1995) "Technology and Masculinity—The Case of the Computer," *European Journal of Women's Studies* 2(3): 379–394.

———. (2003) "Gender and ICT—new connections," in *He, She and IT Revisited: New Perspectives on Gender in the Information Society*, ed. M. Lie, 9–33. Oslo: Gyldendal.

Lie, M., and K.H. Sørensen (1996) "Making Technology Our Own? Domesticating Technology into Everyday Life," in *Making Technology Our Own? Domesticating Technology into Everyday Life*, eds. M. Lie and K.H. Sørensen, 1–30. Oslo: Scandinavian University Press.

Lipsky, M. (1980) *Street-level bureaucracy: Dilemmas of the Individual in Public Services.* New York: Russell Sage Foundation.

Listhaug, O. and M. Wiberg (1995) "Confidence in Political and Private Institutions" in *Citizens and the State*, eds. H.-D. Klingemann and D. Fuchs, 298–322. New York: Oxford University Press.

Loader, I., and A. Mulcahy (2003) *Policing and the Condition of England: Memory, Politics and Culture.* Oxford, UK: Oxford University Press.
Lynch, M. (2000). "Rehabilitation and Rhetoric: The Ideal of Reformation in Contemporary Parole Discourse and Practices," *Punishment and Society*, 2(1): 40–65.
Maguire, M. (2000) "Policing by Risks and Targets: Some Dimensions and Implications of Intelligence-Led Crime Control," *Policing and Society* 9: 315–36.
Mazerolle, L.G., C. Bellucci, and F. Gajewskij (1998) "Crime Mapping in Police Departments: The Challenges of Building a Mapping System," in *Crime Mapping Crime Prevention*, eds. D. Weisburd and T. McEwen, 131–155. New York: Criminal Justice Press.
O'Malley, P. (2000) "Risk Societies and the Government of Crime," in *Dangerous offenders. Punishment and social order*, eds. M. Brown and J. Pratt, 17–33. London: Routledge.
———. (2001). "Risk, Crime and Prudentialism Revisited," in *Crime, Risk and Justice: The Politics of Crime Control in Liberal Democracies*, eds. K. Stenson and R.R. Sullivan, 89–103. Cullompton: Willan.
———. (2004) *Risk, Uncertainty and Government.* London: Glasshouse Press.
Oslo Politidistrikt (2000) *Prosjektrapport—POP 2000*, Oslo: Sentrum politistasjon.
Politidirektoratet (2002) *Strategiplan for forebyggende arbeid 2002–2005*, Oslo: Politidirektoratet.
———. (2007) *Nasjonal strategi for etterretning og analyse*, Oslo: Politidirektoratet.
Reiner, R. (2000) *The Politics of the Police.* Oxford, UK: Oxford University Press.
Shearing, C., and R. Ericson (1991) "Culture as Figurative Action," *British Journal of Sociology* 42(4): 481–506.
Sherman, L., D.P. Farrington, B.C. Welsh, and D. L. MacKenzie (2002) *Evidence-Based Crime Prevention*, London: Routledge.
Silverstone, R., E. Hirsch, and D. Morley (1992) "Information and Communication Technologies and the Moral Economy of the Household," in *Consuming technologies media and information in domestic spaces*, R. Silverstone and E. Hirsch 15–31. London: Routledge.
Smith, M.R., and L. Marx, eds. (1994) *Does Technology Drive History? The Dilemma of Technological Determinism.* Cambridge, MA: MIT Press.
Smith, M., and Tilley, N., eds. (2005) *Crime Science, New Approaches to Preventing and Detecting Crime.* Portland: Willan Publishing.
Stoe, D. (2003) "Using Geographic Information Systems to Map Crime Victim Providers Services: A Guide for State Victims of Crime Act Administrators and Victim Service," http://www.ojp.usdoj.gov/ovc/publications/infores/geoinfosys2003/191877.pdf (accessed February 5, 2009).
Tolloczko, P., and A. Rimstad (1999) *GIS 2000*, Oslo: Sentrum Politistasjon.
Wright, A. (2002) *Policing: An Introduction to Concepts and Practice.* Cullompton: Willan Publishing.
Zedner, L. (2009) "Epilogue: The Inescapable Insecurity of Security Technologies?" in *Technologies of (In)security: The Surveillance of Everyday Life*, eds. K. F. Aas, H. O. Gundhus and H. M. Lomell, 257–270. Abingdon: Routledge-Cavendish.

Contributors

Roland Bal is a Professor of Healthcare Governance in the Department of Health Policy and Management, Erasmus University, Rotterdam, The Netherlands. His research interests include the building and workings of information infrastructures and the evaluation of quality programs in healthcare.

Ellen Balka is a Professor in Simon Fraser University's School of Communication (Canada), where she is also the Director of the Assessment of Technology in Context Design Lab. For over a decade, her work has focused on social aspects of information technology in the health sector. Her publications about social aspects of information technology have appeared in numerous journals and books in computer science, health, communication, and science, technology and society studies. She currently holds a Michael Smith Foundation for Health Research Senior Scholar's Award and is working on a manuscript about the role of technology in the social construction of health indicators.

Alain Desrosières is a member of the Institut National de la Statistique et des Etudes Economiques (INSEE), and of the Centre Alexandre Koyré d'Histoire des Sciences, both in Paris. He is a statistician and a specialist in the history and sociology of statistics. He is the author of *The Politics of Large Numbers: A History of Statistical Reasoning* (Harvard University Press 1998), and, more recently, "Comparing the Incomparable: The Sociology of Statistics" in *Augustin Cournot: Modelling Economics* (Edward Elgar 2007).

Helene I. Gundhus, having previously held a post in Criminology at the University in Oslo, is currently an Associate Professor at the Research Department of the Norwegian Police University College. She obtained her PhD in Criminology on the theme: ICT, knowledge work, and occupational cultures in the police. Her research interests include policing, technology, and crime prevention. Together with K.F. Aas and H.M.

Lomell, she is coeditor of *Technologies of InSecurity: The Surveillance of Everyday Life* (Routledge Cavendish 2009).

Svein Hammer held a post-doc at the Norwegian University of Science and Technology at the time that he wrote his work in this book. He is now working at the Norwegian State Housing Bank, where he is responsible for the strategic development of housing research in Norway. He is a sociologist with a PhD on governmental discourses in Norway, and has written articles about Foucault and (especially) the use of the discourse concept as a tool for analyzing processes of production and reproduction of social reality.

Jon Hovland was a PhD fellow at the Norwegian University of Science and Technology when he wrote his chapter in this book. He currently holds a position as adviser at the Norwegian Ministry of Government Administration, Reform and Church Affairs. His PhD dissertation is in Sociology, to be defended Spring 2010 at the Department of Sociology and Political Science, Norwegian University of Science and Technology. Its title (translated from Norwegian) is "Numbers Talk: Measurement and Accountable Governing in Municipal Administration."

Sonja Jerak-Zuiderent is a PhD researcher at the Healthcare Governance Research Group in the Department of Health Policy and Management, Erasmus University, Rotterdam, The Netherlands. Her research interests include ethnographies of science and technology, standards and situated healthcare practices, particularly on issues of accountabilities and uncertainties.

Christopher Kullenberg is a PhD candidate in Theory of Science at Gothenburg University. His dissertation concerns the history and epistemology of the Swedish social sciences in the postwar era with special concern to the process of quantification.

Heidi Mork Lomell is Senior Research Fellow at the University of Oslo, Faculty of Law. She has written on various topics relating to surveillance, crime control and human rights, most recently in *Technologies of InSecurity: The Surveillance of Everyday Life* (coedited with Katja Franko Aas and Helene I. Gundhus; Routledge-Cavendish, 2009). She is also contributing to the new *International Handbook of Criminology* (Taylor & Francis, 2010) with the chapter "The Politics of Numbers: Crime Statistics as a Source of Knowledge and a Tool of Governance."

Kjetil Rodje is a PhD candidate at the School of Communication, Simon Fraser University, Canada, and holds a cand.polit. degree in Sociology from the University of Oslo, Norway. He is the coeditor (together with Casper Bruun Jensen) of *Deleuzian Intersections: Science, Technology, Anthropology*.

Asunción Lera St. Clair is a Professor of Sociology at the University of Bergen, and Scientific Director of the Comparative Research Programme on Poverty (CROP), International Social Science Council (ISSC), and University of Bergen. Her most recent publications are *Global Poverty, Ethics and Human Rights: The Role of Multilateral Organizations* (with Desmond McNeill; Routledge 2009) and *Ethics, Human Security and Climate Change* (with Karen O'Brien and Berit Kristoffersson; Cambridge University Press 2010).

Ann Rudinow Sætnan is a Professor of Sociology at Norwegian University of Science and Technology. This volume is among the results of her project "For Whom the Bell Curves," which was funded by the Norwegian Research Council. Other publications concerning statistics practices are "All Fetuses Created Equal?" in *Gender, Bodies and Work* (Ashgate 2005) and "Nothing to Hide, Nothing to Fear?" in *International Criminal Justice Review* (2007).

Luisa Farah Schwartzman is an Assistant Professor in the Department of Sociology at the University of Toronto. Her work is concerned with how subjective understandings of social categories are affected by—and also help reproduce—social structures, especially in the Brazilian context. Her recent work includes a paper entitled "Does Money Whiten: Intergenerational Changes in Racial Classification in Brazil," *American Sociological Review* (December 2007).

Sigrunn Tvedten is a PhD fellow in Sociology at the Graduation School of Educational Governance and Social Inequality/Department of Sociology and Political Science, Norwegian University of Science and Technology. Her field of interest is rationalities of educational governance—in particular intersections of discourses on school accountability, and social equity and exclusion/inclusion in education.

Gunhild Tøndel is a PhD fellow working at the Department of Sociology and Political Science, Norwegian University of Technology and Science and NTNU Social Research AS. Her MA thesis was a study of diagnosis coding practices in an internal medicine department. She is currently working on her PhD thesis on statistics construction and use in municipal health and care services.

Chris Williams is a Lecturer in History at the Open University, with an interest in the history of crime and policing. He has published on a number of topics on the history of the criminal justice system in the United Kingdom and its empire, and is currently working on a book exploring the evolution of the sociotechnical systems which have been used over the last two centuries to control policing. On the topic of criminal statistics, he has published "Counting Crimes or Counting People: Some Implications of Mid-Nineteenth Century British Police Returns" *Crime, Histoire & Sociétés/Crime, History & Societies* 4(2), November 2000: 77–93, and "Catégorisation et stigmatisation policières á Sheffield au milieu du XIXe siècle" *Revue d'Histoire Moderne et Contemporaine* 50(1), Jan–March 2003: 104–125.

Name Index

A
Aas, I. H. M. 245, 247
Aas, K. F. 280
Abbott, A. 31
Adam, B. 5
Adams, S 240
Agar, J. 158, 168
Akrich, M. 7
Alkire, S. 145
Anand, S. S. 106
Anderson, M. J. 1, 2, 12
Armatte, M. 59
Aspinall, P. J. 102

B
Bal, R. 14, 15, 224, 279
Bailey, S. 119, 131
Baldersheim, H. 209
Balka, E. 14, 37, 99, 172, 173, 179, 180, 187, 279
Bartrip, P. W. J. 157
Bayatrizi, Z. 201
Bayley, D. H. 203–204
Beck, U. 5
Becker, H. 2
Beirne, P. 191
Bell, D. 195
Bellucci, C. 269
Benjamin 121
Benzecri, J.-P. 54
Berg, M. 99, 173, 178, 186, 227, 248, 251, 256, 261
Berge, K. L. 215
Bernouilli, J. 49
Beveridge, R. B. 176
Bhopal, R. 102, 105, 106–107
Bijker, W. E. 6, 7, 8
Bingham, J. W. 188
Bjorn, P. 172

Black, D. J. 193, 202
Blalock, H. M. 31
Blasius, J. 52
Blau, P. M. 31
Bloomfield, B. P. 245, 248
Boltanski, L. 226, 240
Booth, C. 47, 48, 49
Bottomley, K. 202
Bourdieu, P. 55
Bourhis, R. Y. 113
Bowden, G. 6
Bowker, G. C. 2, 3, 8, 25, 173, 176, 177, 178, 179, 186, 187, 235, 246
Bowley, A. 47, 48, 49
Bradby, H. 102
Brante, T. 76
Broadfoot, P. M. 209, 214
Brown, D. 152
Brown, J. S. 196
Bukve, O. 84
Burrell, G. 76
Byrkjeflot, H. 246, 247
Bøås, M. 139, 152

C
Callon, M. 6, 67, 74, 75, 229, 231, 237
Carlström, A. K. 273, 276
Carney, S. 209
Carter, N. 21
Chantraine, A. 12
Chan, B. 181
Chan, J. 273
Chen, J. R. 101
Chester, N. 159, 164
Christensen, T. 199, 209
Christie, N. 195
Clarke, R. V. 268, 276
Colbert, J.-B. 44, 45
Cole, J. 15

Cole, S. 157
Coleman, C. 201, 202
Comstock, R. D. 104
Corbin, J. M. 229
Cournot, A 24, 279
Cowan, R. S. 6, 13
Crawford, N. S. 106
Crocker, D. 152
Cussins, C. 260
Czarniawska, B. 198

D

Dahler-Larsen, P. 216
Dandeker, C. 159, 168
Daniel, R. 119
Darwin, C. 35, 47, 48
David, C. 193
Davis, J. 157
De Gaulle, C. 45
Deleuze, G. 65, 66
De Michelis, A. 12
De Moivre, A. 41
Desrosières, A. 3, 4, 5, 11, 13, 15,
    22, 24, 28, 33, 41, 62, 64, 65,
    66, 79. 84, 85, 86, 192, 202,
    208–209, 233, 279
Dewey, J. 152
Dixon, K. 193
Dos Santos, D. (Frei David) 121, 123, 125
Dowswell, G. 240
Doyle, J. M. 107
Du Cane, E. 158, 167
Duguid, P. 196
Duncan, O. D. 31
Dunworth, T. 266
Durojaye, L. 175

E

Eck, J. 268, 276
Eco, U. 23
Eide, K. 210
Einstein, A. 1
Emmerij, L. 152
Emsley, C. 159
Engel, E. 44, 50
Engelsen, B. U. 216
Ericson, R. 266, 272
Ericsson, K. 193
Eyerman, R. 67

F

Farah Schwartzman, L. xiii, 14, 15,
    117, 133, 279
Fausto-Sterling, A. 107

Feeley, M. 267
Felson, M. 203
Feuer, L. S. 251
Fienberg, S. E. 1, 2, 12
Fillitz, T. 22
Finkler, K. 245
Fisher, R. 49, 53
Flegg, G. 27
Foucault, M. 8, 10, 13, 15, 37, 64,
    66–67, 79, 80, 81, 82, 83, 86,
    87, 89, 93–94, 240, 280
Freedman, D. A. 31
Freeman, T. 224, 225, 226, 227
Fridjónsdóttir, K. 66, 67, 68–69
Friedrichs, R. W. 76
Frisch, R. 46, 49, 59
Fry, P. 119, 129
Fukuda-Parr, S. 142

G

Gaebler, T. 199
Gajewskij, F. 269
Galton, F. 41, 47–48, 52
Garland, D. 195, 199, 203, 265
Gasper, D. 140, 141, 145, 146
Gatrell, V. A. C. 159
Gauss, C. 41, 47, 49
Geest, S. v. d. 245
Giddens, A. 118, 275
Giere, R. 30
Gieryn, T. 76, 137, 138
Gillard, V. 27, 28
Gilling, D. 202
Giuffre, M. 181
Gjøen, H. 7
Glomsaker, T. 247
Godfrey, B. S. 167, 192
Goldstein, H. 197
Goodhart, C. 201, 204
Gordon, C. 83, 87, 94
Gouin, S. 175, 176
Granheim, M. 221
Greenacre, M. 52
Griesemer, J. R. 9, 23, 76, 138, 173,
    174, 175, 200
Grinten, T. E. D. 241
Guattari, F. 65, 66
Guimarães, A. S. A. 119
Gundhus, H. I. 15, 79, 263, 280
Guston, D. 139, 152
Guy, W. 201

H

Haavelmo, T. 49

Name Index   285

Hacking, I. 3, 4–5, 11, 15, 191–192
Hadden, T. 159
Hafstad, A. 247
Hägerström, A. 70
Haggerty, K. D. 12, 192, 272
Hammer, S. xiii, 1, 10, 14, 76, 79, 82, 86, 94, 100, 207, 280
Haq, K. 143
Haq, M. 140, 142, 143, 144, 148, 151
Haraway, D. 1, 192, 230
Harcourt, B. E. 168, 197, 267
Harper, R. 32
Harrison, S. 240
Hasenbalg, C 119, 125
Heisenberg, W. 201
Hempel, C. G. 69
Henderson, E. 158, 165, 167
Hennebry, J. 103
Hennock, E. P. 49
Henriques, R. 120
Herrnstein, R. J. 29, 31
Hirsch, E. 6, 276
Holmberg, S. 73, 75
Hood, C. 24, 26, 34, 37, 199
Hope, T. 275
Hope, Ø. B. 246, 248
Hopmann, S. 209
Horn, D. G. 65
Hoskin, K. 201, 202, 204
House, E. 31
Hovland, J. xiii, 13, 21, 31, 38, 82, 280
Htun, M. 120
Hughes, D. 245
Hwang, H. 198
Hård, M. 7

I
Igo, S. E. 12
Inglehart, R. 72
Innes, M. 270, 275
Irby, K. 100
Ireland, R. W. 157

J
Jamison, A. 67
Jasanoff, S. 66, 137, 148
Jerak-Zuiderent, S. 14, 15, 224, 280
Johnsen, A. 84
Johnston, L. 197, 265
Jolly, R. 152
Jordan, B. 173
Flavius Josephus 28
Jørgenvåg, R. 246, 248

K
Kafka, F. 27
Kahn, J. 38
Karlsen, G. 210, 214
Karlsen, S. 105
Kaul, I. 142–143,
Kemshall, H. 273, 274, 275
Keymer, E. S. 164
Keynes, J. M. 43, 44, 45, 50
Kirdar, U. 143
Kirkey, S. 181
Kiær, A. 44, 49, 50–52
Kjærnsli, M. 211
Klein, L. 46
Kovalevski, A. 49
Krieger, N. 105, 107, 144
Kuhn, T. 76
Kula, W. 28
Kullenberg, C. 13–14, 64, 86, 94, 280

L
Lampland, M. 24, 27, 37
Laplace, P.-S. 41, 47–48, 49
Lascoumes, P. 41
Latour, B. 6, 7, 8, 23, 66, 75, 198, 226, 228, 229, 239, 257
Le Galès, P. 41
Legendre, A.-M. 49
Le Play, P. 50
Lepore, S. J. 100
Levias, D. 62
Levitas, R. 201
Lie, A. 209
Lie, E. 12, 44
Lie, M. 6, 263, 274, 276
Lie, S. 211
Liesman, S. 22
Lincoln, A. 10
Lipsky, M 273
Listhaug, O. 277
Loader, I. 275
Lomell, H. M. xiii, 1, 14, 79, 100, 191, 280
Long, M. 199
Luhmann, N. 26, 28, 34–35
Lundgren, U. P. 221
Lushington, G. 167
Lynch, M. 273
Lægreid, P. 199, 209

M
Machado, E. A. 120
MacKenzie, D. A. 3
Maggie, Y. 126, 133
Maguire, M. 192, 263, 265

March, L. 44
Marx, K. 70
Marx, L. 276
May, E. 166, 167
Mayston, D.J. 21
Mazerolle, L. G. 269
McIntosh, T. 181
McIntrye, M. 201
McKenzie, K. 106
McLaughlin, E. 199, 203
McNamara, C. 21
McNamara, R. 140
McNeill, D. 139, 140, 152, 281
Medeiros, C. A. 121
Mediås, O. A. 209, 210, 211, 213
Mesman, J. 240
Mills, C. W. 30
Moe, O. 221
Mol, A. 240, 241
Moos, L. 209, 221
Moran, M. 158, 168
Morgan, G. 76
Morgan, M. S. 30, 49
Morris, R. M. 169
Morrison, M. 30
Moynihan, J. 201
Mulcahy, A. 275
Munro, R. 193
Murray, C. A. 29, 31
Mussolini, B. 65
Myrdal, G. 77

N
Naschold, F. 209
Nazroo, J. Y. 105
Neyman, J. 49
Nivière, D. 58
Nobles, M. 119
Nussbaum, M. 145

O
Occam (see Seach, W.)
Offerdal, A. 84
Ohlsson, A. 73, 74, 75
O'Malley, P. 11, 12–13, 91, 158, 273, 274, 275
O'Meara, M. 175
Omi, M. 117, 119
Opedal, S. H. 84
Osborne, D. 199
Oudshoorn, N. 1, 2, 6, 7, 15

P
Painter, C. 203

Parson, T. 73, 75
Patriarca, S. 65
Pearson, E. 49
Pearson, K. 41, 47, 48, 49, 52, 53, 72
Pendakur, R. 103
Peria, M. 120, 121, 125, 133
Petersson, O. 73
Pinch, T. 1, 2, 6, 7, 15
Pogge, T. 146–147, 149, 150
Polanyi, K. 48
Pollitt, C. 240
Pols, J. 240
Pont, B. 209
Porter, T. M. 3, 4, 5, 11, 15, 25, 66, 118
Posner, K. L. 245
Power, M. 23, 34, 227

Q
Quetelet, A. 5, 24, 41, 47, 48, 52, 72, 191–192

R
Ranis, G. 142
Ramos, C. 120, 129
Rawson, R. W. 159
Reddy, S. 149
Reiner, R. 204, 265, 275
Reuter, P. 202
Risch, N. 107
Rimstad, A. 267
Rodje, K. 14, 99, 281
Rokeach, M. 72
Roll-Hansen, N. 12
Rose, L. 209
Rose, N. 90
Rowntree, S. 47, 48, 49
Ruhleder, K. 177

S
Sacco, V. F. 192
Sansone, L. 126, 127
Sasson, T. 194
Schjerve, H. 247
Scholten, G. R. M. 241
Schuurman, N. 172, 187
Schull, M. 181
Schultz, S. 181
Scott, J. C. 118
Seach, W. (William of Occam) 251
Segerstedt, T. T. 69, 77
Sekula, A. 157
Sen, A. K. 142, 143, 144, 145, 150, 152
Sending, O. J. 86

Name Index

Senior, P. A. 105, 106
Sevón, G. 198
Sharif, N. A. 101
Sharman, Z. 172
Shearing, C. 266
Sheriff, R. E. 126
Sherman, L. 276
Sheth, T. C. 101
Shore, J. 162–163
Silva, G. M. D. 119, 120, 129
Silva, N. V. 119
Silverstone, R. 6, 276
Simiand, F. 53
Simon, J. 195, 267
Sin, L. 181
Sindall, R. 157, 162, 204
Sjøberg, S. 210
Skidmore, T. 119
Skolbekken, J.-A. 5
Slåttebrekk, O. V. 246
Smith, D 185
Smith, M. 276
Smith, M. R. 276
Smith, P. 202, 203
Sparks, R. 275
Spearman, C. 54, 72
Spurkeland, J. 84
St. Clair, A. L. 14, 136, 137, 138, 140, 142, 143, 144, 149, 152, 281
Stanford, T. G. 157
Stanley, E. H. (Lord Derby) 164
Star, S. L. xiii, 2, 3, 8, 9, 23, 24, 25, 37, 76, 138, 173, 174, 175, 176, 177, 179, 187, 200, 226, 235, 239, 246
Starr, P. 201
Sterling, T. 105
Stevenson, S. 157
Steyn, M. 181
Stoe, D. 266
Storstein, A. & M. 25, 26, 37
Strathern, M. 26, 204, 227
Streeten, P. 140, 142, 143
Strauss, A. L. 226, 229, 234, 239, 251, 252, 256, 260
Suarez, D. 198
Suchman, L. 173, 226, 239
Suppes, P. 30
Swidler, A. 132, 133
Sætnan, A. R. xiii, 1, 100, 104, 204, 261, 276, 281
Søgnen, A. 211
Sørensen, K. H. 6, 274, 276
Sørhaug, T. 216

T

Teixeira, M. P. 126
Telhaug, A. O. 209, 210, 211, 213
Telles, E. E. 117, 119, 120, 127, 131
Thévenot, L. 226, 240
Thomassen, Ø. 5
Thoreau, H. D. 191, 192
Thurstone, L. 54
Tilley, N. 276
Timmermans, S. 99
Tinbergen, J. 46, 59
Tobias, J. J. 161
Tolloczko, P 267
Torjesen, D. O. 246, 247
Trevelyan, G. M. 164
Tukey, J. 54
Tvedten, S. 14, 207, 281
Tøndel, G. xiii, 15, 245, 248, 260, 281

V

Vandermoortele, J. 140
Vanoli, A. 44
Vissandjee, B. 101
Von Otter, C. 209

W

Wade, R. 140, 149
Wagner, I. 173
Waring, M. 1, 2
Weber, M. 42, 158
Weibull, L. 75
Weiss, T. 152
Wiberg, M. 277
Whitehouse, S. 172, 180
Wiener, C. 99, 177, 227, 228, 229
Wiener, M. J. 161, 157
Winant, H. 117, 118–199
Westerståhl, J. 70, 73, 77
Williams, C. 14, 79, 157, 161, 163, 281
Williams, D. R. 105
Woolf, S. J. 163
Wright, A. 273
Wright, C. 44
Wu, Z. 100, 104

Y

Young, J. 203
Young, M. 181
Young, Ma. 202
Yule, U. 47, 48, 49, 52–53

Z

Zedner, L. 168, 267
Zuiderent-Jerak, T. 240

Subject Index

A
ABF (*See* Activity Based Financing)
abilities: 41, 47, 48, 54
Aboriginal peoples (*See* ethnic categories)
academia/academic/academics: 26, 67, 70, 71, 73, 125, 139, 141, 142, 143, 147, 149, 201
access domain: 180, 181–182, 183, 186
accountable/accountability (*See also* transparency): 10, 11, 14, 21, 35, 99, 111, 136, 149–151, 177, 207, 209, 210–211, 212–214, 216, 217, 220, 225, 226–227, 230, 231, 238–239
accountable communication: 24, 27, 32–25, 36, 37
accountability domain, 180, 182–184, 186
accoutants/accounting: 22, 25, 26, 32, 33, 34, 37, 59, 60–61
activism/activist(s): 117–118, 119, 120–121, 125, 129
Activity Based Financing: 245, 246, 248, 252, 260, 261
actor(s): 2, 3, 6, 8, 13, 25, 26, 34, 35, 36, 54, 73, 75, 84, 94, 117, 123, 129, 132, 137, 138, 139, 141, 145, 149, 172–173, 174, 178, 198, 201, 204, 210, 216, 218, 219; non-human actors/actants, 6–7, 67, 75, 183, 228; rational actors, 208, 217–218, 220
Actor-Network Theory (also called Translation Theory): 6–7, 67, 73
administration/administrative/administrators: 3, 9, 11, 12, 15, 33, 45, 48, 54, 55, 58, 59, 65, 82, 84, 86, 90, 91, 120, 159, 163, 164, 184, 194, 195, 212, 229–230, 234, 235, 246, 247, 248, 250–251, 253, 259, 266
adult (*See* age categories)
affirmative action:14, 117–118, 120, 121, 125, 128, 130, 131, 132–133
affluent (*See* class categories)
Afro-Brazilian (*See* racial categories)
age categories: adult 74, 132–133, 14, 176, 177, 178; child(ren)/pediatric, 49, 164, 176, 177, 178, 186, 187; elderly, 49, 233; infant, 164; youth(s), 74, 196–197, 269
agencies: 5, 12, 14, 45, 46, 68, 69, 108, 119, 136, 137, 139, 140, 141, 142, 143, 145, 148, 152, 158, 159, 163, 201, 225, 269, 271, 274
agricultural surveys: 44
amarela (*See* racial categories)
Anglo-American (*See* racial categories)
annual reports/annual returns: 14, 29, 159–161, 163, 183, 191–204
ANOVA: 30–31
anti-program: 7, 9, 74, 75
assistance unions: 47, 48
audit: 22, 24, 26, 28, 157, 167, 199, 203, 265, 273; audit cultures, 227
Australia: 162, 175, 177

B
Balanced Scorecard: 21, 32–33, 34, 38, 79, 89, 91, 92
basic needs: 109, 140, 142, 245, 151
basic skills: 210, 212, 213–216
behavior: as a variable, 41, 64, 66, 69, 73, 106, 108, 112; anti-social/

Subject Index

criminal/deviant, 69, 265, 268, 276; electoral 55; individual, 41, 46, 108; risk, 11, 110, 268; stakeholders', 46, 56, 57; and stereotyping, 106, 107–108; strategic, 47, 57, 137, 202; organizational, 21
bell curve (also called "normal law"): 41, 47, 48
Bell Curve, The: 29, 31
benchmark(ing): 29, 45, 46, 55–60, 89, 94, 209
"black" (*See* racial categories)
black box(es): 42, 61, 108
Black Movement(s): 117–121, 125, 126, 129, 131, 132
blue collar (*See* class categories)
boundary object(s): xiii, 23, 26, 34, 35, 76, 136, 138, 141–152, 173, 174–175, 178, 186, 200–201; definition of 9, 138–139, 174, 200–201
boundary organization(s): 136–137, 139, 143, 144, 148, 150
boundary work: 76, 136, 137, 138, 139, 141, 142, 143, 145, 146, 147, 148, 149, 150
branco (*See* racial categories)
Brazil: 14, 117–135
budget(s): 44, 45, 50, 56, 58, 92, 184, 245, 247, 248, 249, 250, 255, 265
bureaucracy(ies)/bureaucrat(s): 21, 27, 42, 84, 93, 118, 126, 128, 132, 158–159, 163, 164, 176, 199, 273

C

Canada: 99–116, 172–190
capability (*See also* HD-CA): 14, 136, 144–147, 151
carento (*See* class categories)
categories/categorization/categorizing (*see also* age categories, class categories, etchnic categories, racial categories): 1, 2, 3, 6–7, 8, 14, 24, 25, 47, 48, 53, 54, 55, 57, 64, 66, 79, 80, 81–82, 85, 86, 99–112, 117–135, 145, 146, 147, 160, 164, 175, 176, 185, 187, 193, 202, 219, 237, 248, 250–251, 253, 254, 257
caucasian (*See* racial categories)
cause/causal(ity): 6, 14, 24, 30, 31, 36, 41, 55, 69, 101, 102, 104, 108, 148, 191–192, 194, 196, 197, 199, 237, 238, 251, 264; chicken/egg question, 35
census: 12, 44, 51, 65, 67, 103–104, 133
child(ren) (*See* age categories)
Chinese (*See* ethnic categories)
citizen(s)/citizenship: 2, 12, 13, 14,26, 49, 93–94, 103, 113–114, 118, 121, 131, 137, 141, 200, 225, 266, 268, 269–270
civil servants/civil service: 158, 159, 162, 163, 164, 165
class: 14, 47, 48, 49, 50, 51, 53, 54, 55, 66, 77, 105, 124, 126, 127, 128, 129, 132, 133, 157, 162, 164, 168, 276
class categories: affluent, 51; blue collar, 46; *carente* 121, 133; criminal, 162, 168; dangerous, 168; lower, 48; lower-middle, 133; middle, 50, 77, 127; poor, 48, 49–50, 77; upper, 48, 51; well-to-do, 50; working, 47, 49, 50, 51, 276
classification/classification system(s): 26, 27, 45, 46, 48, 54–55, 57, 58, 99, 102, 104, 106, 111, 117–118, 119, 121–133, 164, 173, 175–177, 179, 180–181, 185–187, 202, 245, 246–248, 253, 256, 257, 259, 261, 275
"club" government: 158, 168
co-construction (*See* mutual construction)
code(s)/coding: 3, 10, 22, 34–35, 58, 85, 164, 209, 228, 234, 235, 238, 246, 248. 249–260, 271; communicative codes (*see* under communication)
Colbertism: 44, 45
communication: 21, 24, 25–26, 27, 28, 30, 32–35, 36, 37, 58, 61, 73–74, 101, 117, 133, 160, 164, 177, 180, 181, 201, 240, 251, 256, 258, 267, 268–269; communication codes/communicative systems: 26, 28, 34–35, 36
communitarians: 145
competition: 10, 43, 89, 93, 94, 209, 219, 229
consensus, 33, 139, 141, 143, 149, 150, 218; lack of, 129
constructivism/constructionism: 11, 26, 66–67, 192–193

consumer(s): 6, 13, 44, 67, 75, 233; consumerism/consumption, 5, 46, 50, 55, 66, 140, 144, 200; consumer price index, 44, 45, 46
contingency table: 48, 54
convention(s): 4, 22, 24, 27–28, 35, 36, 49, 53, 54–55, 58, 178, 260
convicts (see criminals)
coordinate(d)/coordination: 11, 41, 45, 56, 57, 144, 172, 173, 186, 201, 236, 251, 259, 260; uncoordinated, 199
cooperate/cooperation/cooperative: 7, 37, 71, 132, 138, 159, 174, 268, 269, 271
coproduction (*See* mutual construction)
correlate/correlation: 24, 48, 49, 52, 53, 61, 68, 104, 105, 107, 108, 111, 112, 161, 196
correspondence analysis: 52–55
corroborating analysis: 54, 55
count/coutable/counted/counting: 1–2, 8, 15, 23, 36, 103, 104, 112, 157, 158, 160, 161–162, 191, 203, 246–249, 253, 255–256, 259, 270, 273
crime, crime maps/rates/statistics, criminals: 5, 14, 15, 66, 79, 157–169, 191–204, 263–276
crime victims: 197, 263, 265, 268
criminal (*See* class categories)
culture: 60, 72, 90, 99, 108, 109, 110, 133, 139, 142, 145, 150, 152, 202, 203, 209, 211, 227, 266, 271, 272, 273, 274, 276

D

dangerous (*see* class categories)
Darwinism: 35, 47, 48
data/databank(s)/data register(s)/data warehouses: 3, 5, 8, 12, 14, 15, 22, 25, 30–31, 37, 43, 46, 54, 56, 59, 60, 61, 62, 67, 70, 71, 75, 76, 85, 100, 102, 103, 104, 106, 107, 108, 111, 112, 119, 121, 125, 146, 150, 159, 164, 168, 172, 173, 175, 176, 177, 179, 180, 181, 182, 183, 184, 185, 186, 187, 192, 194, 196, 201, 202, 203, 207, 208, 211, 212, 213, 215, 216, 217, 218, 219, 220, 221, 224, 226, 228, 229, 230–233, 236, 237, 238, 239, 241, 245, 246, 247, 248, 261, 263, 264, 265, 266, 268, 269, 270, 271, 273, 275
data analysis: 54, 61
data coders: 228, 234, 235, 238
Definition of Selected Competencies (DeSeCo): 215
deliberative participation: 136, 149-152; definition of, 149
deviance: 69, 76, 191; statistical, 80
determinism/determinists: 5, 6, 21, 267, 269, 276
Diagnosis Related Groups (DRG): 235, 245–261
discourse(s): 2, 3, 10, 11,13, 15, 42, 55, 80, 81–82, 84, 86, 87, 88, 89, 93, 119, 128, 131, 141, 147, 151, 192, 193, 194, 197, 200, 204, 208, 209, 218–219, 224, 240, 263–265, 266, 267, 270, 271, 273, 274
discrimination: 14, 105, 106, 107, 110, 117, 118, 123, 124, 126, 127, 128, 129, 130–131, 132, 133, 146
disease: 5, 100–101, 106, 107, 191, 233, 234, 236, 238, 248, 261
diversity: 43, 59, 99, 108–109, 110–111, 112, 174
domain of care: 178, 180—181, 186
DRG (*See* Diagnosis Related Group)
Durban conference: 120

E

East Asian (*See* ethnic categories)
econometrics/econometricians: 43, 46, 49, 50, 52, 53, 55–60, 61
economic policies: 43, 50
economics: 42, 43, 46, 67, 136, 139, 140, 141, 144, 145, 146, 147
economists/economistic: 52, 59, 60, 61, 90, 120, 144, 146, 151, 172, 173
economy: 10, 32, 34, 35, 38, 42, 43, 44, 45, 46, 59, 68, 81, 209, 224; political 146
education/educational: 14, 29, 31, 53, 57, 125, 128, 130, 133, 159, 194, 200, 207, 208, 209–210, 211, 213, 214, 215, 216, 217, 218, 220, 224, 263
efficiency: 56, 85, 139, 140, 144, 163, 164, 165, 199, 209, 210, 224, 246, 247, 267
egalitarianism: 14, 49
elderly (*See* age categories)

(un)employment: 3, 36, 44, 45, 57, 58, 59, 123, 130, 140, 196, 200
engineer(s)/(social) engineering: 6, 7, 42, 49, 58, 67, 68, 69 70, 73, 74,75, 76, 80
engineer state: 44, 45
England: 42, 47–50, 68, 157–169, 275
Enterprise Master Person Index (EMPI): 180–181
entrepeneur: 6, 59, 87, 90
environment(al): 106, 108, 127, 139, 143, 152, 173, 181, 186, 210, 211, 212, 213, 214, 215, 240
episteme(s)/epistemic/epistemology: 47, 55, 64, 66–67, 70–75, 76, 137, 191, 204
equity: 51, 112–113, 136, 140, 149, 150, 152, 182–184, 186
escurinha (*See* racial categories)
ethnic(ity): 14, 99–114, 121, 126
ethnic categories: Aboriginal peoples, 40, 103, 110, 138, 141; Chinese, 100–101; East Asian, 100–101; European, 100–101; Native Americans, 107; Pacific Islanders, 107; Romani, 69
ethics/ethical: 2, 94, 99, 136–137, 140, 142, 143, 144, 147–151 227, 266, 275
eugenic(s)/eugenicist(s): 41, 47, 48, 49, 52
European (*See* ethnic/racial categories)
European Open Method of Coordination (OMC): 45, 56, 57–58, 59
European Union (EU): 56, 57, 58
Eurostat: 12, 58
experiment(s)/experimental: 25, 31, 46, 53, 54, 55, 226, 229
expert(s)/expertise: 52, 59, 138, 139, 142, 143, 146, 148, 149, 150, 226, 231, 274

F

fact(s): 22, 32, 64, 70, 71, 72, 73, 85, 88, 90, 124, 140, 162, 197, 198, 217, 236, 257, 266, 270
factor analysis: 72
fallacy: 21, 203–204
family/families/familial: 14, 44, 50, 54, 81, 105, 117, 121, 124, 125, 126, 128, 129, 130, 131
federal equity domain (*See* accountability domain)
(in)fertile/fertility: 48–49, 66

Financial Accounting Standards Board: 22
formalism(s): 41–42, 49, 52
France/French: 5, 44, 45, 54, 55, 56, 60, 67, 69, 103, 202
fraud: 27, 117, 118, 121, 122, 123, 124, 130, 133
free trade: 45
French Law Relating to Financial Laws (LOLF): 56, 57, 58, 59
fund(s)/funding: 51, 56, 68, 69, 71, 137, 138, 159, 161, 199

G

Gallup Institute: 12, 73
GEA (*See* Global Environmental Assessment Project)
genetic(al)/genetics: 101, 102, 105, 106, 107–108, 110, 112
geography/geographical: 3, 65, 77, 102, 103, 159, 263
German(y): 44, 56, 69
geographical information systems (GIS): 263–277
Global Environmental Assessment Project (GEA): 139
globalization: 139, 150
GNP (*See* gross national product)
goal(s): 21, 51, 53–54, 58, 112, 125, 129, 136, 138, 139, 140, 141, 144, 146, 149, 151, 152, 175, 197, 210–211, 212–214, 215, 216, 217, 218, 219, 220, 240, 260, 266
Goodhart's Law: 201, 204
governance: 4, 5, 6, 9–11, 12, 13, 15, 21, 25, 31, 33, 41, 71, 83, 88, 89, 93, 139, 161, 164, 199, 200, 204, 211, 216, 217, 241, 245; definition of, 15
government(s)/governmental: 2, 3, 5. 9–11, 12, 13, 14, 15, 25, 51, 58, 65, 67, 68, 69, 79, 80–82, 83, 84, 85–89, 90, 91, 93, 94, 100, 103, 110, 119, 120, 125, 157–169, 182, 183, 193, 196, 199, 201, 208, 209, 210, 211, 212, 213, 220, 225, 263; definition of, 15
governmentality(ies): 10, 14, 79–94, 207, 208, 211–220
gross domestic product (GDP): 58, 146
gross national product (GNP): 144

Subject Index 293

group(s): population (citizen, ethnic, social) groups, 3, 9, 14, 29, 47, 54–55, 64, 69, 99, 100, 101, 102, 103, 104, 105, 106, 107, 108, 109, 110, 111, 112–113, 119, 127, 137, 139, 141, 144, 149, 157, 163, 173, 174, 180, 182, 186, 207, 208, 212, 261; project group, 228–229, 230–231, 236, 237, 239

H

Habitual Criminals Act (HCA)/Habitual Criminals Register: 157–169
HbA1c value: 233–236, 241
health: 5, 14, 56, 81, 84, 90, 99, 100–113, 114, 173, 178, 180, 181, 182, 183, 184, 185, 186, 187, 224, 225, 226, 228, 229, 230, 233, 234, 236, 237–239, 240, 245, 246, 247, 248, 260, 261
HD-CA (See Human Development-Capability)
HDI (See Human Development Index)
HDRO (See Human Development Report Office)
heredity: 48
hispanic(s) (See ethnic categories)
home assistance: 47, 49
Home Office (UK): 159–169, 265
homo economicus: 87
hospital(s): 14, 106, 172, 174, 175, 176, 178, 179, 180, 181, 183, 184, 185, 225, 227, 228, 229, 230, 233, 234, 235, 236, 237, 238, 239. 241, 245, 246–249, 250, 251, 252, 258, 259
Human Development-Capability (HD-CA): 136–152
Human Development Index (HDI): 142, 146–147
Human Development Report Office (HDRO): 136, 141, 142, 143, 144, 146, 147, 150
human rights: 151
Human Security Now: 148

I

ICD (see International Classification of Diseases)
identity(ies): 14, 74, 99, 102, 104, 118, 119, 121, 123, 124, 126, 127, 138, 142, 157, 162, 166, 188, 246, 250, 260, 266

ideology: 5, 131, 132, 133, 140, 193
immigrant(s): 101, 102, 103, 104, 113–114
immigration: 102, 103, 119
incentive(s): 45, 46, 58, 59, 122, 148, 159, 202, 224
income: 29, 31, 44, 50, 51, 53, 58, 101, 105, 121, 129, 147, 149
indicator(s): 10, 11, 12, 14, 29, 33, 36, 41, 42, 43, 45, 54, 56, 57, 58–59, 60, 61, 79–94, 101, 125, 176, 181, 199, 200, 202, 203, 207, 208, 210, 211, 212, 214, 215, 216, 217, 218, 220, 221, 224–241, 265, 270, 271, 276
índio (See racial categories)
infant (See age categories)
International Classification of Diseases (ICD): 246, 248, 252–255, 260, 261
International Labor Organization (ILO): 140
International Monetary Fund (IMF): 32, 34
International Statistical Institute (ISI): 50, 51, 52
invisible hand: 86–87
IPLOS *Individbasert pleie- og omsorgsstatistikk*: 25, 26, 37
Ireland: 165, 167
Islamabad North South Roundtable: 143
Italy: 65

J

Jacobinism: 60
journal publications: 22
(in)justice(s): 33, 37, 117, 132, 136, 140, 144, 146, 147, 148, 149, 150, 151, 165, 168, 199, 201, 260, 163, 265, 274, 275

K

Keynesianism: 43, 44, 45, 50
knowledge: 1, 3, 4, 5–9, 10, 12, 13, 15, 22, 23, 25, 26, 27–29, 32, 36, 43, 64, 65, 66, 67–70, 71, 82, 83, 84, 86, 80–82, 83, 85, 88, 90, 91, 92, 93, 94, 107, 110, 111, 117, 118, 136–137, 138, 139, 140, 141, 142–143, 145, 147, 148, 149, 150, 151, 152, 161, 167, 168, 174, 176–177, 178, 186, 195–196, 197, 200,

294 Subject Index

208, 209, 211, 216–218, 220, 225, 228, 230, 236, 240, 245, 253, 254, 258, 259, 260, 263–265, 266, 267, 268, 269–274, 275, 276

L

Landelijke Medische Registratie (LMR): 231, 234, 235, 238
language: 3, 4–5, 21–22, 25, 37, 53, 55, 56, 58, 66, 76, 85, 99, 101, 102, 103, 104, 108, 109, 112, 113, 119, 144, 151–152, 158, 164, 196, 240, 247
Latin America: 103
law(s): 5, 21, 23, 28, 35, 41, 45, 48, 49, 50, 53, 54, 56, 61, 66, 81, 82, 114, 120, 121, 122, 125, 128, 129, 130, 131, 133, 160, 162, 163, 191, 219, 258, 264, 266, 269–270, 275; County and Borough Police Act (1857), 159; Goodhart's Law, 201, 204; Habitual Criminals Act, (1869) 162, 163, 164, 165; Laplace-Gauss Law, 41; law of large numbers, 41, 61; LOLF, 56, 57, 58, 59; "normal law", 41, 48; Poor Law (1835), 47, 48, 49; The Prevention of Crimes Act (1871), 165; Speenhamland Laws (1795), 49; Three Strikes Law, 27
learning: 2, 6–7, 8, 128–129, 130, 210, 211, 212, 213, 214, 215, 216, 220, 221, 225, 226, 233, 237, 239, 273
legitimacy: 10, 55, 71, 120, 125, 150, 151, 193
liberal(ism)/neoliberalism: 5, 11, 14, 37, 43, 44, 45, 46, 79–94, 140, 209, 211, 218, 219, 220, 224
life expectancy: 146–147
linear regression: 49- 50
LMR (See Landelijke Medische Registratie)
living conditions: 47, 101, 105
local accounting and quality domain: 180, 184, 186
logic(al): 22, 26, 28, 30, 35, 36, 57, 67, 70, 75, 82, 85, 87, 91, 111, 123, 224–225, 226, 235, 275
logistic regression: 46, 52–55
LOLF (See French Law Relating to Financial Laws)

love: 28
lower/lower-middle (See class categories)

M

macro-level: 11–13, 14, 41, 42, 44–46, 47, 50, 59, 60, 61, 118, 157, 209, 248
managener(ial)/management: 9, 10, 11, 15, 21, 24, 34, 36, 37, 42, 48, 49, 55–56, 57, 59, 60, 61, 65, 82, 91, 148, 158, 162, 163, 173, 174, 176, 181–182, 183, 184, 185, 186, 187, 193, 194, 195, 199, 202–203, 204, 207, 209, 210, 211, 213, 216, 218, 220, 225, 226, 228, 229–230, 231, 232, 233, 234, 235, 237, 238, 245, 250, 252, 255, 259, 260, 261, 265, 267, 268, 270, 271–273, 274, 275, 276
market(s): 10, 13, 42–47, 56, 67, 86, 87, 130, 136, 140, 144, 172, 193, 224–226, 227
mathematic(al/ians/s): 4–5, 15, 22, 31, 41, 43, 49, 54, 210
Marxism: 70
(im)measurable/measurability: 34, 35, 88, 89, 91, 92, 140, 163, 203, 211, 216, 229
measure(ment): 1, 3, 8, 10, 11, 15, 21, 22, 23, 24, 27, 30, 34, 35, 36, 38, 41, 45, 46, 48, 51, 57, 61, 64, 65, 67, 71, 72, 73, 79, 81, 82, 84, 85, 87, 88, 89, 91, 92, 93, 94, 99–113, 119, 125, 140, 147, 157, 159, 163, 176, 181, 183, 199, 200, 201–203, 209, 210, 211, 214, 215, 216, 218, 219–220, 224–225, 226, 227, 231, 233, 234, 235, 236, 240, 246, 247, 261, 265, 266, 276
means-ends scheme, 251
media, 35, 99, 120, 201, 250
mediators: 30, 32
medical diagnosis: 15, 237, 238, 245–260
mercantilism: 43, 44
meritocracy: 164
meso-level: 11–13
methodology(ies): 4, 22, 31, 42, 43, 60, 68, 71, 73, 74, 76, 89, 139, 147, 149, 151, 202, 218, 226
metrology/metrological: 22, 54, 59, 61

Subject Index 295

medical record(s)/patient record(s): 172, 173, 178, 179, 180, 183, 186, 238, 254, 256

micro-level: 11–13, 15, 46, 56, 157, 169, 209, 275

migration: 106

minority group(s): 3, 14, 29, 69, 101, 113

model(s)/modeling: 6, 21, 24–35, 36, 37, 42, 43, 44, 46, 47, 50, 53, 54, 55, 59, 60, 62, 69, 72, 73–74, 75, 82–83, 84, 89, 94, 137, 148, 168, 199, 207, 209, 219, 248, 264, 268

monitor(ing): 12, 24, 56, 58, 79, 86, 113, 118, 163, 182, 183, 184, 199, 209, 213, 219, 246, 261, 273

monographic surveys: 50

moral(ity): 37, 48, 51, 67, 70, 72, 79, 127, 129, 136, 140, 142, 144, 145, 146, 147, 151, 179, 187, 260

moral panic: 157, 162

moreninha/moreno (*See* racial categories)

mortality: 104–105, 125, 237–239

mulato (*See* racial categories)

municipal(ity): 25, 27, 28, 29, 30, 31, 32, 33, 34, 53, 79, 84–85, 87–89, 90, 91, 92, 93, 94, 207, 211, 212, 217, 268, 280, 281

mutual construction (also co-construction, coproduction): 1, 2, 3, 6, 7–9, 10, 11, 14, 15, 43, 50, 64, 65–67, 71, 79, 82, 93, 137, 143, 148, 150, 151, 152, 157, 254, 257, 263, 266, 267, 269, 270, 272, 275, 276

N

não-branco (*See* racial categories)

national accounting: 32, 44, 45

National Audit Office (Netherlands): 225

Native Americans (*See* ethnic categories)

natural sciences: 4, 23, 53, 54, 66

negro (*See* racial categories)

neoliberalism: 11, 14, 43–46, 79–94, 139, 140, 209, 224

New Public Management (NPM): 21, 37, 58–59, 60, 199, 209

Netherlands: 225, 233, 241

NGO(s): 121, 139, 141

9–11: 22

non-white (*See* racial categories)

norm(s)/normative: 24, 28, 65, 69, 70, 81, 103, 136, 138, 140, 142, 144, 147, 149, 151, 231, 234, 259, 273

normal distribution/normal law/statistical normality: 30, 41, 48, 61, 80

Norway: 12, 14, 25, 17, 29, 32, 44, 49, 50, 84, 86, 87, 89, 93, 94, 207–220, 245, 246, 247, 248, 252, 261, 264, 267, 268, 276–277

NPM (*See* New Public Management)

number(s): 1, 4, 8, 10–11, 12, 13, 14–15, 21–38, 41, 57, 59, 61, 64, 65, 66, 70, 72, 73, 76, 79, 80, 81, 82, 84, 85–86, 88–89, 90–92, 94, 157–169, 185, 187, 191–204, 207–220, 227, 229, 233, 234, 235, 236, 237, 239, 240, 246, 253, 259, 260, 275

O

objectification: 3, 24, 45, 56, 85, 233

observation(s): about those of others, 1, 8, 23, 24, 27, 35, 41, 43, 47, 52, 61, 107, 130, 160, 192, 251, 264, 268; about our own, 23, 33, 36, 79, 87, 93, 215, 227, 228, 236, 246, 249, 253, 266, 267, 269

Occam's razor: 251

OECD (*See* Organization for Economic Co-operation and Development)

OMC (*See* European Open Method of Coordination)

operationalize(ation): 85, 99, 104, 108, 112, 139, 145, 147, 149, 216, 219, 231

opinion polls: 59, 73

organization(s)/organizational: 3, 8, 21–37, 76, 79, 84, 88, 89, 90–91, 93, 118–119, 120, 121, 125, 126, 131, 132, 136, 137, 139, 142, 143, 144, 147, 148, 149, 150, 151, 159, 173, 174, 178, 179, 186, 187, 191, 192, 193, 194, 195, 201, 202, 208, 211, 212, 214, 216, 217, 220, 226, 233, 241, 247, 264, 266, 267, 268, 270, 271, 272, 273, 274, 276

296 Subject Index

Organization for Economic Co-operation and Development (OECD), 210, 215
"other", the: 102
outdoor relief: 47, 48, 49

P

Pacific Islanders (*See* ethnic categories)
pardo (*See* racial cateogories)
parliament(s)/parliamentary: 12, 50, 51, 56, 57, 159, 162, 201, 221
path analysis: 31
patient(s): 33, 110, 129, 172, 173, 175, 176–177, 178–180, 181, 182, 183–184, 185, 186, 187, 188, 224, 225, 233–234, 235, 236, 237, 238, 239, 240, 245–261
pediatric (*See* age categories)
pediatric triage: 172, 176, 177, 178, 186, 187
performance indicators: 14, 33, 42, 55–56, 59, 199, 200, 202–203, 224–241, 265, 271, 276
photographer(s)/photography: 6, 157, 165, 166
PISA (*See* Programme for International Student Assessment)
plan(ning): 5, 10, 11, 12, 13, 15, 21, 43, 44, 45, 57, 59, 69, 70, 71, 80, 85, 86, 92, 106, 111, 149, 207, 209, 210, 230–231, 247, 249, 252, 264, 265, 268–269, 273, 276
pluralism: 144
police returns (*See also* crime statistics): 159–160, 161–163
policing: 79, 159, 194, 195, 197, 198, 200, 201, 203–204, 263–276
policy(ies): 2, 5, 14, 29, 32, 42, 43, 45, 46, 47, 48, 49, 50, 56, 57, 58, 59, 60, 68, 104, 107, 108, 112, 113, 117, 118, 119, 120, 121, 123, 125, 126, 128, 129, 132, 137, 138, 139, 140, 141, 142, 143, 145, 146, 149, 150, 151, 157, 159, 162, 168, 172–187, 194, 198, 201, 224, 260, 263, 273, 275, 276
policy design(ers)/policy maker(s) / policy making: 5, 106, 117, 118–121, 123, 125, 126, 129, 130, 132, 133, 136–137, 138, 139, 140, 141, 142, 145, 146, 148, 149, 151, 181, 182, 225, 274
political arithmetic: 64, 66, 76
political science: 41, 54, 61, 68, 141
politician(s)/politicization/politics: 2, 3, 4, 5, 11, 12, 13, 14, 15, 26, 33, 35, 43, 47, 48, 49, 50, 58, 60, 66, 71, 73, 74, 84, 86, 87, 106, 107, 118–119, 130, 131, 136, 137–141, 142–143, 144, 145, 146, 147, 148, 149, 150, 151–152, 157, 158, 162, 178, 186, 187, 192, 200, 208–209, 210, 215, 227, 247, 251, 259, 272, 274
poor (*See* class categories)
POP (*See* Problem-Oriented policing)
population(s): 29, 41, 46, 47, 50, 65, 66, 68, 80–82, 86, 99, 100, 101, 102–103, 106, 107, 108, 110, 111–112, 119, 157, 159, 160, 161, 162, 163, 164, 177, 178, 191, 196, 218, 261
positivism: 67, 69, 70, 209, 219
poverty: 3, 14, 42, 47–50, 124, 136, 137, 139, 140, 141, 142, 144, 146, 147, 148, 149, 151
power(ful)/empower(ment): 2, 4–5, 6, 7, 8, 10, 13, 21–37, 65, 80–82, 83, 84, 86, 87, 90, 91, 93, 100, 137, 140, 141, 149, 151, 162, 168, 196, 220, 227, 238, 256, 269
practice(s)/praxis: 10, 12, 13, 15, 24, 27, 28, 33, 36, 37, 43, 46, 55, 64–76, 79, 80, 81, 82, 83, 84, 85, 86, 87, 89, 90, 91, 93, 94, 109, 110, 118, 158, 159, 163, 166, 167, 168, 169, 172, 173, 175, 177, 178, 180, 184, 185, 186, 202–203, 208, 211, 212, 219, 220, 226, 227, 228, 230, 233, 234, 236, 238, 239, 240, 245, 246, 248, 252, 253, 254, 255, 256, 257, 258, 260, 263–276; statistical practice(s), 2, 3, 5, 7–8, 9, 10, 11, 12, 13, 15, 22, 23, 24, 35, 46, 84–85, 86, 87, 93, 94, 157, 161, 220
preto (*See* racial categories)
price index: 44, 45, 46, 50, 58
privatize(ation): 187, 224
probability: 4–5, 30, 41, 61, 82

Problem Oriented Policing (POP): 197, 264, 265, 267–269, 271, 272, 274, 275
problem owner(s): 229, 230, 231, 234, 235, 237, 239, 241, 269
production index: 29–30
production output/productivity: 21, 246, 247
profession(al/s): 10, 11, 58, 60, 61, 74, 88, 138, 168, 180, 194, 202, 225, 234, 241, 247, 249, 250, 260, 269, 272, 274, 276
Programme for International Student Assessment (PISA): 210, 211, 220

Q

qualitative categories/measures/methods: 31, 66, 76, 218, 246
quality: 89, 92, 94, 174, 179, 208, 209, 212–214, 224, 228, 229, 230, 235, 238, 250, 261; quality assurance/assessment/indicators/measurement, 14, 22, 38, 81, 85–86, 88, 180, 184, 186, 207, 208–211, 212–214, 215, 216, 217, 218, 219, 224–226, 227, 228, 246, 273; of life, 144; of numbers, 4, 61, 79
quality domain (*See* local accounting domain)
quality manager/management team: 228–238
quantify/quantification: 21, 24, 27–29, 30, 31, 33, 35, 36, 37, 42, 46, 50, 53, 56, 57, 58, 59, 61, 64, 66, 67–75, 76, 140, 142, 144, 203, 216, 246
quantitative methodology: 4, 10, 43, 52, 57, 59, 60, 64–65, 67–68, 69, 75, 76, 80, 92, 118, 119, 142, 207, 209, 215, 216, 218
questionnaire(s): 71, 121, 122, 131, 133, 212

R

race: 14, 99, 100, 102–103, 104, 105, 106, 107–108, 109, 110, 113, 117–133
racial categories: Afro-Brazilian, 119, 126, 133; *amarela*, 122; Anglo-American, 102; black, 107, 119; *branco*, 119, 121, 122, 123, 124, 125, 126, 127, 128, 129, 131, 133; caucasian, 102, 107, 108; *escurinha*, 127; European, 119; *índio*, 119, 127, 128; *moreninha*, 127; *moreno*, 126, 127, 131; *mulato*, 123, 124, 126, 129; *não-branco*, 119, 120, 125; *negro*, 119–122, 123, 124, 125, 126, 127, 128, 129–130, 131, 132, 133; non-white, 103, 119, 122; *pardo*, 119, 120–123, 124, 125, 126, 128, 129, 130, 131, 133; *preto*, 119, 125, 126, 127, 130–131, 132, 133; white/whiteness, 102, 103, 105, 107, 119
"Racial Project": 117, 118–120, 131 132
racism/racial discrimination/prejudice: 14, 101, 102, 105, 106, 107, 108, 110, 114, 117, 118, 123, 124, 126, 127, 128, 129, 130–131, 132, 133
random sampling: 49, 51, 76
rational choice: 73, 75, 218
rational expectations theory: 46
rationalize(ation): 42–43, 49, 71, 85, 133, 208, 219
realism: 61
reductionism: 23, 144
register(s) (*See* databanks/data registers)
regression: 29–30, 31, 46, 48–50, 52–55
relativism: 8, 11, 42, 69
representation(s): 24, 26, 36, 52, 72, 109, 138, 187, 202, 208, 247, 250, 252–255, 259, 260, 261
representative method: 50, 51, 52
research(ers): 5, 6, 13, 21, 22, 25, 26, 29, 43, 50, 56, 60, 64, 65, 66, 67, 68, 70–75, 76, 99, 100, 102, 104, 105, 106–108, 111, 112, 126, 130, 131, 137, 138, 139, 143, 149, 152, 175, 176–177, 216, 229, 246, 248, 276
research domain: 180, 184–185
resource(s): 2, 21, 26, 57, 59, 69, 81, 108, 111, 117, 146, 147, 174, 176, 181, 182, 184, 185, 187, 198–199, 200, 207, 211, 212, 213, 214, 217, 218, 219, 220, 246, 247, 250, 261, 265, 269, 270, 271, 272
risk: 5, 11, 12, 51, 79, 91, 93, 100, 101, 104, 107, 108, 109, 110,

298 Subject Index

111, 112, 158, 168, 182, 203, 225, 231, 233, 238, 257, 263, 264, 265, 266, 267, 268, 269, 270, 271, 272, 274, 275
Rokeach Value Survey: 72
Romani (*See* ethnic categories)

S

sampling: 42, 44, 45, 47, 48, 49, 50–52, 60, 61, 76, 167
Scanning, Analysis, Response and Assessment (SARA): 264, 268
school(s): 14, 27, 29, 34, 54, 56, 84, 88, 90, 120–121, 122–123, 127–128, 129, 130, 133, 146, 147, 207–221
science(s)/scientific: 3, 5–6, 7–8, 15, 23, 30, 31, 33, 42, 43, 53, 54, 55, 66, 64, 65, 66–67, 69, 70, 71, 72, 73, 85, 102, 105, 106, 107, 120, 136, 137–141, 142, 145, 150, 152, 174, 192, 209, 210, 241, 245, 251, 266, 272, 273, 276
Science and Technology Studies (STS): 5, 6, 7–8, 11, 13, 66, 69, 121, 136, 137, 139, 145
scientist(s): 3, 67, 76, 117–118, 119–120, 125, 130, 137, 138, 141, 148, 174
Scotland/Scottish: 103, 165, 167
SES (*See* socioeconomic status)
skin color (*See* race)
social class (*See* class)
social engineer(ing): 6, 7, 44, 45, 49, 65, 67, 68, 69, 70, 73, 74, 75, 76, 80
socialism: 44, 57, 59
social science: 4, 13, 21, 30, 41–43, 52, 53, 54, 55, 61, 64–77, 118, 119, 120, 125, 130, 139, 141, 209
social scientist(s): 21, 41, 42, 52, 67, 69, 73, 76, 117, 119, 125, 130
social world(s): 1, 2, 9, 42, 49, 55, 71, 138, 149, 174, 175, 178, 201, 260
society: 1, 4, 6, 33, 42, 43, 44, 45, 46–47, 51, 64, 65, 66, 67, 69, 70, 71, 72, 73, 74, 75, 76, 80–81, 86–87, 88, 100, 101, 104, 112, 113, 118, 119, 120, 125, 128, 139, 164, 167, 191–192, 194, 195, 200, 201, 203, 209, 215, 220, 224, 265, 268

socio-economic status (SES): 104, 105, 119, 121, 122
sociologist(s)/sociology: 6, 30, 31, 41, 42–43, 52, 55, 64–77, 79, 120, 137, 141, 145, 146–147, 275
stakeholder(s): 2, 31, 44, 46, 56, 57, 141, 149, 198, 200–201, 225, 227
standard(s)/standardization: 2, 3, 10, 12, 24, 25–27, 30, 32, 33, 34, 35–36, 37, 42, 58, 61, 65, 71, 72, 73, 75, 76, 80, 82, 85, 86, 88, 91–92, 101, 102, 106, 107, 109, 143, 158, 161, 164, 165, 175, 176, 183, 208, 209–210, 219, 220, 224, 230, 231, 234, 236, 241, 250, 253, 254, 261, 264, 271, 273, 274
standard of living: 50, 149
state, (the): 2, 5, 9, 10, 11, 12, 13–14, 15, 27, 42–47, 50, 56, 58, 60, 64–77, 80, 81, 84, 85, 86, 87, 118, 119, 120, 121, 125, 129, 132, 129, 157, 164, 168, 169, 187, 209, 224, 226, 227, 250, 252, 260
statistical inference: 30, 41, 49, 54
statistician(s): 51, 58, 59, 60, 61, 65
statistics: This entire book is about statistics. Our working definition of statistics can be found on page 3 and in footnotes 4 and 5 on page 15. An overview over the chapters can be found on pages 13–15 and in the table of contents. The introduction (pages 1–15) and index reflect the many ways in which we discuss statistics.
stereotypes: 106, 108, 110, 111;
structural functionalism: 65, 67, 70, 72, 74, 75–76
STS (*See* Science and Technology Studies)
survey(s): 3, 42, 44–45, 47, 48, 49, 50–52, 58, 59, 60, 65, 67–68, 70, 71, 72, 73–74, 75, 76, 77, 99, 103–104, 113, 131, 210, 211, 212, 215
surveillance: 5, 80, 118, 157, 159, 168, 169, 270, 273
Sweden/Swedish: 13, 64–65, 67–70, 71, 75, 76, 77, 210, 238

T

target(s)/targeting/target groups: 24, 34, 37, 46, 48, 94, 108–109, 110, 111, 129, 133, 197, 199, 200, 201–202, 203, 207, 208, 212, 264, 265, 271, 272, 273–274

technocrat(s)/technocracy: 23, 45, 142, 149, 151, 158, 168

technology(ies): 5–11, 12, 13, 23–24, 42, 56, 60, 67, 79, 82, 83, 85, 90, 91, 93–94, 157, 158, 168, 173, 177, 186, 208, 216, 219, 220, 240, 241, 245, 246, 247, 252, 260, 263, 265–267, 268, 269, 272, 273, 274–275, 276

tool(s): 2, 3, 4, 5–6, 8, 9, 12, 21, 31, 32, 33, 34, 36, 37, 41–42, 43, 44–46, 47, 48, 50, 53, 56, 57, 60, 65, 73, 76, 80, 83, 84, 88, 93, 109, 129, 136, 138, 147, 148, 176, 178, 193, 207, 211, 213, 215, 216, 217, 218–219, 225, 245, 256, 257, 258, 261, 265, 267, 268, 270, 271, 272, 273, 274–275, 276

Translation Theory (*See* Actor Network Theory)

transparency: 10, 14, 24, 25, 26, 33, 34, 38, 45, 148, 177, 224, 226, 227, 239–240, 248

truth claim(s)/knowledge claim(s): 5, 8, 13, 100, 149

U

UNDP (*See* United Nations)

Union of Soviet Socialist Republics (USSR): 45, 49

United Kingdom (*See also* England, Scotland, Wales): 157, 158

United Nations (UN): 136, 137, 138, 139–140, 141, 143, 146, 147, 150, 151, 152

United Nations Development Programme (UNDP): 136, 141, 142–143, 144, 146, 149, 150

United States of America (USA): 29, 44, 45, 67, 68, 69, 70, 71, 72, 89, 102, 107, 113, 119, 246, 265

utilitarianism: 144

V

values: cultural/ethical/social etc. 2, 3, 5, 61, 69, 70, 72, 106, 112, 113, 136–137, 138, 140, 141, 142, 144, 148, 149, 150, 151, 229, 230, 233–234, 236, 240, 250, 259, 271, 273; statistical, 1, 20–21, 183, 226, 228, 231, 235, 240, 241

variable(s): 22, 31, 46, 50, 52, 53, 54, 55, 60, 61, 72, 102, 103, 104, 105, 106, 107, 108, 111–112, 130, 160, 197, 203, 261, 268

verification/verify: 32, 34, 51, 55, 56, 59, 60, 251

victim(s)/victimization: 27, 29, 126, 127, 197, 263, 265, 268

W

Wales: 157–169, 275

Welfare/welfare state: 13–14, 29, 43, 44, 45, 50, 65, 67–70, 76, 80, 86, 147, 224, 226, 269

"white"/"whiteness" (*See* racial categories)

World Bank: 136, 140, 142, 145, 146

World Health Organization (WHO): 248, 261

worldview: 117, 118, 132, 136, 251

Y

Yale University: 246

youth(s) (*See* age categories)

#0232 - 060716 - C0 - 229/152/17 - PB - 9780415811057